Technological Change and Mature Industrial Regions

Series Editor: Philip McCann, *Professor of Economics, University of Waikato, New Zealand and Professor of Urban and Regional Economics, University of Reading, UK*

Regional science analyses important issues surrounding the growth and development of urban and regional systems and is emerging as a major social science discipline. This series provides an invaluable forum for the publication of high quality scholarly work on urban and regional studies, industrial location economics, transport systems, economic geography and networks.

New Horizons in Regional Science aims to publish the best work by economists, geographers, urban and regional planners and other researchers from throughout the world. It is intended to serve a wide readership including academics, students and policymakers.

Titles in the series include:

Regional Knowledge Economies
Markets, Clusters and Innovation
Philip Cooke, Carla De Laurentis, Franz Tödtling and Michaela Trippl

Entrepreneurship, Industrial Location and Economic Growth
Edited by Josep Maria Arauzo-Carod and Miguel Carlos Manjón-Antolín

Creative Cities, Cultural Clusters and Local Economic Development
Edited by Philip Cooke and Luciana Lazzeretti

The Economics of Regional Clusters
Networks, Technology and Policy
Edited by Uwe Blien and Gunther Maier

Firm Mobility and Organizational Networks
Innovation, Embeddedness and Economic Geography
Joris Knoben

Innovation, Agglomeration and Regional Competition
Edited by Charlie Karlsson, Börje Johansson and Roger R. Stough

Technological Change and Mature Industrial Regions
Firms, Knowledge and Policy
Edited by Mahtab A. Farshchi, Odile E. M. Janne and Philip McCann

Technological Change and Mature Industrial Regions

Firms, Knowledge and Policy

Edited by

Mahtab A. Farshchi

London South Bank University, UK

Odile E. M. Janne

Birkbeck College, University of London, UK

Philip McCann

University of Waikato, NZ; University of Reading, UK

NEW HORIZONS IN REGIONAL SCIENCE

Edward Elgar

Cheltenham, UK • Northampton, MA, USA

Published by
Edward Elgar Publishing Limited
The Lypiatts
15 Lansdown Road
Cheltenham
Glos GL50 2JA
UK

Edward Elgar Publishing, Inc.
William Pratt House
9 Dewey Court
Northampton
Massachusetts 01060
USA

A catalogue record for this book
is available from the British Library

Library of Congress Control Number: 2008937413

PEFC
PEFC/16-33-111
CATG-PEFC-052
www.pefc.org

ISBN 978 1 84720 089 1

Printed and bound in Great Britain by MPG Books Ltd, Bodmin, Cornwall

Contents

PART II MATURE REGION–INDUSTRY CASE STUDIES

PART III REGIONAL POLICY AND MATURE INDUSTRIAL REGIONS

Contents

PART II MATURE REGION–INDUSTRY CASE STUDIES

PART III REGIONAL POLICY AND MATURE INDUSTRIAL REGIONS

Contributors

Josep-Maria Arauzo-Carod, Grup de Recerca d'Indústria i Territori, Department of Economics, Universitat Rovira i Virgili, Reus, Spain.

David Bailey, Birmingham Business School, University of Birmingham, Birmingham, United Kingdom.

Andrew Beer, School of Geography, Population and Environmental Management, Flinders University, Adelaide, Australia.

Michael H. Best, Center for Industrial Competitiveness (CIC), University of Massachusetts Lowell, Lowell, USA.

Peter Clark, School of Business and Management, Queen Mary and Westfield College, University of London, London, United Kingdom.

Edson Paulo Domingues, Centre for Development and Regional Planning, Department of Economics, Federal University of Minas Gerais, Brazil.

Mahtab Akhavan Farshchi, Faculty of Engineering, Science and Built Environment, London South Bank University, London, United Kingdom.

Eduardo Amaral Haddad, Department of Economics, University of Sao Paulo; São Paulo, Brazil; and Regional Economics Applications Laboratory, University of Illinois, USA.

Geoffrey J. D. Hewings, Regional Economics Applications Laboratory, University of Illinois, Urbana, Illinois, USA.

Simona Iammarino, SPRU, University of Sussex, The Freeman Centre, Brighton, United Kingdom.

Marko Jaklič, Faculty of Economics, University of Ljubljana, Ljubljana, Slovenia.

Odile E. M. Janne, Department of Management, School of Management and Organizational Psychology, Birkbeck College, University of London, London, United Kingdom.

Jonathan Jones, Economics, Business School, University of Newcastle upon Tyne, Newcastle, United Kingdom.

Declan Jordan, Department of Economics, University College Cork, Cork, Ireland.

Seiji Kobayashi, Nihon Daigaku Kaikan (Nihon Graduate Business School), Nihon University, Tokyo, Japan.

Ilaria Mariotti, Dipartimento di Ingegneria, Politecnico di Milano, Milan, Italy.

Philip McCann, Department of Economics, University of Waikato, New Zealand; and Department of Economics, University of Reading, Reading, United Kingdom.

Eoin O'Leary, Department of Economics, University College Cork, Cork, Ireland.

Ivana Paniccia, Faculty of Economics, Luiss G. Carli University, Rome, Italy.

John B. Parr, Department of Urban Studies, University of Glasgow, Glasgow, United Kingdom.

Fernando Salgueiro Perobelli, Department of Economics, Federal University of Juiz de Fora, Brazil.

Yannis Psycharis, Department of Regional Planning, University of Thessaly, Volos, Greece.

George Petrakos, Department of Regional Planning, University of Thessaly, Volos, Greece.

Anja Cotič Svetina, Faculty of Economics, University of Ljubljana, Ljubljana, Slovenia.

J. Kim Swales, Fraser of Allander Institute and Department of Economics, University of Strathclyde, Glasgow, United Kingdom.

Holli Thomas, School of Geography, Population and Environmental Management, Flinders University, Adelaide, Australia.

Elisabet Viladecans-Marsal, Department of Econometrics, Statistics and Spanish Economy, Universitat de Barcelona, Barcelona, Spain.

Nick von Tunzelmann, SPRU, University of Sussex, The Freeman Centre, Brighton, United Kingdom.

Colin Wren, Economics, Business School, University of Newcastle upon Tyne, Newcastle, United Kingdom.

Hugo Zagorsek, Faculty of Economics, University of Ljubljana, Ljubljana, Slovenia.

During the writing of this book we tragically lost our colleague Hugo Zagorsek (1977–2008), a fellow traveller whose intellect, wisdom, enthusiasm and scholarship were an inspiration to us all.

PART I

MATURE REGIONS: TECHNOLOGY, SOCIETY AND INDUSTRIAL STRUCTURE

1. The Problems of Mature Regions: An Introduction and Overview

Mahtab Akhavan Farshchi, Odile E. M. Janne and Philip McCann

1.1 INTRODUCTION

This chapter provides an overview of the basic themes and insights of the book, and outlines the problems involved in developing policies to promote mature industrial regions. The complex and specific nature of regional issues in mature regions is considered in the context of globalisation and technological changes. In particular, the focus of this book is on the issues relating to the technology, knowledge and industrial structure of mature regions, and the challenges that these pose for policies designed to promote such regions.

Many of the basic characteristics of mature regions are simply very different from those of buoyant or emerging regions. Prosperous and growing regions tend to be the regions which are typically the focus of widespread analysis and observation. As such, many of the insights gained from buoyant or emerging regions may not be appropriate or relevant for the development of mature regions. While mature regions experience industrial restructuring and transformation, they demonstrate very different features and behaviour in response to processes of technological change. So, for example, the labelling of an industrial cluster as being 'mature' is insufficient to understanding its specific characteristics and evolutionary trajectory, and potentially misleads policy making. Yet, while buoyant or emerging regions are the focus of much research, observation and policy analysis, mature regions are largely overlooked. Ironically, however, these mature regions tend to be the very regions which require policy intervention. As such, there is something of a mismatch between the origins from where policy insights arise and the situations in which they need to be applied. As such, while the need for regeneration may be very clear, opinions diverge on the likely prospects for mature regions as well as appropriate policy responses. This is particularly so when policies are based on naïve analyses which employ simple analytical constructs. All too often such naïve and short-term analyses then trigger reactive responses to the perceived structural crisis and industrial decline. Such responses often lead to protectionist policies and 'regional champion' approaches, in contrast to more nuanced and proactive policies which support and enhance existing regional capabilities, and emphasise technological and organisational transformation.

As a response to this situation, the arguments in this book consider the challenges facing policy makers dealing with mature regions, and the ways in which a detailed understanding of transformation processes in mature regions may foster more realistic and nuanced policy responses. Challenging arguments revolve around the suitable allocation of policy funds and policy responses to globalisation. This book examines these difficult issues.

1.2 INDUSTRIAL RESTRUCTURING AND MATURE REGIONS

What are the prospects for the growth and development of mature industrial areas in the context of increased internationalisation and globalisation?

The reasons why mature regions are often overlooked in the research literature is that in the 'typical' path of development, the nature of mature regions is usually assumed to be one of decline. In particular, a simple product cycle view of the world assumes that once a region is specialising in the production of goods or services which have reached a later mature stage in the cycle, then the region will tend to exhibit certain problematic characteristics. These characteristics are poor opportunities for innovation, limited employment growth, standardised technologies and declining real wages relative to other regions. Various terminologies are often applied to mature regions, such as *structural decline*, *industrial crisis* or *hollowing out*, and this perceived negative terminology often prevents policy makers and academics from engaging in a proactive analysis of the origins, nature and appropriate policy responses to such situations. However, as we will see in this book, the industrial restructuring associated with mature regions is far more complex than these simple terms imply. The structural transformations of these regions involve detailed understanding of the links between economics, history, geography and technological change, and the analysis of these transformations required an inter-disciplinary approach to these issues.

The book explicitly adopts an inter-disciplinary approach to analysing structural transformation of mature regions. The contributions include sociological analyses, business history approaches, various different economic analyses, geographical approaches, and critically, technological analyses. The chapters in the book were all specially commissioned for the book from a range of leading scholars working all over the world and specialising in these difficult issues. The themes that are dealt with examine the nature of a region's technological capabilities and how these can be developed and transformed. We examine these issues from the perspective of internal and external trade linkages, from the perspective of internal and external investment drivers, from the perspective of internal and external technological drivers, and from the perspectives of economic history and economic geography. We explore a range of case studies from different parts of the world and different region-industry groupings in order to provide evidence-based insights into how these various internal and external links apply to different region-industry types. What becomes clear is that there is no such thing as a 'typical' mature region; regional capabilities are clearly context-specific. The detailed local knowledge of the policy maker therefore becomes essential. However, there are also commonalities across mature transforming regions, and in order to aid policy makers in their decision-making, an understanding of such commonalities is essential. Yet, the commonalities across mature regions can only be uncovered by a careful analysis of the trade and technological structures of regions. Only by detailed trade and technological analyses will we better understand the complexity and dynamics of the structural adjustment taking place in mature regions. Such analyses also hold out the prospect of the development of more appropriate, more nuanced, and also more parsimonious typologies for aiding the decision-making of policy makers. In sum, mature regions are changing, and not necessarily declining. In order to identify long-run growth dynamics of these regions, academics and policy makers need to focus on the trade and technological transformation of the region, rather than on short-term issues and policy reactions.

1.3 THE ORGANISATION OF THE BOOK

The book is divided into three parts. In the first part, (Chapters 1–6), the complex and specific nature of regional issues in mature industrial areas is considered in the context of changes due to globalisation, trade and technology.

A differentiation perspective is adopted in the chapters by Nick von Tunzelmann (Chapter 2) and Simona Iammarino, Odile Janne and Philip McCann (Chapter 3). These chapters analyse regional capabilities and the complex multi-level interactions of local, national and global actors, as a means of improving our understanding of the technological nature of industrial transformation in mature regions and clusters. It is argued in these chapters that the nature and complexity of regional capabilities need to be addressed by policy makers. Von Tunzelmann examines the concepts of capabilities used in various literatures, and offers a framework of analysis for policy making in the development and regeneration of regional economies. Iammarino, Janne and McCann review and discuss the literature on the co-evolving processes of the globalisation of innovation by multinational enterprises (MNEs) and regional innovation, and its implication for the analysis of industrial transformation and renewal. Local industrial specificity and sophisticated multi-level industrial and technological interactions through MNEs are at the centre of industrial transformation and restructuring in mature regions and clusters for innovation, and their analysis offers insights for regional policies (Chapter 3).

The chapters by Fernando Perobelli, Eduardo Haddad and Edson Domingues (Chapter 4) and Geoffrey Hewings and John Parr (Chapter 5) analyse the highly sophisticated networks of linkages and interpendencies that define the economic environment of industrial regions in Brazil and the US Midwest respectively. Perobelli et al. (Chapter 4) build a hierarchy of the Brazilian states, showing that the economically weaker North, North-East and Centre-West states present a higher degree of dependence on the performance of the other Brazilian states than their own economy. Hewings and Parr (Chapter 5) similarly use input–output analysis in the context of the US Midwest, and argue that intra-sectoral trade and more traditional agglomeration economies, especially in manufacturing, tend to be more spatially dispersed at the level of the Midwest rather than the individual state or metropole. These two chapters emphasise the challenging nature of policy intervention at the state level when existing structures of economic interactions have to be taken into account. The last chapter in this part by Josep-Maria Arauzo-Carod and Elisabet Viladecans-Marsal (Chapter 6) discusses the economic disparities and manufacturing restructuring of Spanish regions. The authors present the recent regional dynamics in Spain and highlight the role of both industrial specialisation and regional policy in the success (or otherwise) of industrial restructuring and economic transformation of carefully defined mature regions.

The second part of the book (Chapters 7–12), looks at specific cases from a range of different countries, in order to distinguish what aspects of mature regions' technology, knowledge or structure are location- or region-specific from what are more generally applicable to mature industrial regions in all countries.

The chapters by Peter Clark (Chapter 7) and David Bailey and Seiji Kobayashi (Chapter 8) focus on the UK automotive region of the West Midlands. While Clark critically discusses the Marshallian origins and geo-historical competitive constraints of the region, Bailey and Kobayashi examine the specific factors contributing to the recent collapse of the region's MG Rover plant, and the role played by regional policies in the transformation and response of the region. Clark explains the decline and hollowing out of the automotive industry in Birmingham by carefully considering the regional specificity of the sector's social structure, cultural conventions of coordination and consequent problematic ability to embrace radical innovation-design cycles. He argues that the historical inheritance of the Birmingham region and trajectory has thwarted the long-term production of cars in the region. As Bailey and Kobayashi demonstrate, while policy at the national UK level actually exacerbated the problems, in contrast, the local regional regeneration policies, which focused on helping local supplier firms to diversify into other industries, proved to be a major success story.

In Chapter 9, Michael Best returns to some of the themes originally developed by von Tunzelmann (Chapter 2). Best argues the case for a differentiation across mature regions according to their distinctive regional capabilities, providing a detailed profile of finely grained products, technologies and organisational changes associated with the successful emergence of the Massachusetts' medical device cluster. The account of the ways in which the Massachusetts region managed to build a medical devices cluster underlines the importance of its ability to change and transform existing technologies and structures, rather than merely prolonging its inherited specialisation in traditional and minicomputer industries.

Andrew Beer and Holli Thomas (Chapter 10) discuss the impacts of the shock of recent plants closure to Australia's Southern Adelaide region, and evaluate the subsequent policy responses to industrial restructuring in the region. They argue that regional plans have fallen short of integrating the wider strategic role of universities and the education sector into the long-term restructuring agenda for the region. The emergence and consolidation of other sectors such as the education sector and other knowledge industries must be taken into consideration when devising long-term regeneration strategies for the region. Beer and Thomas, along with Bailey, Kobayashi and Best, warn against the failure of too narrowly defined policy making in mature regions. In particular, policies with only a limited consideration of wider regional stakeholders, an over-emphasis on unsuitable support for existing regional industrial 'champions', or the inability to integrate changing sectoral activities with organisational changes, will only fail in the long run.

The development of taxonomies provides the tools for deriving some generalities about diversity that a differentiation perspective requires, and this approach is adopted in the specific contexts of the Italian industrial districts by Ivana Paniccia (Chapter 11) and the West Midlands (UK) and Saxony

(Germany) automotive clusters by Odile Janne and Mahtab Akhavan Farshchi (Chapter 12) .

The chapter by Paniccia provides an original classification of Italian industrial districts based on their observed structural characteristics, and discusses the issues of the industrial structural changes or decline for each type (Chapter 11). The capacity of regional industries for technological upgrading in traditional products and production methods, and the role of technological diversity and integrated processes, are also themes discussed in previous chapters, and Paniccia develops these arguments on the basis of careful observations of the behavioural, structural, geographical and historical properties of the industries. Paniccia develops the argument for more horizontal (national) policies, as opposed to district (or cluster) specific measures, in Italy. These insights are complemented by the evidence-based research of Janne and Farshchi (Chapter 12), who argue that important insights for policy making may be gained from considering how different types of externalities, and particularly localised knowledge spillovers, influence innovation and growth in the industrial transformation of clusters. Their work of comparing and contrasting the automotive clusters of the West Midlands and Saxony demonstrates how the nature of the local externalities affecting innovation helps to explain the different growth prospects of certain types of regional mature industries.

What emerges from the two first parts of this book is that the ability of mature regions or clusters to undertake technological changes, supply-chain (trade) changes and organisational changes, is the key to understanding a region's restructuring and its long-run industrial transformation. Each of these three types changes are actually different forms of innovation, and one form of innovation without the other two forms of innovation is not sufficient to engender a fundamental and successful transformation of a mature region. Facilitating all three types of changes is essential and should be the focus of regional policy.

The third part of the book, containing Chapters 13–18, focuses on the development of specific tools for policy making and existing policies employed in various mature region cases. The identification and development of regional capabilities and regional policy initiatives are critically considered.

In Chapter 13, Kim Swales explains the case for regional policies from the perspective of welfare economics. Starting from the fundamental theorems of welfare economics, Swales examines the various cases for regional policy, highlighting how arguments regarding the provision of public goods and the problems associated with market failure relate to regions specifically. The criteria for regional redistribution are examined in the light of a formal cost benefit analysis. The emphasis here is on understanding the specific nature of the shadow prices for labour, capital and taxation in a regional policy framework. Swales argues that the transparency and accountability of such a methodology should always form the basis of regional policy, whereas in reality this is hardly ever the case.

In Chapter 14 Ilaria Mariotti examines the actual practical, as opposed to the theoretical, basis for the regional policies which have previously been employed in The Netherlands, United Kingdom and Italy. Mariotti adopts an historical approach to understanding how the policy frameworks developed in each of the countries. What becomes clear is that the main aspects of the evolution of each of the national-regional policy frameworks are common to each of the countries. In particular, over recent years there has been a movement away from reducing regional inequalities and towards improving regional competitiveness, at least in principle. However, the specific ways by which this is supposed to be are rather different, and still very much dependent on each country's traditional approach to regional policy. In Italy, the focus is primarily on traditional industries, in The Netherlands it is on firms' relocation, while in the UK the focus has been more on attracting inward FDI.

Declan Jordan and Eoin O'Leary in Chapter 15 look at the challenges facing regional policy in Ireland from the perspective of innovation promotion. Their results indicate the absence of any strong innovation interaction between locally- or regionally-based concentrations of suppliers, customers or competitors, or between firms and tertiary education institutions, or government agencies. The absence of such linkages therefore provides a major problem for the design of Irish regional policy, as much current thinking regarding clusters and competitiveness is based on the assumed existence of such interactions. The reasons for this can be clearly understood in that the role of inward FDI in Ireland has been such a significant reason for the rapid growth of the Irish economy. Yet, at the same time, the backward linkages of the immigrant FDI have been at a national, rather than a regional scale. Regional policy based on much contemporary thinking has only very limited leverage there.

The themes highlighted in both of the previous two chapters are picked up in Chapter 16 by Colin Wren and Jonathan Jones who examine the role of regional policy in determining the location behaviour of inward FDI. Although regional policy plays a significant role in relocation decisions, there are major difficulties associated with identifying exactly how effective these schemes really are. One issue which emerges is a scepticism regarding the importance of agglomeration economies for many types of activities. The observed location and relocation behaviour of FDI could arise for a range of reasons such as supply-chain issues, acquisition behaviour and signalling effects, all of which are discussed in the various chapters of this book, as well as agglomeration effects. A better understanding of the role of each these alternative location mechanisms is critical for evaluating regional policy.

Anja Svetina, Marko Jaklič and Hugo Zagorsek in Chapter 17 analyse the ways in which properly implemented policy learning is likely to bring successful outcomes in terms of the quality and coordination of policy for supporting industrial transformation. They investigate how appropriate policy learning necessitates the use of evaluation criteria that are both qualitative and quantitative, and also require the involvement of stakeholders. While their arguments are equally applicable to all countries, they discuss the specific experience of implementing cluster policy in the transition economy of Slovenia in which these best-practice principles were not adopted. Using a policy life cycle framework, the authors explain how the failure of the Slovenian government to understand and adopt a policy-learning approach into their policy management practices, the inheritance of a rigid industrial structure and mindset as well as short-term political changes, led the government to recently stop the Slovene cluster programme despite its promising early impact.

In Chapter 18 Yannis Psycharis and George Petrakos examine the role and performance of regional policy in Greece. The major issue which emerges here is that the nature and role of regional policy cannot be divorced from the institutional structure of the national economy. The level of centralisation–decentralisation determines the allocation of power and resources at the various administrative levels. As such, although regional policy is mildly redistributive, in addition to questions of physical geography, the effectiveness of regional policy over the long term is demonstrated to be subject to much wider institutional and governance issues. As such, their analysis uncovers an additional level of issues which need to be considered in relation to the design and effectiveness of regional policy, over and above the industrial, technological and geographical levels.

2. Regional Capabilities and Industrial Regeneration

Nick von Tunzelmann

2.1 INTRODUCTION

There has been widespread resort to the language of 'capabilities' in recent times, used variously across different levels of systems from individual to global, to describe a large variety of processes (e.g. 'social capabilities' for growth), and a variety of functions (e.g. 'production capabilities', 'technological capabilities', 'user capabilities'). This chapter attempts a unifying framework for defining 'capabilities' applicable at any required level, including the regional. The main extensions to orthodox static notions involve both interactive and dynamic 'capabilities'. This allows us to portray the regional (re-)generation of innovation system capabilities. A substantial literature already exists on 'regional systems of innovation' (and much more on 'clusters') but less on how to *start up* or *revive* innovative regional systems (Feldman, 2001). The latter involves a 'blending' of existing and new strengths: in products (high-tech, medium-tech or low-tech?), processes (advanced or traditional?), and indeed policies/strategies (mission-oriented or diffusion-oriented?). 'Systems integration' at the regional level involves coordinating a diverse structure of 'value networks'. In this light, the chapter will explore interactive and dynamic capabilities in policy-making at the regional level. It will show the need to 'align' objectives of all types of actors, both public and private, simultaneously with the need to 'align' micro, meso, and macro levels. Efforts to do so are inherently limited by the 'capabilities' of the state/authority to link its policy-making to client needs.

Section 2.2 of this chapter will outline a conceptualisation of capabilities, including dynamic capabilities, and Section 2.3 develops the concepts in a setting of regional regeneration. Section 2.4 analyses the key concept of 'network alignment', while Section 2.5 assesses capabilities at the policy level. Section 2.6 concludes.

2.2 DEFINING CAPABILITIES

The notion of 'capabilities' has taken hold in recent times to indicate the elements seen as necessary to the successful development of each and every unit of analysis.

a) Capabilities have been used variously to describe the whole range of levels of systems, from the individual up to the global, thus spanning capabilities of firms (micro systems), of industries and/or regions (meso systems), of countries (macro systems) and so forth.
b) Capabilities have been used to describe a variety of processes, both static and dynamic – one thinks, for instance, of the 'social capabilities' for growth as flagged by Abramovitz (1986, 1994), which might be thought of as a dynamic version of the popular notion of 'social capital'.
c) Capabilities are used to refer to a wide variety of functions, e.g. 'production capabilities', 'technological capabilities', 'user capabilities', and so on.

Other terminology is used in what appears to be somewhat similar ways, e.g. language such as 'competencies' or 'capacities'. A number of authors have attempted to draw distinctions between these concepts, but with little consensus or standardisation as yet (e.g. Foss and Knudsen, 1995; Dodgson and Bessant, 1996). So a key question is whether all these terms and all these applications of the term 'capabilities' are describing the same sort of thing, or instead whether there are useful distinctions that might be drawn, perhaps of relevance to fine-tuning policy. As it is, there is at present a mass of confusion both theoretically and empirically about this clearly important set of concepts. It does not help that formal theoretical approaches such as that from economics largely avoid probing this whole area; in effect a new mode of analysis has to be developed to make much headway.

The approach to be adopted here draws primarily on the work of Sen (1985) on consumer capabilities and on Penrose (1959) on the capabilities of firms (producers). The former begins with a classification of product 'characteristics', drawing on the work of Lancaster (1966) and others. This arrays a certain type of product (e.g. a motor car) by a range of such characteristics (e.g. top speed, fuel consumption, safety level, number of passengers, etc.). Better products come higher up the 'quality ladder' and new products will insert themselves at an appropriate point on the consumer's scale of wants. However, as Sen (1985) stresses, simply having a better product says little or nothing about the individual consumer's capacity to benefit from those improved characteristics. First, consumers may find themselves in different *circumstances* that affect their extent of benefit (this includes, for instance, the impact of regulation, such as speed or pollution limits on cars). Second, individual consumers may have different *abilities* to reap any such benefits (e.g. to drive a car safely). This combination of circumstances and abilities represents their *capabilities*. Some factors influencing capabilities may be (wholly or partly) under their own control as individuals, while others may not be. We can think of different levels of *endowments* with regard to such capabilities, which include the individual's *competencies* (the level of relevant training etc., attained) and their *capital* (accumulated wealth, tangible or intangible). We can also think of *enhancements* to their capabilities, most often from on-going learning and knowledge acquisition, and especially through raising one's level of pertinent skills. Finally, we need to note that, as welfare economists emphasise, individuals may obtain different levels of *utility* (satisfaction) from the same level of capabilities.

The definition of capabilities of producers follows analogously. Here we think of firms as the agent responsible for transforming technologies into products. That which is the output for the firm – the product – is of course the input for the consumer. In orthodox economics approaches, all firms need to be at the frontier in terms of their ability to transform technologies into products, otherwise they will be competed out of existence, but plainly this does not happen in reality. In practice, firms – like consumers – confront a range of resources or possibilities, this time in technologies (very broadly defined) rather than products. These are the characteristics of the resources they utilise. Again, they have differing capabilities to effect the production of their products out of these resources and technologies, given by differences in circumstances and differences in abilities. These differences in capabilities underlie, but do not fully determine, the *profitability* (equivalent to utility in the consumer case) they obtain from their production activities, e.g. because other functions are required to get the best out of the technologies. The theory of the growth of the firm as developed by Penrose (1959) and modified by Richardson (1972) (who used the language of capabilities) is consonant with this view of the capabilities of producers.

This tells us something about where the products consumed by the consumers come from, but not where the technologies to produce them come from. To see that, we must move further upstream and look at the supply of the technologies. We can initially suppose they stem from another type of organisation that we label 'laboratories'. In laboratories, people with various scientific and technical competencies are employed to design and develop such technologies. Again, their degree of success in doing so is conditioned by their capabilities; in this case to produce not products but technologies. The profitability of the laboratory is given in part by the extent to which it can *appropriate* the benefits of the technologies it produces, e.g. through

Intellectual Property Rights (IPRs). In practice, much of the twentieth century saw the movement of laboratories 'in house', i.e. within the firm, because of the need for close interactivity between the development of the technologies and the development of the products that used those technologies – and also to overcome IPR problems by internalising them. This altered the organisational location of laboratories but not their function, and often not their geographical location.

Where do the scientists and technicians come from with their competencies? They come out of another institution, which we can label 'universities', producing a certain type of human capital with varying levels of capability. The procedure can be replicated successively – eventually we can build up not just a capability chain but a whole capability network, drawing on the fact that producers (like firms) need a range of inputs, and not just technologies in a narrow sense. Apart from universities and educational systems (for human capital), they need supply chains (for materials and working capital), banks (for physical capital), governments (for public infrastructural capital), and so on. In practice, the main issue for management is aligning all these value networks with the varying capabilities of the agents concerned into a functioning productive system.

This hints at where regional systems come into the picture, in terms of aligning the multiplicity of interactions in a specific regional setting, but there are important complications to the picture that underline any such role. The first problem is that in this complex structure of interrelations, there is no guarantee that the way in which the capabilities of one type of agent (say producers) evolve is in the same direction as those of another type of agent (say consumers). First, what economists refer to as *asymmetries of information* may take them in different directions, i.e. one agent may conceal from another what it does or believes. Second, what we can term *asymmetries of knowledge* may prevent them from going in the same directions, even if they perfectly understand each other.

Thus interactions among types of agents may be necessary:

a) *ex ante*, through information and knowledge exchanges;
b) *ex post*, through market exchanges.

Clearly, the more that is exchanged *ex ante* using exchanges of knowledge, the less the mismatch that markets will have to sort out *ex post*. From a regional or national perspective, asymmetries represent a waste of resources that may be costly to put right.

Among such agents, different actors (e.g. different firms in an industry or different regions in a country) may have differing capabilities for the interaction process itself, in the form of different abilities to learn and to absorb knowledge generated elsewhere. To see this we can make use of a taxonomy of forms of learning, based on empirical work on US data originally carried out by Malerba (1992).

Table 2.1 Internal and external forms of learning for the firm

Source	Internal	External
Production	Learning by doing	Learning from spillovers
Consumption	Learning by using	Learning from interacting
Technology supply	Learning from R&D	Learning from S&T

The rows of this table conform to the tripartite structure of production, consumption and (technology) supply already developed in relation to capabilities – each corresponds to a different form of learning. It should be stressed that while we focus here primarily on the supply of technologies, in practice learning can take place in all functions of the firm and not just in 'technology' in the narrow sense, e.g. in marketing, in organisation or in finance. Second, and particularly important, are the columns that distinguish learning taking place inside the firm (etc.) from learning which is external to it and that will need to be absorbed in various ways. The potential here is evidently vastly greater than what the firm can do internally; in practice the degree of absorption (absorptive capacity) is somewhat limited. Firms can learn from spillovers from other producers, i.e. horizontally, or by interacting with upstream suppliers and downstream users, i.e. vertically, as well as from independent research carried out in the regional or national (or international) science and technology system. The main point is that the scope for interaction is potentially huge, but resources have to be devoted to interaction, from internal or external (e.g. regional) sources, if that potential is to be harnessed. Managing interaction is not a free lunch.

These complex interactions are taking place within an ever-changing environment. This is a context famously described by Schumpeter (1911) as one of 'dynamic competition'. We can distinguish 'high-velocity' industries (Eisenhardt and Martin, 2000) from established industries; in the former (mostly high-tech industries) both technologies and markets are changing at particularly high speeds. We can also distinguish periods of 'fast history' from those of 'slow history' (Klein, 1977), although things are constantly changing even under slow history. In this sense, decisions have to be made in 'real time'.

This leads to our definition of 'dynamic capabilities' (cf. Teece et al., 1997). The extent to which interactive capabilities of producers are dynamic depends upon the extent to which the resources or products are not only attuned to the capabilities of their suppliers or consumers, but are so when the latter capabilities are constantly undergoing change. That is, as the circumstances and abilities of suppliers and consumers alter, which we take to be occurring so frequently it might as well be supposed to be continuous, the producer has to update its own capabilities in order to handle the new resources and new products in 'real time', i.e. within a time frame that is appropriate to that changing environment. The latter time interval, in order to be appropriate, will have to take into account 'time-based competition', i.e. the fact that competitors will be endeavouring to beat the producer in question to the market or the resource, and this of course is Schumpeter's notion.

More formally, 'dynamic capabilities' of firms represent the extent to which the changes in their own capabilities (in production) influence or are influenced by changes in the capabilities of consumers and/or suppliers, in real time. This adds a dimension of causality to the above description; a causality which can flow in both directions, in line with Schumpeterian views. The dynamic capabilities of all other agents – laboratories, universities, banks, governments, etc., as well as consumers – can be formulated in analogous ways.

2.3 REGIONAL CAPABILITIES AND REGIONAL SYSTEMS OF INNOVATION

The same applies to regions, even though these represent a different way of aggregating agents – by geographical space rather than by function. Regions can of course be considered as particular congregations of suppliers, producers, consumers, etc., each with their own unique level of capabilities. While much government policy at national and supranational as well as regional levels has been directed in recent years at promoting clusters, there is no necessary reason why clusters should be progressive – indeed it is all too easy for them to cut themselves off and become pockets of traditional values, especially in eras of 'fast history' which may somewhat pass them by.

For a cluster to be progressive its capabilities need to be interactive, i.e. in tune with those of its suppliers and its customers, and to be dynamic, i.e. interactive in real time in a context where suppliers' and customers' needs and abilities are constantly changing. These are the basic requirements of a 'regional system of innovation' (e.g. Cantwell and Iammarino, 2003), where the emphasis is shifted from mere co-location to co-evolution. However, even the burgeoning literature on regional systems of innovation says comparatively little about how to start up or regenerate such a regional system, i.e. establishing or re-establishing a dynamic cluster. This is the primary perspective of the present chapter.

At one level, starting or restarting a regional system could be conceived of as replacing a set of more or less traditional routines with a revolutionary set of changes; in which case, there is much to be learnt from the experience of past industrial revolutions (von Tunzelmann, 1997). However, that past experience indicates that even revolutions are built in part from patterns that are more evolutionary in nature. In reality, there would seem to be greater chances of success from adopting a strategy of *blending* new sources of strength with existing strengths. Abilities are cumulated rather than acquired through market or other mechanisms; the more a region can harness extant accumulations of capabilities but harness them to new purposes, the more likely it would appear to have a chance of establishing a competitive advantage.

We can trace this potential for blending in regard to products or industries, to processes, and to policies or strategies.

2.3.1 In products

Many regions and countries have taken to heart international benchmarks for success through striving to increase the proportion of high-tech products or industries in their region, as measured via either production or exports. It is common to see this benchmark quoted as a guide for how well or badly the region is doing. As we have argued extensively elsewhere (e.g. von Tunzelmann and Acha, 2005), this can be a very misleading indicator. It is of course the case that Japanese and some other East Asian countries became 'tigers' through raising their exports of high-tech products, but that does not necessarily mean that this is an optimal path for later comers, or indeed for regions seeking regeneration.

There are several reasons for advocating caution on this score. In the first place, the definition of a high-tech industry, which conventionally relies on Organisation for Economic Co-operation and Development (OECD) gradings, is suspect. The boundaries of industries tend to span a range of activities, some of which are indeed high-tech by any reasonable standard, but many of which are not. There are very few (if any) industries that are fully high-tech by a modest range of criteria (Peneder, 2001). The East Asian experience was dominated by entry into low-tech levels of high-tech industries, using undoubtedly high market potential to upgrade their technological capabilities over time, but it is less obvious that it continues to apply outside of a few cases such as China.

The point is that each region must make some (realistic) assessment of its potential capabilities in the face of market opportunities and technological opportunities that are subject to constant change (dynamic competition). We draw a sharp distinction between 'high-tech industries' and 'high-tech activities'. All regions aiming at accelerating growth must consider the potential of high-tech activities for what they do well, but these exist across many industries. Just as there are few true high-tech industries, so nowadays there are few true low-tech industries. Industries such as food processing or textiles, which score low on OECD rankings for technology inputs, have many pockets of high-tech development that we would label as 'high-tech activities'. The issue is to draw on the latter in order to boost existing or growing strengths in areas where it is possible to appropriate adequate returns from these new advances. For example, many regions are internationally identified with a particular food or textile product. The issue then becomes one of blending high-tech activities into these purportedly low-tech products. For regenerating regions, there is clearly a major decision to be made as between reviving sagging industries (which may be of the 'rust-belt' type) and implanting new ones; but in either case the question comes down to one of assessing the region's dynamic capabilities in the light of: (a) the resources it can draw upon (endowments) or is able to enhance; (b) the profitability of the strategy in a context of national and international competition.

2.3.2 In processes

Although often equated with a flood of exports of cheap, technology-based products, the Japanese and East Asian pushes were in fact more deeply based on a growing command of production processes. These originated in well-known forms such as quality control, zero defects or just-in-time manufacture, but came to take hold in a dynamic sense (von Tunzelmann, 1995). Without these dynamic environments, however, simple 'static' imitation by other regions did not prove so effective. Because of the high dependence on context-specific systems (including reward and incentive systems, business-to-business relations, macro climates, etc.) and tacit knowledge, these progressive forms of production processes are probably inherently more difficult to imitate, and therefore to act as a pathway to later comers.

As with products, there is a lowered emphasis on technology for its own sake, and a heightened emphasis on cumulativeness and appropriability. The stress in the literature has shifted to agility and flexibility rather than dedicated high-tech process equipment, and rightly so. Flexibility can encompass traditional as well as advanced processes. However, even flexibility has its limits, and compromises have to be sought to balance the dynamic requirements of being abreast of, or even ahead of, the market, against the existing capabilities at hand, including accumulated tacit knowledge of how to produce; in other words, an adequate level of dynamic capabilities.

This has been abetted by a shift of emphasis from 'static' economies of scale that have to do with larger capacity (e.g. larger plant or equipment sizes), to more 'dynamic' economies of scale. The latter come in two main forms. The first refers to scale economies in real time, namely those that derive from doing things faster. Time-saving changes involve not just greater throughput but less downtime and more systemic operation of equipment, plant and productive systems. Flexibility is a modern expression of some of these values. There is an obvious link between this perception of scale economies in real time and our notion of dynamic capabilities as interactive in real time.

The second form of dynamic scale economies consists of a more historical process of learning in a cumulative fashion over time. Modern growth theory stresses the increasing returns that follow from what Isaac Newton called 'standing on the shoulders of giants', i.e. capitalising on the work of one's predecessors, but this should not be taken to downgrade the amount of additional effort that Newton, or for that matter present-day corporations, have to apply – a point we will revert to in a moment.

2.3.3 In policies and strategies

Firms have to build their strategies on these bases, combining technological, production and market capabilities with other functions such as financial resources. Regions have to consider the interactions among the firms that constitute them. Just as firms have to promote dynamic capabilities in production and suppliers in technology, so policy-makers at a regional or national level have to promote dynamic capabilities in respect of the institutional and economic environment in which the firms can best flourish.

The traditional approach of policy-makers to promoting high-tech activities was through 'mission-oriented' programmes (Ergas, 1987). These drew on the notion of a 'linear model', according to which more expenditures on scientific research fed through into more invention, which when commercialised resulted in a greater diffusion of innovations and ultimately higher rates of economic growth. This linear model has since been supplanted by demand-pull and subsequently 'coupling' models, that reverse or counterbalance the lines of causation. Policy emphasis on letting the market decide has further promulgated 'diffusion-oriented' programmes, in which the main emphasis is placed on dissemination and bottom-up determination of policies, rather than top-down, science-led structures. In 'coupling' approaches, rather than top-down vs. bottom-up, there needed to be a fruitful dialogue between the two.

Growing out of these views about 'blending' technology in high-tech and low-tech industries, 'blending' traditional and advanced processes, and 'blending' mission-oriented and diffusion-oriented policies, concern has been drawn again to the old issue of 'make versus buy'. The regenerating region evidently has to 'buy' high technology, advanced processes and mission science to a large degree, in the sense of acquiring it externally. However the blending process will be very much internally inspired, even if external agents are brought in to try to smooth the process. This raises the issue of the absorptive capacity of the region in question. The seminal studies of absorptive capacity by Cohen and Levinthal (1989, 1990) showed that its extent is highly dependent on the activity carried out by the actor in question. Thus, as they demonstrated, much Research and Development (R&D) carried out by firms is not intended to extend the technological frontier so much as to increase the firm's awareness of the vicinity below the frontier – a vicinity that other firms may be occupying but in which the firm in question is still feeling its way. The same applies to a (regenerating) region, implying that it will need to carry out a fair amount of research just to absorb developments originating in other places. That is, even before the blending process proper begins, the region must commit itself to expenditures that represent 'make' in the domain of technology. The pay-off is that without these expenditures the blending is likely to be much more problematic and perhaps undoable.

The converse, it may be pointed out, applies to an external agent seeking to export its technology (processes, etc.) to the region in question. The multinational corporation (MNC) will have undertaken the necessary expenditures already to develop the technology in its headquarter country. It then faces the reverse type of blending by which it seeks to adapt its accumulated expertise to a new spatial context.

The need for blending carries a range of implications for changes in modes of governance and forms of organisation. Typical instances – though by no means an exhaustive list – include the following:

a) Absorption may require the promotion of greater mobility of the personnel involved, particularly to absorb the more tacit or currently uncodified aspects of the new products, processes, policies, etc. There has been much debate about whether this ought to be done by bringing in experts from outside to guide the process or instead sending one's own experts elsewhere to learn from the experience of others. The issue is basically the same as that just discussed in respect of regions versus MNCs, but in microcosm. The general consensus is that experts from outside will aim to teach the region's firms what to do, whereas sending one's own people to leader regions may be a better way to learn, and the latter is evidently preferable. However intuition suggests that some 'blending' of the two may be desirable. Either way, organisations will have to build such mechanisms into their regular procedures, possibly including formal external linkages.

b) Another likely form of blending may be that in modes of governance, for example as between competition and collaboration. While these are sharply contrasted in most texts on industrial organisation, the evidence at hand indicates that the reality of the capitalist enterprise is one in which these modes intertwine in ways that almost defy analysis. When it comes to questions of the kind we are posing here, of the blending old and new ways of doing things, such collaboration may go beyond the mere 'helping out' to a more creative form of collaboration among firms that will be aiming to succeed in more competitive markets. On the other side, policy-makers may have to promote both competition and collaboration in the strategies – again a delicate balancing act.

c) In order to match the old and the new, and especially where the new may seem somewhat exotic, specialist institutions devoted to bridging may need to be set up. The cognitively complex areas of science on which many of these advances are likely to be based implies that there may be one or more intermediate stages of translation between upstream developments and downstream marketing, for instance in emerging areas such as nanobiotechnology. The kinds of dissemination involved in diffusion-oriented strategies may also necessitate brokerage, as Cooke and Morgan (1994) illustrated for the case of Denmark.

d) Institutional change may have to be carefully handled and phased so as to facilitate the blending process. Existing institutions are likely to be geared toward existing and thus older ways of doing things. These may have to be modified or bypassed. There then emerges another delicate issue in the guise of whether more appropriate institutions can be set up 'ahead of demand'. The experience in many cases of trying to do so has been to create 'orphans' that atrophy from lack of connection in either direction. Again, a blending of institutions, oriented towards the phasing of a broader developmental process, appears warranted.

2.4 NETWORK ALIGNMENT

The blending of differing new streams and also between old and new streams can be carried out through any one of the classic modes of governance – markets, hierarchies (public or private) or networks. Fashion in this respect has changed over the last half century, with hierarchies (both public and private) being the norm accepted from the 1950s to the 1970s, at a time when markets were seen as failing, then markets becoming ascendant – in theory if not in practice – in the 1980s, when conversely governments were seen as failing through being out of touch with commercial reality and crowding out resources for financially responsible investment. Neither position however came to terms with the key criticism of the opposing perspective – that governments did not possess the capabilities to override commercial instincts in the 1950s–70s, but nor did markets suddenly become long-sighted in the 1980s.

Networks therefore came into fashion in the 1990s, somewhat earlier in academic as opposed to governmental circles. Networks of many different kinds were alighted upon as a solution to key developmental problems. They were constituted between pairs or groupings of large firms, between large firms (as hubs) and a constellation of small-firm suppliers or distributors, between small high-tech firms (like the celebrated Silicon Valley phenomenon), or between small non-tech firms (like the 'Third Italy'). They came in many different forms, e.g. formal or informal, strategic alliances or joint ventures, etc.

The networks that are of interest to us are those in which capabilities and knowledge are decentralised, so that even in a hub-type of network, where the central corporation clearly exercises a high degree of power, the surrounding constellation of suppliers and distributors do contribute some knowledge-based input to the functioning of the network. Such networks – and a large number are probably of this kind – are built up from these decentralised capabilities.

Networks however come with problems of their own. Somehow, the differing elements of knowledge, spread across the decentralised units of the network, have to coalesce into a coherent, functioning productive system. This has led many of the hub firms to take on systems integration roles, in the course of which they frequently hollow out many of their pre-existing functions (including sometimes actual production) through outsourcing, etc., in order to coordinate the value network. Management of the network to a degree supersedes management of the corporation, and in view of the decentralisation of power, responsibility and especially knowledge, is often yet more complex. Moreover, the structure of the system depends on one's individual vantage point. It is not uncommon these days to observe industry structures in which there are multiple and overlapping systems integrators, each integrating a system at a different level or in a different guise. What we then see are systems of systems or networks of networks. In such a complex world in organisational terms, network failure takes over from market failure, government failure and corporate failure, as in the older modes of governance, and to a degree comprehends all of them.

To see this, we can depict the structure of networks commonly encountered at multiple levels and in multiple forms.

a) Networks form firstly to amalgamate different *functions*, e.g. technology, production, marketing, finance, management. The traditional corporate hierarchy combines these internally within the individual firm, and even small firms usually carry out all of these functions to a greater or lesser degree. But other forms of organisation split different functions between different enterprises and establishments, and it is becoming increasingly common to outsource one or more of these functions, even from traditional firms.
b) In the second place, networks form to combine different *resources*, particularly what economists describe as the factors of production ('production' here covers all activities, not just manufacturing). This proposition is a very familiar one, except that the diversity of resources and their origins for modern productive activity is often underplayed. As knowledge capital increases relative to physical capital alongside the expansion of the knowledge-based economy, this diversity also increases. The firm or other unit responsible for organising production, in the broad sense we use that term, has somehow to bring them all together. With rising complexity it is not surprising to find delegation of responsibilities in this regard, often to networks.
c) Thirdly, we have the *spatial* networks that arise in regional space, for example international and global networks, national, regional and local networks. What is often nowadays termed multi-level governance tends to address this territorial level, based on varying degrees of propinquity, and often adding the proposition that modern information and communication technologies overcome some of the limitations of having to be nearby (the so-called 'death of distance' argument), so again complexity can be thought of as rising.

The key issue for any mode of governance is taken here as that of alignment. Alignment involves orienting heterogeneous functions, resources and spaces in order to achieve mutually compatible outcomes. Networks have been adopted to obviate some of the problems of market and corporate failures. They are intended to permit a more adaptive balance between specialisation (as in market governance) and integration (as in corporate governance). Networks however raise coordination problems of their own, in view of the very decentralisation they espouse. They are intended to speed up development processes by permitting a degree of specialisation, but often lack adequate management structures for proper integration of the activities in real time.

In order to overcome network failures of these types, the constituents of the network will need interactive and dynamic capabilities of the kinds outlined earlier. Misalignments at the spatial level between regions, countries and localities typically require policy capabilities coming to take on a critical role, to restore alignment.

2.5 POLICY CAPABILITIES

Governments require similar sorts of static, interactive and dynamic capabilities to firms, for the process of producing policies. Policies to engage with market failure have been fostered for many years. However, the context of possible network failure takes administrations into somewhat uncharted waters.

Governments can be thought of as meeting objectives for *regulation*, *development* and *distribution* (McGowan et al., 2004). Trade-offs need to be sought between these targets, and different countries and different regions establish such trade-offs in differing ways. While East Asian governments have tended to privilege development, if need be at the cost of regulation and distribution,[1] Western European governments have given greater emphasis to regulation, while Eastern Europe in the Soviet era stressed distribution.[2] Over time there has perhaps emerged a greater consensus favouring regulatory and distributional objectives that remain compatible with sustained development instead of explicitly or inadvertently obstructing it.

Achieving such a balance between development, regulation and equity involves an appropriate blending of capabilities in the formulation and implementation of policies, and therefore in the policy-making bodies concerned. There does not necessarily have to be a close relationship between the government and business, so long as government is able to establish a context within which business can suitably frame development-oriented strategies (e.g. Kim and von Tunzelmann, 1997, for the case of Taiwan).

At the regional level, the horizontal policies for combining resources and functions have to be associated with a vertical process at the spatial level, allowing policies to mesh smoothly with national and international policies set at higher levels and maybe with local governments at lower levels. The network governance context implies no great master plan to roll everything together, but rather that the loopholes for network failure of the types noted above are adequately addressed. Alignment of policies is a matter of constructing mutually compatible targets for the outputs and inputs involved, and assuring that the instruments to achieve them actually do so.

There are emerging but as yet somewhat limited guidelines for how a lack of joined-up thinking can be overcome at a regional level. It is clear in the first place that regional policy-makers need, like their national counterparts, to operate through the demand side as well as on the supply side. It is hard to see how cautious economic policies can provide the right indirect incentives for risk-taking and new developments. Here conflict with the national level may become serious. Our suggested policy framework of blending, for example in relation to technologies and products, may also be defended as an attempt at a compromise between economic progress and economic precaution. It suggests moderating the excesses of enthusiasm for technological leaps on the supply side as well as promoting an environment for change on the demand side.

In the second place, there is a need to promote interactivity, to ensure that not just the actions of government but those of all actors are compatible. How this is done is a very open question. In the UK, governmental attention has shifted to clusters as the units for integration of systems at a regional level, following the advice of Porter (1990) and others. What the clusters should consist of is often left obscure in this literature. Should they be based on product or process affinities? Should they be based around resources or around institutions? The adoption of instruments to grow clusters hinges on the choices made, and the room for manoeuvre of regional and national governments is a moot point.

In particular, the cluster approach may not help very much in terms of what endowments are needed to be in place and what can be enhanced through the development process itself – in other words which regional capabilities are truly dynamic and will yield dynamic economies of scale, and which are possible dead ends. This brings us back to the call for instituting and promoting dynamic capabilities. Again, our suggestion of blending old and new is an attempt to forge a workable compromise.

More formally, the dynamic capabilities dimension constructs a case for instituting regional systems of innovation (Asheim and Gertler, 2005) rather than merely clusters. From our earlier discussion, the difference lies in the extent of looking ahead to opportunities that are commensurate with the region's supply capacities on the one side but also with the future demands of global markets on the other side. Global competition implies that the only way that many regions might succeed in what are often understood as high-tech industries would be through a 'race to the bottom' in terms of low wages or other factor payments. This in turn almost guarantees that the region concerned will be trapped in the lowest-tech end of the knowledge spectrum, and that even if successful for a time, that time will surely be limited as it is in turn undercut by an emerging region with even lower wages. The outcome will thus probably be rather short-lived and unsustainable, and possibly deleterious for its distributional considerations. The main argument favouring the cultivation of regional systems of innovation on the negative side is to avoid such a low-level trap.

Balancing developmental, regulatory and distributional issues calls for the meshing of many talents in regional government and institutions as well as in the welter of private-sector activities involved in economic activity. A prime function of the former is to align the objectives of these very various types of actors, in particular aligning the public sector with the private sector. This is a tall order, since neither the public sector itself nor the very diverse private sector appears in most cases to be well integrated. Even more problematically, the tools for such public–private alignment – i.e. the instruments – are poorly established in the literature, and still more so in practice. Addressing this deficiency is perhaps the most demanding task confronting regional development in the present day. Moreover, the levels of government relevant to decision-making at the regional level also overlap with national and sometimes international government above and with local governments below.

2.6 CONCLUSIONS

Industrial regeneration at the regional level is seen in this chapter as dependent on the cultivation and enhancement of capabilities and their alignment, within what will almost certainly be a quite heterogeneous regional system. The argument is that capabilities involve orienting 'production' (very broadly defined so as to include the production of services both public and private, etc.) to the demands of users and the supplies of providers. The kinds of capabilities required are not just static, in the sense of comparative advantage, but both interactive and dynamic. The dynamic element involves reaction in real time and over historical time. The chapter suggests ways in which increasing returns might arise in these terms.

The analysis here lead to an argument for a blending of the old and the new, with the old drawing on the leading sources of existing comparative advantage and the new reflecting emerging technological (supply) and market (demand) opportunities. This strategy is much more likely to succeed in most regions than the prevailing tendency to abandon the old and clamber on a bandwagon of what is seen as the new. This is what is distinctive about regeneration.

This diversified 'system' may be promoted through a variety of modes of governance, but the network mode is coming to be omnipresent, for reasons touched on in the chapter. Networks – loosely defined as all modes of governance that do not centrally involve either the exercise of power (hierarchies) or market exchange – themselves exist in diversified forms, variously covering functional matters, resource matters and spatial matters. Organisations and institutions typically arise to forge systems integration, but the boundaries of such integration are likely to be ragged. The wider networks involved at the regional level are likely to be systems of systems, with the proviso that many of these systems will be very loosely constructed.

By definition, the alignment of such loose systems cannot be brought about simply by the exercise of power – indeed, any such power might be to the detriment of balanced development. In practice, it is usually a combination of the more powerful public-sector and private-sector agents that generate some kind of alignment here, but network failure of the kinds detailed above can easily prevail. Regional administrations have to breed the policy capabilities of both interactive and dynamic kinds to cope with this often rather confused context. It has to be said, on a final note of warning, that the instruments currently developed to meet such targets are at present poorly established and even more poorly implemented.

NOTES

1. This is an evident oversimplification. Many scholars have, for instance, traced take-offs in East Asia to distributional measures such as land reform and to a continued low level of inequality (e.g. World Bank, 1993).
2. Similarly, the five-year planning mechanisms aimed at development as well as social equity, though with usually little attention to regulation.

REFERENCES

Abramovitz, M. (1986), 'Catching up, forging ahead, and falling behind', *Journal of Economic History*, **46**, 385–406.

Abramovitz, M. (1994), 'The origins of the postwar catch-up and convergence boom', in J. Fagerberg, B. Verspagen and N. von Tunzelmann (eds), *The Dynamics of Technology, Trade and Growth*, Aldershot. UK and Brookfield, US: Edward Elgar, pp. 21–52.

Asheim, B. and Gertler, M. S. (2005), 'The geography of innovation: regional innovation systems', in J. Fagerberg, D. C. Mowery and R. R. Nelson (eds), *The Oxford Handbook of Innovation*, Oxford: Oxford University Press, pp. 291–317.

Cantwell, J. and Iammarino, S. (2003), *Multinational Corporations and European Regional Systems of Innovation*, London: Routledge.

Cohen, W. M. and Levinthal, D. A. (1989), 'Innovation and learning: the two faces of R&D', *Economic Journal*, 99, 569–96.

Cohen, W. M. and Levinthal, D. A. (1990), 'Absorptive capacity: a new perspective on learning and innovation', *Administrative Science Quarterly*, **35**, 128–51.

Cooke, P. and Morgan, K. (1994), 'The creative milieu: a regional perspective on innovation', in M. Dodgson and R. Rothwell (eds), *The Handbook of Industrial Innovation*, Aldershot. UK and Brookfield, US: Edward Elgar, pp. 25–32.

Dodgson, M. and Bessant, J. (1996), *Effective Innovation Policy*, London: International Thomson Business Press.

Eisenhardt, K. M. and Martin, J. A. (2000), 'Dynamic capabilities: what are they?', *Strategic Management Journal*, **21**, 1105–21.

Ergas, H. (1987), 'The importance of technology policy', in P. Dasgupta and P. Stoneman (eds), *Economic Policy and Technological Performance*, Cambridge: CEPR/Cambridge University Press, pp. 51–96.

Feldman, M. P. (2001), 'The entrepreneurial event revisited: firm formation in a regional context', *Industrial & Corporate Change*, **10**, 861–91.

Foss, N. J. and Knudsen, C. (1995), *Towards a Competence Theory of the Firm*, London: Routledge.

Kim, S.-R. and von Tunzelmann, N. (1997), 'Aligning internal and external networks: Taiwan's specialization in IT', SPRU Electronics Working Paper Series, **17**.

Klein, B. (1977), *Dynamic Economics*, Cambridge MA: Harvard University Press.

Lancaster, K. J. (1966), 'A new approach to consumer theory', *Journal of Political Economy*, **74**, 132–57.

Malerba, F. (1992), 'Learning by firms and incremental technical change', *Economic Journal*, **102**, 845–59.

McGowan, F., Radosevic, S. and von Tunzelmann, N. (eds) (2004), *The Emerging Industrial Structure of the Wider Europe*, London: Routledge.

Peneder, M. (2001), *Entrepreneurial Competition and Industrial Location*, Cheltenham, UK and Northampton, MA, USA: Edward Elgar.

Penrose, E. T. (1959), *The Theory of the Growth of the Firm*, Oxford: Blackwell.

Porter, M. E. (1990), *The Competitive Advantage of Nations*, New York: Free Press.

Richardson, G. B. (1972), 'The organisation of industry', *Economic Journal*, **82**, 88–96.

Schumpeter, J. A. (1911), *The Theory of Economic Development*, transl. R. Opie, Cambridge MA: Harvard University Press, 1934 edn.

Sen, A. K. (1985), *Commodities and Capabilities*, Amsterdam: North-Holland.

Teece, D. J., Pisano, G. and Shuen, A. (1997), 'Dynamic capabilities and strategic management', *Strategic Management Journal*, **18**, 509–33.

von Tunzelmann, G. N. (1995), *Technology and Industrial Progress: the foundations of economic growth*, Aldershot, UK and Brookfield, US: Edward Elgar.

von Tunzelmann, N. (1997), 'Innovation and industrialization: a long-term comparison', *Technological Forecasting and Social Change*, **56**, 1–23.

von Tunzelmann, N. and Acha, V. (2005), 'Innovation in "low-tech" industries', in J. Fagerberg, D. C. Mowery and R. R. Nelson (eds), *The Oxford Handbook of Innovation*, Oxford: Oxford University Press, pp. 407–32.

World Bank (1993), *The East Asian Miracle: economic growth and public policy*, Washington DC: World Bank.

3. Multinational Firms and Technological Innovation: The 'Global versus Local' Challenge

Simona Iammarino, Odile E. M. Janne and Philip McCann

3.1 INTRODUCTION

Regional innovation processes and policies can only be understood with reference to a broad multi-level analysis of the interactions between local, national and global actors. As a consequence, in discussions of the links between globalisation and geography in technological innovation, the common use of simple global versus local dichotomies in terms of the location decisions and strategies of multinational enterprises (MNEs) is often misleading. This chapter provides insights for a clearer understanding of the concepts and processes linking technology, organisational forms and institutions. Rather than being in a global–local dichotomy, MNEs are key drivers of the globalisation of innovation.

The chapter is organised as follows. Section 3.2 reviews the links between the globalisation of innovation and MNEs. The global versus local challenges for MNEs and regions are discussed in Section 3.3. Section 3.4 discusses the research challenges of the co-evolution of global and local issues. Section 3.5 highlights the challenges of policy making for mature regions in that context. Conclusions and directions for future research are summarised in the last section.

3.2 MULTINATIONAL ENTERPRISES AND THE GLOBALISATION OF INNOVATION

The concept of the globalisation of innovation is the link between the increased international integration of economic activities and the increasing importance of knowledge in socio-economic processes (Archibugi and Iammarino, 1999). Yet, the concept of the *globalisation of innovation* has been interpreted in many different ways causing some confusion and inconsistency of its use as an analytical construct. Attempts have been made (Archibugi and Michie, 1995) to clarify these issues by identifying three main features of the concept which are seen as being complementary rather than mutually exclusive of each other. These are: the international exploitation of technology which is produced on a national basis; the global generation of innovations; and global technological collaborations. The first feature category reflects innovators' efforts to gain economic advantages through the exploitation of their own technological competencies in non-domestic markets. The second category accounts for innovations conceived on a global scale from the moment they are generated, i.e. those created by MNEs. The third category encompasses national and international agreements, involving both private and public actors, for the joint development of specific technological discoveries. The categories are strongly interdependent and they encompass both competitive and collaborative arrangements (Archibugi and Iammarino, 1999).

Of all types of firms, MNEs play the most important role in terms of promoting the global integration of economic activities, the creation of new technical knowledge and the integration of different socio-economic regimes (Giddens, 1990; Cantwell and Iammarino, 1998). Yet, the links between multinationality, economic integration and innovation have been interpreted in different ways. One interpretation views innovation as primarily a function of research and development (R&D). R&D functions gain in importance as technological progress becomes more complex, and higher levels of skills and organisational expertise are associated with larger stocks of knowledge. Therefore, MNEs tend to be more research-intensive than other domestically-oriented firms (Iammarino et al., 2007). Yet, within this perspective, two approaches can be distinguished. On the one hand, there is the industrial organisation view, where, on the basis of the linear model of technological processes, investment in R&D leads directly to cost reductions, higher product quality, increased corporate competitiveness and larger international markets (e.g. Dunning, 1970; Markusen, 1984). On the other hand, transaction cost theories of the firm propose that R&D activities generate more intensive knowledge flows and a greater complexity in transactions which, in turn, lead to a greater degree of vertical integration, industrial concentration, and the expansion of firms across national boundaries (e.g. Buckley and Casson, 1976; Hennart, 1977; Rugman, 1981).

Another interpretation emphasises the two-way relationship between multinational expansion and innovation. Following also the developments of evolutionary theories of the firm (Nelson and Winter, 1977, 1982), the core of the analysis has shifted from a uni-directional link from innovation to internationalisation, to a multi-way view of the relationship between these phenomena. These arguments are based on Schumpeterian approaches such as the resource- and competence-based theories of firm growth, in which high R&D intensity and internationalisation can be viewed as complements (e.g. Cantwell, 1989, 1992; Patel and Pavitt, 1991; Kuemmerle, 1999; Petit and Sanna-Randaccio, 2000; Castellani and Zanfei, 2002; Frenz and Ietto-Gillies, 2007). In this view, technological competence provides firms with capabilities for expansion through continued learning in production, while more learning generates greater competence, hence rising market shares. As such, MNEs are increasingly regarded as evolving organisations which interact with socio-economic contexts in both the home and host locations (Teece, 1977; Dunning, 2000; Janne, 2002a&b; Frenz and Ietto-Gillies, 2007). These interactions are of different kinds, and the evolution of the MNE organisational features is central to explaining the multi-way relationship between MNE expansion and global innovation processes. On the one hand, the technology characteristics of the home environment are argued to play a key role in determining an MNE's capability (Cantwell and Janne, 1999; Frost, 2001; Le Bas and Siera, 2002; Noorderhaven and Harzing, 2003). On the other hand, in terms of host locations, those environments which generate local spillovers are an important pull factor. Therefore, as well as its technological and organisational competencies, the innovative activity of an MNE also depends on the geography of knowledge (Pearce and Papanastasiou, 1999; Cooke, 2001; Cantwell and Piscitello, 2002; Cantwell and Iammarino, 2001, 2002; Janne, 2002a&b; Sanna-Randaccio, 2002; Simmie, 2003; Iammarino, 2005) and this issue itself has various features to it.

Recently, attention has shifted from viewing the MNE primarily as a vehicle of technology transfer towards the concept of the 'network firm' (Bartlett and Ghoshal, 1989; Bartlett and Ghoshal, 1990; Hedlund, 1994), which exploits both internal and external networks for promoting innovation. One major aspect of this is that MNEs play a crucial role in the cross-border creation and integration of knowledge (e.g. Chesnais, 1988; Pearce, 1989; Cantwell, 1989, 1992, 1994; Granstrand et al., 1992; Birkinshaw, 1996; Fors, 1998; Niosi, 1999). They play such roles by establishing integrated networks of affiliates (Malmberg et al., 1996; Zanfei, 2000; Frost, 2001; Ietto-Gillies, 2001; Veugelers and Cassiman, 2004; Narula and Zanfei, 2005). MNEs may adopt strategies of technological differentiation and product diversification across geographical locations (Cantwell and Janne, 1999), which lead to complex networks of intra-firm cross-country specialisations in innovative activity (Cantwell and Iammarino, 2003). As well as intra-firm networks there are also inter-firm networks and broader interactions between affiliates and their external environment (Almeida, 1996; Almeida and Kogut, 1999). Dunning and Wymbs (1999) demonstrate that the degree of multinationality is associated with the perception that global technological advantage has foreign sources which can be tapped into. Hence, this emphasis on the globalisation of innovation has resulted in a questioning of the rather narrow role conferred upon foreign subsidiaries by Vernon (1966), with attention now focused on the important position assumed by subsidiaries in MNE innovation activities (Birkinshaw and Hood, 1998; Birkinshaw, 2000; Frost et al., 2002; Pearce, 1999).

3.3 THE 'GLOBAL VERSUS LOCAL' CHALLENGE IN THE CREATION OF INNOVATION

The global versus local challenge in the creation of innovation is often expressed from the firm perspective in terms of the need to 'globalise' or 'localise' innovative activities, or combine the two within the MNE. However, in terms of innovation, as we have seen, the interaction between the MNE and its locations is more complex than this (Dicken, 1994; Buckley and Ghauri, 2004). Technological diversity in geographically dispersed locations generates a pressure on the MNE to organise innovative activities into globally integrated networks, while the effects of this on the individual region are complex and ambiguous (Janne, 2002b).

3.3.1 The Firm's Challenge

MNEs are perceived to have moved from the 'local' model of individual market servicing towards a 'global' and network model of strategies for sourcing specific skills and knowledge (Harzing, 1998). In this context, conventional assumptions regarding the global versus local dichotomy can be misleading, as the 'global' will not always be represented by centralisation and headquarters, and the 'local' will not always be by decentralisation and subsidiaries (Buckley and Carter, 2002; Noorderhaven and Harzing, 2003). Network-based organisations entail complex sets of global and local ('glocal') interactions and strategies (Buckley and Carter 2002). Hence, subsidiaries may represent 'global' nodes of corporate knowledge (Zander, 1999; Cantwell and Mudambi, 2005), while the home country headquarters may be regarded as representing the 'local' dimension (Cantwell and Janne, 1999; Noorderhaven and Harzing, 2003). As such, these arguments largely invert the traditional understanding. However, not all subsidiaries will be involved in an MNE's innovative efforts, as it will depend on the extent to which those subsidiaries are themselves engaged in knowledge creation and on their relationships with the host environment (Almeida and Kogut, 1997; Janne, 2002a). A deeper understanding of this requires a shift of analytical focus from the broad industry level towards more finely defined technologies (Cantwell and Iammarino, 2000; Janne, 2002a&b; von Tunzelmann and Acha, 2005), functional activities in the value-chain of MNEs (Ernst, 2002; Ernst and Kim, 2002) or product quality differentials (Storper et al., 2001). For analysing the innovation of MNEs, simple distinctions between high-/low-tech, new/mature and sunrise/sunset industries are not very useful because they underestimate the range and complexity of corporate innovation processes.

3.3.2 The Region's Challenge

Evolutionary and systemic approaches to technological change have pointed out that while what happens inside the firm is indeed fundamental, the immediate socio-economic environment around the firm also plays a critical role in a firm's innovation (Foray and Freeman, 1993; Cantwell and Janne, 2000; Cantwell and Iammarino, 2003). According to this argument, while globalisation is driven by the search of MNEs for exploitable knowledge in different locations, localised knowledge is also a critical determinant of firm's behaviour. The region or cluster is seen as being very important in the globalisation process rather than simply playing a largely passive role as in early MNE theories (Cooke, 2005). The analysis of local advantages for innovation (Storper and Venables, 2004) has focused on the interaction between innovation and agglomeration economies. Various typologies linking innovation to spatial configurations have been proposed (McCann and Mudambi, 2005; Iammarino and McCann, 2006; Paniccia, in Chapter 11 of this book). These show that simple core-periphery discussions, or constructs based on simple analytical dichotomies, are largely fruitless. For example, the idea that codified knowledge represents the global dimension of innovation while tacit knowledge represents only the local dimension of innovation is not correct (Breschi and Lissoni, 2001; Morgan, 2004). Therefore cluster-policy frameworks linked to innovation and MNEs must be constructed on more sophisticated and nuanced understandings (Asheim and Coenen, 2005; Cooke and Leydesdorff, 2006; Moodysson et al., 2006). For clusters specialised in an industry mainly drawing on an analytical knowledge base (e.g. science-based), such as biotechnology and genetics, innovation policy should focus on the knowledge infrastructure, university–industry interactions, spin-off firms and global networking. For clusters specialised in an industry which mainly draw on a synthetic knowledge base (e.g. engineering-based), such as automobile manufacturing and industrial machinery, innovation policies should focus primarily on inter-firm localised learning, and knowledge transfers mediated via the user–producer interactions of existing firms.

3.4 GLOBAL LOCAL CO-EVOLUTION: THE CHALLENGES FOR RESEARCH

In the 'global versus local' debate we need a much richer understanding of the links between geography, MNEs and innovation that the simple dichotomy implies. The use of simple concepts and labels such as 'clusters', 'networks', 'centres of excellence' and 'knowledge-based economy', is problematic because, while setting the policy agenda, they provide limited theoretical underpinnings and ambiguous explanatory and predictive power (Gordon and McCann, 2000; Martin and Sunley, 2003; Godin, 2006). In contrast, many insights are being gained from a growing interdisciplinary literature. This has also resulted in the elaboration of typologies and hierarchies as analytical tools for guiding further research. In particular, the analysis of the firm's relationship with its environment for innovation has resulted in the generation of multidisciplinary terminologies converging around a knowledge-based approach to the firm and the region.

As a way of opening the 'black box' of complex interactions between innovation, clusters and MNEs, taxonomies based on multidisciplinary approaches have been developed. These taxonomies provide a useful tool to better understand how geographical characteristics relate to aspects of the behaviour of firms (including MNEs), and vice versa. They offer analytical frameworks and arguments for regional policy to tackle the specific activities that are involved in localised innovation processes. A recent taxonomy combines transaction costs-based and evolutionary knowledge-based classifications of industrial clusters to explain their nature and evolution and help uncover the specific innovation processes underlying particular industrial or regional configurations (Iammarino and McCann, 2006). From a transaction costs perspective, MNEs in traditional (and mature) oligopolistic industries are more likely to fit (and invest) in clusters of the 'industrial complex' type that minimise the possibility of unintentional spillovers (Gordon and McCann, 2000; McCann and Mudambi, 2004, 2005). However, firms have different means of avoiding rivalry by using barriers to entry and controlling for pure spillovers. This may be achieved by contracting for knowledge, developing alliances (Narula and Santangelo, 2007) or choosing different modes of entry into clusters (Lorenzen and Mahnke, 2002). Although originally developed from a somewhat different perspective, the industry-innovation distinction employed in this innovation-geography taxonomy (Iammarino and McCann, 2006) closely resembles that of the Pavitt's original industry-innovation taxonomy (1984).

A different approach to developing taxonomies has focused on the types of specific benefits geographical proximity confers to firms for innovation and competitiveness. A classification of the knowledge base of firms and industries has been used to explain the role, type and relative importance of clusters and regional innovation systems (Asheim and Coenen, 2005; Asheim et al., 2007). One implication is that each industry generates and commercialises a different combination of analytical, synthetic and symbolic knowledge, each of which requires a specific balance of local and global networks for knowledge creation and innovative activities (Moodysson et al., 2006). Malmberg and Maskell (2006) have classified the benefits of spatial proximity between interacting parties into vertical, horizontal and social dimensions of localised learning, as well as distinguished their social and institutional underpinning.

As well as these taxonomies, there are various other examples of this inter-disciplinary type of thinking. For instance, Tallman et al. (2004) use notions from the theory of organisational knowledge in strategic management to explain stocks and flows of knowledge in industrial clusters and their role in understanding cluster-based competitive advantage. Their model helps to explain mechanisms linking territory and firms' interactions and proposes that cluster-specific architectural knowledge, as opposed to component knowledge (Henderson and Clark, 1990), is the key to local sustained competitive advantage. The nature of architectural knowledge of clusters is explained as having more the character of 'knowing' than 'knowledge', as firms need to be involved and collectively knowledgeable of the architecture to be part of it, a theme that is also echoed in the discussion on the nature of knowledge in the theory of the firm and MNE (Barnard and Cantwell, 2006).

Meanwhile, an alternative approach is that of Cooke (2005) who uses concepts related to firms' dynamic capabilities (Penrose, 1959; Teece et al., 1997) and regional innovation systems in order to develop a theory of economic geography based on regional knowledge capabilities. This approach emphasises the pivotal function of network nodes in the multi-level nature of innovation processes. Finally, another interesting example can be found in the notion of national innovative capacity developed by Furman et al. (2002) which combines insights from endogenous growth theory (from the field of economics), the concepts of national innovation system (from innovation and evolutionary economics) and national competitive advantage and clusters (from strategic management) to explain a country's potential for innovation.

While each of these lines of interdisciplinary research offers directions for analysis, there still remains much to be done. In particular, while there has been a shift of analytical emphasis away from (power) hierarchy models of MNE activities across time and space to network models, inequality still prevails. The deepening process of the international division of labour is more complex than initially depicted in the simple core-periphery model, and some concepts such as that of 'concentrated dispersion' (Ernst, 2002; Ernst and Kim, 2002) implicitly entail the idea of hierarchies of locations. As a consequence, more refined and sophisticated locational rankings of interactions between MNEs and locations are required (Cantwell and Janne, 1999, 2000; Cantwell and Iammarino, 2001, 2003; Janne, 2002a&b). Future research may look at a more finely grained definition and explanatory power of such hierarchies which allow for the role of real geographical features (e.g. climate, coastline, soil quality), migration and demographic changes (Buckley and Ghauri, 2004). Another area of potential research development is regarding whether different types of clusters or regions match the characteristics of different hierarchical levels and also phases in the product life cycles. In this context, mature industrial clusters might, for example, be associated more closely with intermediate locations, 'technological' industrial districts with high order locations, and urban regions in the periphery with lower order locations. In this context, both qualitative (Jones and Khanna, 2006) and quantitative (Fortanier and Van Tulder, 2007) work focusing on business and economic history becomes essential (Cantwell and Janne, 1999, 2000).

3.5 POLICY ISSUES

As we have seen here, there are now a wide range of theoretical arguments concerning the links between innovation, economic geography, globalisation and MNEs. However, linking these various theoretical strands to policy frameworks is complicated, and this is particularly the case for mature regions which are attempting to market themselves as attractive locations for MNEs and inward foreign direct investment (FDI). This is because any attempt at policy making for regional industrial transformation and innovation promotion in mature regions involves dealing with issues relating to technology. Moreover, as we have seen in this chapter, these technological issues operate simultaneously on different levels, namely at the local, inter-regional, national and international levels. Therefore, in order to be effective the scale and scope of regional policies need to take into account complex multi-level industrial and technological dimensions. On the basis of the arguments presented here, in mature regions, traditional approaches to regional policies are likely to be ineffective in supporting innovation and regeneration if they build solely on generic and predefined characteristics of mature clusters. Simple core-periphery modes or product cycle arguments are likely to be insufficient templates for appropriate policy. This is also true for policies which build simply on the application of off-the-shelf concepts such as clusters, or focus on fashionable industry such as high technology, if such policies are undertaken without a great deal of careful analysis. Our discussion here provides some insights as to how we might think about appropriate regional policies in mature regions. More effective policy intervention for regional revival and growth is likely to build on a better understanding of the industrial and technological transformation which underlies a region's capabilities. Firstly, there is a need to focus on the specific technological evolution and skills profile of the region's industrial clusters. On the basis of this evaluation, in the case of mature regions in particular, it is then also necessary to explore the extent to which existing technologies, skills and competencies can be used in new ways and in different sectors so as to foster innovation and growth. The notion of knowledge capabilities and technological trajectories implies that regional policy needs to find an appropriate way of linking the region's past and inherited skills base to its potential future opportunities.

3.6 CONCLUSIONS

In the beginning of the chapter we outlined a tripartite categorisation of the features of the globalisation of innovation. In the rest of the chapter we focused primarily on only one particular category, namely that of the global generation of innovation. The other two categories, namely the international exploitation of technology nationally produced on world markets and the global technological collaborations, were not discussed in detail here. Yet, as we pointed out, these three features of innovation globalisation are not mutually exclusive of each other. Therefore, there are likely to be additional interactions between each of these features, which have not been discussed here. Given the complexities of the arguments regarding just one of these features of innovation globalisation, it is likely that there are additional significant challenges to be addressed regarding our understanding of the co-evolution of global actors and local economies. Future research therefore needs to take into account all of the three features of the globalisation of innovation.

The preceding discussion, however, has already provided some clues as to these likely interdependencies. The majority of theories now suggest that corporate dynamic capabilities (Teece et al., 1997) may be associated with the openness of firms to their external knowledge environment (Fontana et al., 2006) as well as with their internal knowledge-generating capabilities. Moreover, as we have seen here, the internal and external knowledge capabilities of the MNE are regarded as being increasingly interrelated, as manifested in new network forms of global organisation. However, there is still only rather limited and fragmented empirical evidence available regarding these issues. In particular, there is a serious lack of suitable large empirical surveys on these issues, relating specifically to MNEs and regions. The limited data available suggest that there are complex MNE patterns of spatial differentiation and inequality, which produce spatial hierarchies of activities. These insights represent a richer and more nuanced picture than the simple core-periphery model of MNE location and growth would provide. Moreover, these unequal spatial distributions are likely to be constructed with respect to particular regions. Not only do MNEs appear to have an unevenly distributed geographical dispersion of activities in each of the main regional blocks of the European Union, North America and Asia-Pacific, but that the relationship between these three regions and the rest of the world is unequal. As such, an increased focus on the regional strategies amongst this triad of macro-regions may be more appropriate than focusing on global strategies (Rugman, 2005). Lastly, the complexities of the links between MNEs and regions calls for more interdisciplinary research on theoretical concepts and taxonomies which better allow for the design of appropriate regional policies in mature regions.

REFERENCES

Almeida, P. (1996), 'Knowledge sourcing by foreign multinationals: Patent citation analysis in the US semiconductor industry', *Strategic Management Journal*, **17**, winter special issue, 155–65.

Almeida, P. and Kogut, B. (1997), 'The exploration of technological diversity and the geographic localization of innovation', *Small Business Economics*, **9**, 21–31.

Almeida, P. and Kogut, B. (1999), 'Localization of knowledge and the mobility of engineers in regional networks', *Management Science*, **45** (7), 905–15.

Archibugi, D. and Michie, J. (1995), 'The globalisation of technology: A new taxonomy', *Cambridge Journal of Economics*, **19** (1), 121–40.

Archibugi, D. and Iammarino, S. (1999), 'The policy implications of the globalisation of technology', *Research Policy*, **28** (2–3), 317–36.

Asheim, B. and Coenen, L. (2005), 'Knowledge bases and regional innovation systems: Comparing Nordic clusters', *Research Policy*, **34**, 1173–90.

Asheim, B., Coenen, L., Moodysson, J. and Vang, J. (2007), 'Constructing knowledge-based regional advantage: Implications for regional innovation policy',

International Journal of Entrepreneurship and Innovation Management, **7** (2-5), 140–55.
Barnard, H. and Cantwell, J. (2006), 'Knowledge in the theory of the firm and MNC: Asset or action? A commentary on Foss N. "Knowledge and organisation in the theory of the MNC"', *Journal of Management Governance*, **10**, 21–7.
Bartlett, C. A. and Ghoshal, S. (1989), *Managing Across Borders: The Transnational Solution,* Boston: Harvard Business School Press.
Bartlet, C.A. and Ghoshal, S. (1990), 'Managing innovation in transnational corporations', in C.A. Bartlett, Y. Doz and G. Hedlund (eds), *Managing the Global Firm*, London: Routledge.
Birkinshaw, J. (1996), 'How multinational susbsidiaries mandates are gained and lost', *Journal of International Business Studies*, **27** (3), 467–96.
Birkinshaw, J. (2000), 'Strategy and Management in MNE Subsidiaries', in T. Brewer and A. Rugman (eds), *The Oxford Handbook of International Business*, Oxford: Oxford University Press.
Birkinshaw, J. and Hood, N. (1998), 'Multinational subsidiary evolution: Capability and charter change in foreign-owned subsidiary companies', *Academy of Management Review*, **23** (4), 773–95.
Birkinshaw, J. and Hood, N. (2000), 'Characteristics of foreign subsidiaries in industrial clusters', *Journal of International Business Studies*, **31** (1), 141–54.
Breschi, S. and Lissoni, F. (2001), 'Localised knowledge spillovers versus innovative milieux: Knowledge 'tacitness' reconsidered', *Papers in Regional Science*, **80**, 255–73.
Buckley, P. J. and Carter, M. J. (2002), 'Process and structure in knowledge management practices of British and US multinational enterprises', *Journal of International Management*, **8**, 29–48.
Buckley, P. J. and Casson, M. C. (1976), *The Future of Multinational Enterprise*, London: Macmillan.
Buckley, P. J. and Ghauri, P. N. (2004), 'Globalisation, economic geography and the strategy of multinational enterprises', *Journal of International Business Studies*, **35**, 81–98.
Cantwell, J. A. (1989), *Technological Innovation and Multinational Corporations*, Oxford: Basil Blackwell.
Cantwell, J. A. (1992), 'The Internationalisation of Technological Activity and its Iimplications for Competitiveness', in O. Granstrand, L. Håkanson and S. Sjölander (eds), *Technology Management and International Business: Internationalisation of R&D and Technology*, Chichester: John Wiley.
Cantwell, J. A. (ed.) (1994), *Transnational Corporations and Innovatory Activities*, London: Routledge.
Cantwell, J. A. and Iammarino, S. (1998), 'MNCs, technological innovation and regional systems in the EU: Some evidence in the Italian case', *International Journal of the Economics of Business*, Special Issue, **5** (3), 383–408.
Cantwell, J. A. and Iammarino, S. (2000), 'Multinational Corporations and the Location of Technological Innovation in the UK Regions', *Regional Studies*, **34** (3), 317–22.
Cantwell, J. A. and Iammarino, S. (2001), 'EU regions and multinational corporations: Change, stability and strengthening of technological comparative advantages', *Industrial and Corporate Change*, Special Issue on Geography of Innovation and Economic Clustering, **10** (4), 1007–37.
Cantwell, J. A. and Iammarino, S. (2002), 'The Technological Relationships between Indigenous Firms and Foreign-owned Multinational Corporations in the European

Regions', in P. McCann (ed.), *Industrial Location Economics*, Cheltenham, UK and Northampton, MA, USA: Edward Elgar.

Cantwell, J. A. and Iammarino, S. (2003), *Multinational Corporations and European Regional Systems of Innovation*, London and New York: Routledge.

Cantwell, J. A. and Janne, O. E. M. (1999), 'Technological globalisation and innovative centres: The role of corporate technological leadership and locational hierarchy', *Research Policy*, **28**, 119–44.

Cantwell, J. A. and Janne, O. E. M. (2000), 'Globalisation of Innovatory Capacity: the Structure of Competence Accumulation in European Home and Host Countries', in F. Chesnais, G. Ietto-Gillies and R. Simonetti (eds), *European Integration and Global Corporate strategies*, London: Routledge.

Cantwell, J. A. and Mudambi, R. (2005), 'MNE competence-creating subsidiary mandates', *Strategic Management Journal*, **26**, 1109–28.

Cantwell, J. A. and Piscitello, L. (2002), 'The location of technological activities of MNCs in European regions: The role of spillovers and local competencies', *Journal of International Management*, **8**, 69–96.

Cantwell, J. A. (1989), *Technological Innovation and Multinational Corporations*, Oxford: Basil Blackwell.

Castellani, D. and Zanfei, A. (2002), 'Multinational experience and the creation of linkages with local firms: Evidence from the electronic industry', *Cambridge Journal of Economics*, **26** (1), 1–25.

Chesnais, F. (1988), 'Technical co-operation agreements between firms', *Science Technology Industry Review*, **4**, 57–119.

Cooke, P. (2001), 'Regional innovation systems, clusters, and the knowledge economy', *Industrial and Corporate Change*, **10** (4), 945–74.

Cooke, P. (2005), 'Regionally asymmetric knowledge capabilities and open innovation. Exploring 'Globalisation 2' – A new model of industry organisation', *Research Policy*, **34**, 1128–49.

Cooke, P. and Leydesdorff, L. (2006), 'Regional development in the knowledge-based economy: The construction of advantage', *Journal of Technology Transfer*, **31**, 5–15.

Dicken, P. (1994), 'The Roepte lecture in economic geography. Global-local tensions: firms and states in the global space-economy', *Economic Geography*, **70** (2), 101–28.

Dunning, J. H. (1970), *Studies in International Investment*, London: Allen & Unwin.

Dunning, J. H., (2000) (ed.), *Regions, Globalization and the Knowledge-Based Economy*, Oxford: Oxford University Press.

Dunning, J. H. and Wymbs, C. (1999), 'The Geographical Sourcing of Technology Based Assets by Multinational Enterprises', in D. Archibugi, J. Howells and J. Michie (eds), *Innovation Policy in a Global Economy*, Cambridge: Cambridge University Press.

Ernst, D. (2002), 'Global production networks and the changing geography of innovation systems. Implications for developing countries', *Economics of Innovation and New Technology*, **11** (6), 497–523.

Ernst, D. and Kim, L. (2002), 'Global production networks, knowledge diffusion, and local capability formation', *Research Policy*, **31**, 1417–29.

Fontana, R.; Geuna, A. and Matt, M. (2006), 'Factors affecting university–industry R&D projects: The importance of searching, screening and signalling', *Research Policy,* **35**, 309–23

Foray, D. and Freeman, C. (eds) (1993), *Technology and the Wealth of Nations*, London: Pinter.

Fors, G. (1998), 'Locating R&D Abroad: The Role of Adaptation and Knowledge-seeking', in P. Braunerhjelm and K. Ekholm (eds), *The Geography of Multinational Firms*, Boston, Dordrecht and London: Kluwer Academic Publishers.

Fortanier, F. and Van Tulder, R. (2007), 'Internationalization Trajectories of Multinational Enterprises: 1990-2004', paper presented at the conference 'Four decades of International Business at Reading: Looking to the future', 16-17 April, Reading.

Frenz, M. and Ietto-Gillies, G. (2007), 'The Impact of multinationality on innovation. An analysis of the UK Community Innovation Survey', *International Review of Applied Economics*, **21** (1), 99–117.

Frost, T. S. (2001), 'The geographic sources of foreign subsidiaries' innovation', *Strategic Management Journal*, **22**, 101–23.

Frost, T. S., Birkinshaw, J. M and Ensign, P. C. (2002), 'Centers of excellence in multinatinational corporations', *Strategic Management Journal*, **23**, 997–1018.

Furman, J. L., Porter, M. E. and Stern, S. (2002), 'The determinants of national innovative capacity', *Research Policy*, **31**, 899–933.

Giddens, A. (1990), *Consequences of Modernity*, Cambridge: Polity Press.

Godin, B. (2006), 'The linear model of innovation. The historical contruction of an analytical framework', *Science, Technology & Human Values*, **31** (6), 639-67.

Gordon, I. R. and McCann, P. (2000), 'Industrial clusters: Complexes, agglomeration and/or social networks?', *Urban Studies*, **37** (3), 513–32.

Granstrand, O., Håkanson, L. and Sjölander, S. (eds) (1992), *Technology Management and International Business. Internationalization of R&D and Technology*, Chichester: Wiley.

Harzing, A.-W. (1998), 'Integration within MNCs: From Multidomestic to Global and Transnational firms?', in H. Mirza (ed.), *Global Competition Strategies in the New World Economy: Multilateralism, Regionalisation and the Transnational Firm*, Cheltenham, UK and Northampton, MA, USA: Edward Elgar.

Hedlund, G. (1994), 'A model of knowledge management and the N-form corporation', *Strategic Management Journal*, Special Issue: Strategy: Search for New Paradigms, **15**, 73–90.

Henderson, R. and Clark, K. B. (1990), 'Architectural Innovation: The Reconfiguration of Existing Product Technologies and the Failure of Established Firms', *Administrative Science Quarterly*, **35**, 9–30.

Hennart, J.-F. (1977), 'A Theory of Foreign Direct Investment', University of Maryland Ph.D. dissertation.

Iammarino, S. (2005), 'An evolutionary integrated view of regional systems of innovation. Concepts, measures and historical perspectives', *European Planning Studies*, **13** (4), 495–517.

Iammarino, S. and McCann, P. (2006), 'The structure and evolution of industrial clusters: Transactions, technology and knowledge spillovers', *Research Policy*, **35**, 1018–36.

Iammarino, S., Sanna-Randaccio, F. and Savona, M. (2007), 'The perception of obstacles to innovation. Multinational and domestic firms in Italy', Working Paper Bureau d'économie théorique and appliquée (BETA) no. 2007-12.

Ietto-Gillies, G. (2001), *Transnational Corporations. Fragmentation Amidst Integration*, London: Routledge.

Janne, O. E. M. (2002a), 'The emergence of corporate integrated innovation systems across regions. The case of the chemical and pharmaceutical industry in Germany, the UK and Belgium', *Journal of International Management*, **8**, 1–23.

Janne, O. E. M. (2002b), 'Agglomeration and Comparative Advantage in Innovative Activities: The Impact of Multinationals in Europe', conference paper presented at the 28th European International Business Academy (EIBA) annual conference, Athens, 8–10 December.

Jones, G. and Khanna, T. (2006), 'Bringing history (back) into international business', *Journal of International Business Studies*, **37** (4), 453–88.

Kuemmerle, W. (1999), 'The drivers of foreign direct investment into research and development: An empirical investigation', *Journal of International Business Studies*, **30** (1), 1–24.

Le Bas, C. and Siera, C. (2002), 'Location versus home country advantages in R&D activities: Some further results on multinationals' locational strategies', *Research Policy*, **31**, 589-609.

Lorenzen, M. and Mahnke, V. (2002), 'Global Strategy and the Acquisition of Local Knowledge: How MNCs enter Regional Knowledge Clusters', paper presented at the 2002 DRUID summer conference, June.

Malmberg, A. and Maskell, P. (2006), 'Localized learning revisited', *Growth & Change*, **37** (1), 1–18.

Malmberg, A., Sölvell, O. and Zander, I. (1996), 'Spatial clustering, local accumulation of knowledge and firm competitiveness', *Geografiska Annaler*, **78B** (2), 85–97.

Markusen, J. R. (1984), 'Multinationals, multi-plant economies and the gains from trade', *Journal of International Economics*, **16**, 205–26.

Martin, R. and Sunley, P. (2003), 'Deconstructing clusters: chaotic concept or policy panacea?', *Journal of Economic Geography*, **3**, 5–35.

McCann, P. and Mudambi, R. (2004), 'The location behaviour of the multinational enterprise: Some analytical issues', *Growth and Change*, **35** (4), 491–524.

McCann, P. and Mudambi, R. (2005), 'Analytical differences in the economics of geography: the case of the multinational firm', *Environment and Planning A*, **37**, 1857–76.

Moodysson, J., Coenen, L. and Asheim, B. (2006), 'Explaining Socio-Spatial Patterns of Innovation: Analytical and Synthetic Modes of Knowledge Creation in The Life Science Industry', paper presented at the fifth Proximity Congress, Bordeaux, France, June 28-30.

Morgan, K. (2004), 'The exaggerated death of geography: Learning, proximity and territorial innovation systems', *Journal of Economic Geography*, **4**, 3–21.

Narula, R. and Santangelo, G. D. (2007), 'Location and R&D alliances in the European ICT industry', UNU-MERIT WORKING Paper Series, 2007–08.

Narula, R. and Zanfei, A. (2005), 'Globalisation of Innovation: The Role of Multinational Enterprises', in J. Fagerberg, D. Mowery and R. R. Nelson (eds).

Nelson, R. R. and Winter, S. G. (1977), 'In search of useful theory of innovation', *Research Policy*, **6** (1), 36–76.

Nelson, R. R. and Winter, S. G. (1982), *An Evolutionary Theory of Economic Change*, Cambridge MA: Harvard University Press.

Niosi, J., (1999), 'The internationalization of industrial R&D: From technology transfer to the learning organization', *Research Policy*, **28** (2–3), 107–118.

Noorderhaven, N. G. and Harzing, A.-W. (2003), 'The "country-of-origin effect" in mulitinational corporations: Sources, mechanisms and moderating conditions', *Management International Review*, **43** (Special Issue 2), 47–66.

Patel, P. and Pavitt, K. L. R. (1991), 'Large firms in the production of the world's technology: an important case of non-globalisation', *Journal of International Business Studies*, **22** (1), 1–21.

Pavitt, K. L. R. (1984), 'Sectoral patterns of technological change: towards a taxomnomy and theory', *Research Policy*, **13**, 343–74.

Pearce, R. D. (1989), *The Internationalisation of Research and Development by Multinational Enterprises*, London: Macmillan.

Pearce, R. D. (1999), 'Decentralised R&D and strategic competitiveness: globalised approaches to generation and use of technology in multinational enterprises (MNEs)', *Research Policy*, **28** (2–3), 157–78.

Pearce, R. D. and Singh, S. (1992), *Globalizing Research and Development*, London: Macmillan.

Pearce, R. D. and Papanastasiou, M. (1999), 'The Technological Competitiveness of Japanese Multinationals. The European Dimension', *Thames Essays on Contemporary International Economic Issues*, Michigan: The University of Michigan Press.

Penrose (1959), *The Theory of the Growth of the Firm*, London: Basil Blackwell.

Petit, M. L. and Sanna-Randaccio, F. (2000), 'Endogenous R&D and foreign direct investment in international oligopolies', *International Journal of Industrial Organization*, **18**, 339–67.

Porter, M. E. (1990), *The Competitive Advantage of Nations*, London: Macmillan.

Rugman, A. M. (1981), *Inside the Multinationals: The Economics of Internal Markets*, New York: Columbia University Press.

Rugman, A. M. (2005), *The Regional Multinationals*, Cambridge University Press.

Sanna-Randaccio, F. (2002), 'The impact of foreign direct investment on home and host countries with endogenous R&D', *Review of International Economics*, **10** (2), 278–98.

Simmie, J. (2003), 'Innovation and urban regions as national and international nodes for the transfer and sharing of knowledge', *Regional Studies*, **37** (6&7), 607–620.

Storper, M. and Venables, A. J. (2004), 'Buzz: face-to-face contact and the urban economy', *Journal of Economic Geography*, **4** (4), 351–70.

Storper, M., Chen, Y. C. and De Paolis, F. (2001), 'Trade and the location of industries in the OECD and European Union', *Journal of Economic Geography*, **2** (1), 75–107.

Tallman, S., Jenkins, M., Henry, N. and Pinch, S. (2004), 'Knowledge, clusters and competitive advantage', *Academy of Management Review*, **29** (2), 258–71.

Teece, D. J. (1977), 'Technology transfer by multinational firms: The resource costs of transferring technological know-how', *Economic Journal*, **87** (2), 242–61.

Teece, D. J., Pisano, G. and Shuen, A. (1997), 'Dynamic capabilities and strategic management', *Strategic Management Journal*, **18** (1), 509–33.

Vernon, R. (1966), 'International investment and international trade in the product cycle', *Quarterly Journal of Economics*, **80**, 190–207.

Veugelers, R. and Cassiman, B. (2004), 'Foreign subsidiaries as a channel of international technology diffusion: Some direct firm level evidence from Belgium', *European Economic Review*, **48**, 455–76.

Von Tunzelmann, N. and Acha, V. (2005), 'Innovation in "Low-tech" Industries', in J. Fagerberg, D. C. Mowery and R. R. Nelson (eds), pp. 407–32.

Zander, I. (1999), 'How do you mean 'global'? An empirical investigation of innovation networks in the multinational corporation', *Research Policy*, **28**, 195–214.

Zanfei, A. (2000), 'Transnational firms and changing organisation of innovative activities', *Cambridge Journal of Economics*, **24** (5), 515–42.

4. Interdependence among the Brazilian States: An Input-Output Approach[1]

Fernando Salgueiro Perobelli, Eduardo Amaral Haddad and Edson Paulo Domingues

4.1 INTRODUCTION

The idea of sectoral dependence, sectoral linkages and regional interdependence is presented in the input–output literature in different ways. Hirschman (1958) analysed sectoral dependence from the demand side (exploring backward effects) while Cella (1984) focused on the supply side (forward effects). Chenery and Watanabe (1958) measured the backward effects by reference to the direct coefficient matrix *A*, only while Rasmussen (1958) and Hirschman (1958) promoted the notion of a key sector. These ideas were further extended by Sonis and Hewings (1994) in the development of the concept of Fields of Influence. This concept is used to verify if the impact of a coefficient change (technology change) was concentrated in one or two other sectors or more broadly diffused throughout the economy.[2] A parallel development (Guilhoto et al., 1994; Sonis et al., 1995) was the notion of pure linkages, an improvement on the ideas of Cella (1984) and Clements (1990).

Another interesting way to compute linkages is by means of the method of hypothetical extraction. The original method of hypothetical extraction (Strassert, 1968) involves calculating the difference in output when an individual sector is removed from an economy. The size of the difference will indicate the importance of the sector that was hypothetically isolated in the economy context (Dietzenbacher et al., 1993). Based on the original method of extraction, it is impossible to discriminate between backward and forward linkage effects.

The literature also presents several different approaches for the extraction method. Cella (1984) proposed an improvement on the original method. Instead of starting with the two types of linkages (backward and forward),

Cella (ibid.) first defined the total linkages' effect of a specific industry and then sought to identify the other two components. The measure of total linkages proposed by Cella (ibid.) has the following characteristics: i) it was constructed based on a consistent input-output model of the economy with a fixed set of technical coefficients, ii) it is possible to split the result into two components (backward and forward linkage) and iii) it does not include the feedback processes that are intrinsic to the selected industry.[3]

However Clements (1990) argued that the decomposition of linkages proposed by Cella (ibid.) overestimated the forward linkages. According to Clements (1990) the second part of Cella's forward linkages measure is really a part of backward linkages. In order to solve (or minimise) this problem, Clements (ibid.) proposed a new disaggregation of total linkages.[4]

The regional extraction method, which will be presented in more detail in the next section,[5] makes some adaptations to Strassert's original method. Instead of extracting a sector, we will implement a regional extraction (one at a time) in the interregional input–output model. Hence, we can examine how the isolation of one region will affect production in the rest of the economy. It also allows for the explicit differentiation between backward[6] and forward[7] linkages. According to Miller and Lahr (2001) there is a place for separate backward and forward linkage indicators in a cross-economy comparison of economic structure (i.e. in this chapter, across regions in a multiregional economy). With the purpose of reaching this aim, the extraction will occur precisely in these linkages. In order to calculate the backward linkages of a sector (or region), all intermediate deliveries that this sector (or region) buys are hypothetically extracted. For the forward linkages, all the intermediate deliveries that a sector (or region) sells are extracted. Based on these steps, it is possible to calculate the backward linkages of the isolated region, and also indicate the dependence of this region upon the inputs from the rest of the economy. The forward linkages are derived in a dual manner. Instead of using the input coefficients matrix (matrix A) we will use the output coefficients (allocation matrix).[8]

Miller and Lahr (2001) examine all possible extractions and also speculate on the plausibility of the economic significance that might underpin them. They point out that a number of alternative extractions produce identical results for certain measures of sectoral importance.

Hence, the framework described earlier, when implemented in a multi-regional input–output matrix, will enable us to analyse, in detail, the structure of the Brazilian states economic interactions. It is important to point out that the interactions in this chapter will be treated as the trade among the Brazilian states. Before presenting the results we will describe the methodology and briefly highlight some points discussed in the literature about the importance of trade for development.

4.2 REGIONAL EXTRACTION METHOD[9]

Consider the general case of an interregional input–output model with *N* regions and *n* productive sectors in each region.[10] The model is given by:

$$x = Ax + f \tag{4.1}$$

where: x – the *nN*-element column output vector.
A – the *nN x nN* matrix of input coefficients.
f – the *nN*-element column vector of final demand.

The solution of equation (4.1) will be:

$$x = (I - A)^{-1} f \quad or \quad Lf$$
$$where \;\; L = (I - A)^{-1} \;\; is\ the\ Leontief\ Inverse$$

The output vector is partitioned as follows:[11]

$$x = (x^{1'},\ldots,x^{I'},\ldots,x^{N'})$$
$$where\ x^{I} = (x_1^{I},\ldots x_i^{I},\ldots x_n^{I})'$$

The coefficient matrix is constructed as follows:

$$A = \begin{bmatrix} A^{11} & \cdots & A^{1N} \\ \vdots & \ddots & \vdots \\ A^{N1} & \cdots & A^{NN} \end{bmatrix} \tag{4.2}$$

The extraction method considers the effect of hypothetically isolating one region on the output of the rest of the economy. Without loss of generality, consider the case where the first region was extracted. Thus, the remaining *N-1* regions will represent the rest of the economy.[12] Hence, we can write $x = (x^{1'}, x^{R'})'\ with\ x^{R} = (x^{2'},\ldots,x^{I'},\ldots,x^{N'})'$ a *n(N-1)* element column vector.

In a similar way, we have:

$$A = \begin{bmatrix} A^{11} & A^{1R} \\ A^{R1} & A^{RR} \end{bmatrix} \tag{4.3}$$

Analogous to the equation (4.3), the Leontief inverse in its partitioned form is given by:

$$L = (I - A)^{-1} = \begin{bmatrix} L^{11} & L^{1R} \\ L^{R1} & L^{RR} \end{bmatrix} \tag{4.4}$$

Based on the equation (4.4) we have:

$$x^1 = L^{11} f^1 + L^{1R} f^R \tag{4.5a}$$
$$x^R = L^{R1} f^1 + L^{RR} f^R \tag{4.5b}$$

With the hypothetical extraction of region 1, the model in equation (4.1) will assume the form:

$$\overline{x}^R = A^{RR} \overline{x}^R + f^R$$

The vector $\overline{x}^R$ represents the production of the rest of the economy with the first region removed. The solution of the reduced equation is:

$$\overline{x}^R = (I - A^{RR})^{-1} f^R \tag{4.6}$$

The difference between x^R (equation #.5b) and $\overline{x}^R$ (equation 4.6) provides the extraction effect of region 1 upon the product of the rest of the economy. In order to interpret the elements of vector $x^R - \overline{x}^R$, we have to calculate the matrix L as the inverse of partitioned matrix[13] as follows:

$$L^{1R} = L^{11} A^{1R} (I - A^{RR})^{-1} \tag{4.7a}$$
$$L^{R1} = (I - A^{RR})^{-1} A^{R1} L^{11} \tag{4.7b}$$
$$L^{RR} = (I - A^{RR})^{-1} + (I - A^{RR})^{-1} A^{R1} L^{11} A^{1R} (I - A^{RR})^{-1} \tag{4.7c}$$

Hence we have:

$$x^R - \overline{x}^R = L^{R1} f^1 + \left[L^{RR} - (I - A^{RR})^{-1} \right] f^R \tag{4.8a}$$
$$= (I - A^{RR})^{-1} A^{R1} L^{11} \left[f^1 + A^{1R} (I - A^{RR})^{-1} f^R \right] \tag{4.8b}$$

The interpretation of the expression $x^R - \overline{x}^R$ can be divided into two parts: the first one $\left(L^{R1} f^1\right)$ describes the production in the rest of the economy that is necessary to satisfy the final demand f^1 in region 1 and the second part, $\left[L^{RR}-(I-A^{RR})^{-1}\right]f^R$, describes the production in the rest of the economy $L^{RR}f^R$ that is necessary to satisfy the final demand in the rest of the economy f^R.

We can observe that the elements of vector $x^R-\bar{x}^R$ show the interdependence between region 1 and the other regions. According to Dietzenbacher *et al.* (1993), these interdependencies are fundamentally backward in their nature. These can be demonstrated using the matrix A^{R1} (whose elements indicate the backward dependence of *1* on *R*) and A^{1R} (whose elements indicate the backward dependence of *R* on *1*).

In order to better understand the expression $x^R-\bar{x}^R$, we will use equation (4.8b) and examine this using the idea of an interregional spillover effect and interregional feedback effects developed by Miller and Blair (1985). In order to satisfy the final demand f^1 in region 1, this region must produce $L^{11}f^1$. Region 1 does not have all the inputs necessary to reach this level of production. So, with the aim of achieving this production, it is necessary that region 1 purchases inputs direct from the other regions. The amount of inputs purchased will be $A^{R1}L^{11}f^1$. To provide these inputs, the production in the rest of the economy that is required is $\left(I-A^{RR}\right)^{-1}A^{R1}L^{11}f^1$. The same analysis can be made for the demand in the rest of the economy f^R.

Applying the traditional idea of interregional feedbacks to region 1, it is possible to demonstrate that the feedbacks for this region will be obtained by comparing the outputs of region 1 within the interregional model to the outputs of region 1, within the single-region model. Essentially we have:

$$x^1-\bar{x}^1 = L^{11}f^1 + L^{1R}f^R - \left(I-A^{11}\right)^{-1}f^1 \tag{4.9}$$

Taking the equations (4.7) and (4.8) and interchanging the superscripts *1* and *R* we will have:

$$x^1-\bar{x}^1 = \left(I-A^{11}\right)^{-1}A^{1R}L^{RR}\left[f^R + A^{R1}\left(I-A^{11}\right)^{-1}f^1\right] \tag{4.10}$$

Based on the regional extraction framework it is possible to show that the vector $x^1-\bar{x}^1$ measures the backward dependence of the rest of the economy on the region 1. In other words, the vector enables us to measure the impact of extracting from the economy, all the *N-1* regions in *R* upon the output of the remaining region 1.

4.2.1 Forward Linkages

Turning to the forward linkage effects if we consider the accounting equation $x = Te + f$, where T – is the matrix of intermediate deliveries, e is the summation column vector, $e = (1,1,...,1)'$, f – is the final demand vector and x – is the vector of total production, it is possible to define $x = Ax + f$, where $A = T\hat{x}^{-1}$.

The matrix B (the allocation matrix) can be defined as follows:

$$B = \hat{x}^{-1}T \tag{4.11}$$

In similar way, the accounting equation $x' = e'T + v'$, where v' – is the row vector of primary inputs imply that:

$$x' = x'B + v' \tag{4.12}$$

which can be rewritten as:

$$x' = v'(I - B)^{-1} = v'G \tag{4.13}$$

The equation (4.1) presents the demand driven input–output model and the equation (4.12) is the dual form of equation (4.1) and can be taken as the supply driven input-output model. The forward linkages can be obtained based on the vector $\left(x - \bar{x}\right)'$.

We can implement the extraction (or isolation) of one region. When the region 1 is extracted we will have:

$$\begin{aligned}\left(x - \bar{x}\right)' &= \left[\left(x^1 - \bar{x}^1\right)', \left(x^R - \bar{x}^R\right)'\right] \\ &= \left(v^{1'}, v^{R'}\right)\left\{\begin{bmatrix} G^{11} & G^{1R} \\ G^{R1} & G^{RR} \end{bmatrix} - \begin{bmatrix} \left(I - B^{11}\right)^{-1} & 0 \\ 0 & \left(I - B^{RR}\right)^{-1} \end{bmatrix}\right\}\end{aligned} \tag{4.14}$$

Hence, the vector $(x^R - x^{-R})'$ will represent the forward linkages of region 1 upon the rest of the economy and the vector $(x^1 - x^{-1})'$ will represent the forward linkages of the rest of the economy upon region 1.

4.3 INTERACTIONS AND DEVELOPMENT: A BRIEF COMMENT

The relevant problem of regional economic development "...revolves around a region's ability to become integrated into the larger markets of the world through exports" (North, 1959). North's ideas will be explored in this chapter through an examination of a region's exports and its linkages with the external regions. Without an extensive time series of data on exports, it would be difficult to identify any causal relationship between exports and development. In the present chapter, we therefore focus on the nature and spatial structure of interdependence among the Brazilian states.[14]

According to North it is possible to consider the idea that trade works as an engine of growth. His idea is linked to the export base theory, whereby the existence of outside demand is a necessary condition for regional growth, allowing for other considerations such as location, comparative advantage, and transfer costs.

According to Heckscher-Ohlin-Samuelson (H-O-S) theorem, trade results from differences in relative factor endowments. Thus, if a country's production is intense in labour it will export labour-intensive goods. The theorem is still able to explain much of the trade between developing and developed countries. However, in the early 1960s the literature pointed out limitations in the capacity of H-O-S theorem to explain some trade patterns. During that period there was the recognition that a growing percentage of expanding world trade was occurring between advanced countries with similar factor endowments, and in particular in the form of intra-industry trade.

In order to explain the intra-industry trade Krugman and Helpman and Krugman (1985) developed a theory that was based on the existence of increasing returns to scale and imperfect competition. They pointed out that intra-firm and intra-industry trade between advanced countries will occur mainly in industries characterised by scale economies and oligopolistic market structures.

Hence, 'countries at early stages of economic development tend to behave according to the H-O-S theorem, i.e., by exporting the goods in which they have comparative advantage. So, it should be expected that the trade between two developing countries would be largely concentrated in some specific goods' (Magalhães *et al.,* 2000), and in particular, in primary or labour intensive goods. In contrast, the share of industrialised goods in the trade between developed countries is high.

There are a small number of papers in the literature dealing with the idea of interregional trade within a single economy. We can highlight Thompson (1965), Okazaki (1989) and Hewings *et al.* (1998). Thompson's (1965) idea of the evolutionary development of urban areas is based on the fact that the internal structure of the economy modifies as the process of growth and development occurs with an increasing intensity of interactions among sectors. Thompson's (1965) hypothesis is that while a region grows, there will be an increase in production intermediation. In other words, greater interaction can represent an infilling process that happens in the structure of interdependence among sectors in a specific economy. This process of interaction can happen through the establishment of direct links among sectors for which there were no previous links, or through the increase in the volume of trade among sectors that had previous linkages. Thus, as the national economy matures, Thompson's (1965) development path would display a growth of intraregional flows at a greater rate than interregional transactions. In addition, it is probable that the interregional flows will be essentially interindustry flows.

Okazaki (1989) proposed another development stage, namely that of 'hollowing out'. Okazaki studied the interactions among sectors within the Japanese economy and verified that the degree of interaction had actually begun to fall. In particular, Okazaki (ibid.) showed evidence that the degree of dependence among local sales and purchases is decreasing within the Japanese economy. This process was classified as a 'hollowing-out' effect. According to Okazaki (ibid.), this process can be explained by the competition from South Korea, China and Indonesia. The idea here is that local Japanese suppliers are replaced by less expensive inputs from other international markets. This hollowing-out effect occurs in a mature economy.

Hewings *et al.* (1998) put together Thompson's and Okazaki's ideas and concluded that during the process of development of an economy, it is possible to observe a process of increasing complexity in the linkages among the industrial sectors. This process can be explained by: i) an increase in per capita income that generates demand for a wide range of goods and as a consequence may increase the number of goods produced, and ii) by an increase in the size of the national market that generates opportunities to introduce new suppliers of intermediate goods. As a consequence, there is an increase in the degree of intra-national intermediation. The whole process can be represented by a logistic curve with a slow period of linkage development followed by relatively rapid deepening and extension of the linkage space. Eventually, however, the hollowing–out process may occur, especially in response to cheaper transportation costs, making it possible for local firms to source materials outside the region and to serve markets in other parts of the country. Therefore, we can infer that the volume of trade between poor regions or in early stage of development tends to be small. As the degree of development increases there is also an increase in trade within the region, and as a consequence, there will be an improvement in the interdependence within the region. On the other hand, in those regions that present a higher degree of development, we can expect the hollowing-out process to occur.

4.4 EMPIRICAL RESULTS FOR THE BRAZILIAN ECONOMY

The empirical results of the extraction method for the Brazilian economy are based on the 1996 interregional input-output table for the 27 Brazilian states.[15] For the present purpose, the Brazilian table was aggregated into 8 sectors. The sectoral classification is as follows: 1 – Agriculture, 2 – Industry, 3 – S.I.U.P, 4 – Construction, 5 – Trade, 6 – Financial services, 7 – Public sector and 8 – Other services.[16]

4.4.1 A Brief Characterisation of the Brazilian Economy

Table 4.1 shows the distribution of GDP by macro region and for some selected states. The result enables us to provide a brief characterisation of the Brazilian economy. We can point out that there is a huge spatial concentration of development at the Southeast region, and the pattern of spatial concentration did not change during the period of analysis. The North, South and Centre-west increased their share of GDP, but the increase was not strong enough to change the pattern of concentration in the Southeast region. This region is still responsible for more than 55% of Brazilian GDP. The increase in the Centre-west's share is due to the expansion of the agricultural sector, and mainly in the export of soybeans.

Despite the relatively de-concentration in the industrial production of São Paulo and Rio de Janeiro City we can observe that there is a tendency of spatial re-concentration in the macro-region of the Southeast and South (Diniz, 1993; Diniz and Crocco, 1996). This tendency can be explained by the development of the second level metropolises and the medium size cities. These represent areas with high economic and industrial growth potential. As a consequence, there could be an increase in their income and production and thus an increase in the trade interactions.

Table 4.1 *Share of Brazilian macro regions and selected states in the Brazilian GDP*

	1985	1990	1995	2000
North	*3.8*	*4.9*	*4.6*	*4.6*
AM	1.5	1.8	1.7	1.7
PA	1.5	2.1	1.9	1.7
Northeast	*14.1*	*12.9*	*12.8*	*13.1*
BA	5.4	4.5	4.1	4.4
CE	1.7	1.6	1.9	1.9
PE	2.6	2.7	2.7	2.6
Southeast	*60.2*	*58.8*	*58.7*	*57.8*
MG	9.6	9.3	9.7	9.6
RJ	12.7	10.9	11.5	12.5
SP	36.1	37.0	35.5	33.7
South	*17.1*	*18.2*	*17.9*	*17.6*
PR	5.9	6.3	5.9	6.0
SC	3.3	3.7	3.6	3.9
RS	7.9	8.1	8.3	7.7
Centre-west	*4.8*	*5.2*	*6.0*	*7.0*
Brazil	100.0	100.0	100.0	100.0

Source: IBGE (2004).

4.4.2 Backward Effects: Analysis Based on the Buying Region Side

The results presented in this section are based on the equations (4.8b) and (4.9). The equations generate both the interdependence between region 1 and the other regions $\left(x^R - x^{-R}\right)$ and the backward dependence of the rest of the economy on the region 1 $\left(x^1 - x^{-1}\right)$.

The application of the extraction method enables us to construct a typology of the Brazilian macro regions in terms of the degree of interdependence within and outside of the macro region. We will represent the results as maps of standard deviation from the mean. The analysis in this chapter shows a significant difference in the level of spatial interaction within the five regions considered.

We can divide the Brazilian macro regions into two groups in terms of backward effects. The North, Northeast and Centre-west regions form group one (Figure 4.1). This group is characterised by: (a) a small degree of internal interactions, which means that internal macro regional interdependence is very small. This can be represented by the ellipses. For Figures 4.1A and 4.1C we can see that when one of the states which is located in either of the regions of North or Centre-west are isolated, the impact within the region is below the mean for all the states; (b) a high degree of dependence towards the Southeast region in terms of acquisition of goods; and (c) there is a weak integration within the group. In other words the flows among North, Northeast and Centre-west are still very small (less than 4%). These results corroborate the arguments in the literature that pointed out that trade among poor regions or regions in early stage of development tend to be small.

Magalhães *et al.* (2000) corroborate the findings for the Northeast region. By implementing a Dendrinos–Sonis model they concluded that, in general terms, there is a weak degree of interaction within the Northeast region.

The second group is formed by the Southeast and South region. The main points are: (a) there is a high degree of intra-regional interdependence, which means that there is a high impact in the product of the states located at the region when one of the other states located at Southeast or South are isolated; and (b) the dependence upon the other Brazilian macro regions is incipient.

The high degree of intra-regional interdependence demonstrated by the states located in the Southeast and South also corroborates the arguments in the literature which suggests that regions with a high degree of development tend to exhibit a high degree of trade within the region. As a consequence, there will be an increase in the level of interdependence within the region. We can observe this pattern in Figure 4.2.

Thus, we can confirm that the patterns of spatial interaction demonstrated by regions in the Southeast and South are such that regions with high level of interconnection within the regions can be related to their level of development and their intra-industry trade.

Another interesting result about the Brazilian states' interconnection is taken from the comparison between BL (backward effects) and IF_b (backward inter-state feedbacks) (Table 4.2). According to the results it is possible to conclude that the backward dependence of each isolated state upon the rest of the Brazilian economy is more important than the backward dependence of the rest of the Brazilian economy upon the isolated state. This is true for every Brazilian state except for São Paulo ($BL > IF_b$). The results presented in Table 4.2 also enable us to conclude that for the states located in the North, Northeast, Centre-west and for Espirito Santo state, the backward dependence upon the rest of the Brazilian economy is much more important than for the others states. For those states BL is greater than 10.00. This result enables us to infer that the economy of those states (regions) is strongly oriented towards other parts of the country, especially the Southeast (São Paulo, Rio de Janeiro and Minas Gerais) and the South (Paraná, Santa Catarina and Rio Grande do Sul).

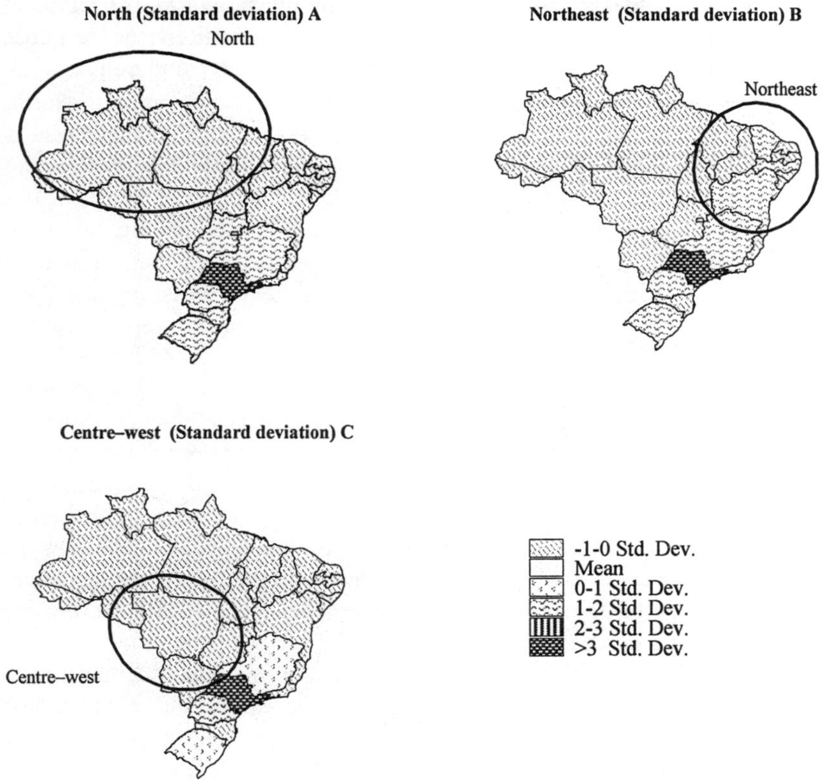

Figure 4.1 *Backward effects: standard deviations*

4.4.3 Forward Effects: An Analysis Based on the Idea of Selling Region

Forward results were calculated based on equation (4.14). The vector $x^R - \bar{x}^R$ measures the dependence of region 1 upon the regions in R (rest of Brazil) with regard to the sale of its output. On the other hand, vector $x^1 - \bar{x}^1$ represents the forward dependence of the regions in R upon region 1 (hypothetically isolated). The value of FL (forward effect) is obtained by summing all off-diagonal elements in each column. IF_f represents the forward dependence of the rest of Brazil upon region 1, which means the forward interstate feedbacks.

The forward effects' results can be taken as a proxy for the amount of products sold by the states that are being examined. Thus, as Figure 4.3A shows, we can confirm that the principal destination of the production of the

North is the Southeast and South region (above 55% and the direction of trade is represented by the arrow). In other words, the majority of the goods produced in the North region are consumed in the Southeast and South.

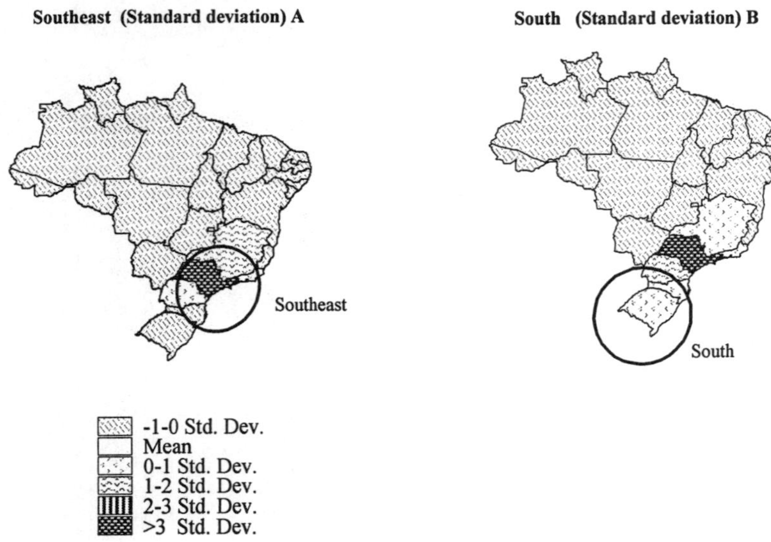

Figure 4.2 Backward effects: standard deviations

The analysis of Figure 4.3A also enables us to confirm that there is a weak interaction within the macro region (below 3%). Thus, the interregional flows are greater than intra-regional flows.

We can confirm that the same pattern occurs in the Northeast and Centre-west, but with a small difference. As we can observe at Figure 4.3B and 4.3C the most important market for the production of those regions is also the Southeast, except for Esprito Santo state and the South region, and also except for Santa Catarina state. Despite the importance of the Southeast and South we can also highlight the role played by Pernambuco, Ceará and Bahia states as markets for the production of the Northeast (Figure 4.3B) and by Mato Grosso and Goiás states as markets for the production of the Centre-west (Figure 4.3C). Thus examining the linkages by the selling region we can confirm that the North region does not exhibit a high degree of intra-interdependence and the Northeast and Centre-west exhibit a degree of intra-regional interdependence which is above the mean.

The pattern exhibited by the interdependence, in terms of the selling region, for the Southeast and the South can be summarised as follows:

Table 4.2 *Backward versus backward inter-state feedbacks*

Isolated State	BL	IFb
North	*24.127*	*0.031*
Acre (AC)	30.265	0.003
Amazonas (AM)	20.321	0.154
Amapá (AP)	19.629	0.001
Para (PA)	21.007	0.042
Rondônia (RO)	20.853	0.012
Roraima (RR)	29.468	0.001
Tocantins (TO)	27.343	0.007
Northeast	*24.873*	*0.090*
Alagoas (AL)	34.680	0.049
Bahia	24.632	0.232
Ceará (CE)	22.735	0.130
Maranhão (MA)	27.502	0.024
Paraíba (PB)	23.312	0.050
Pernambuco (PE)	22.392	0.223
Piauí (PI)	20.550	0.013
Rio Grande do Norte (RN)	26.930	0.040
Sergipe (SE)	21.126	0.050
Southeast	*5.902*	*2.261*
Espírito Santo (ES)	12.024	0.236
Minas Gerais (MG)	5.444	1.305
Rio de Janeiro (RJ)	3.977	0.630
São Paulo (SP)	2.164	6.874
South	*6.821*	*0.590*
Paraná (PR)	9.779	0.780
Santa Catarina (SC)	6.249	0.456
Rio Grande do Sul (RS)	4.434	0.533
Centre-west	*30.806*	*0.110*
Distrito Federak (DF)	37.874	0.021
Goiás (GO)	14.505	0.233
Mato Grosso (MT)	43.916	0.122
Mato Grosso do Sul (MS)	26.927	0.065

Source: Based on the model results.

i) these regions exhibit a high degree of intra-regional interdependence, (see ellipses in Figure 4.4D and 4.4E), and the trade among the states located both in the Southeast and the South is above the mean, which means that the most important market for the products from those regions is the regional market; and ii) the inter-regional interdependence is less intense when compared with

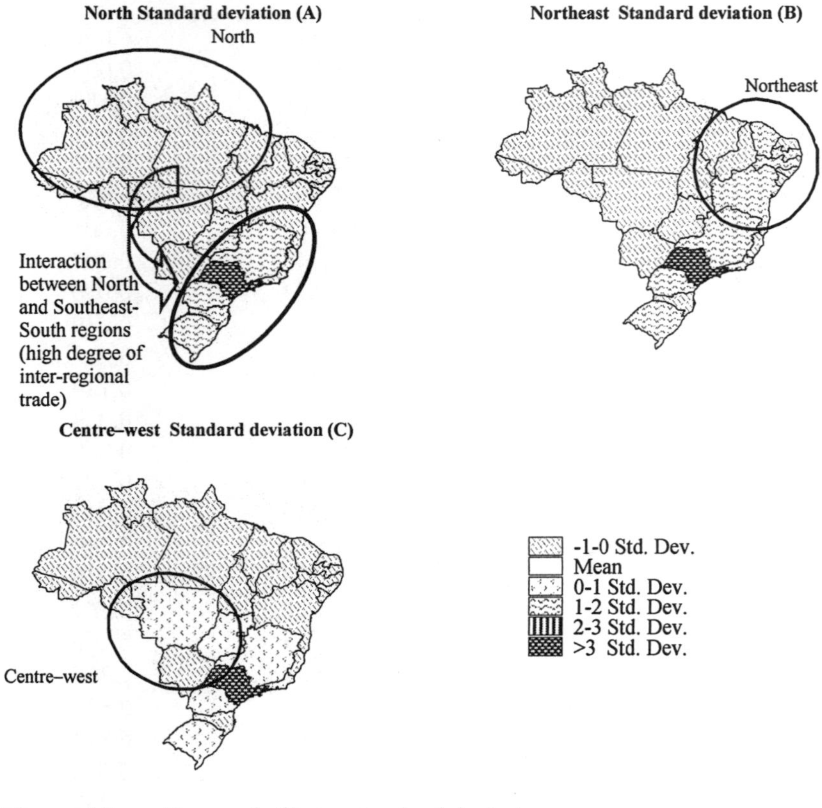

Figure 4.3 Forward effects: standard deviations

the intra-regional interdependence. We can observe from Figure 4.4D and 4.4E that the interaction with the rest of Brazil is below the mean.

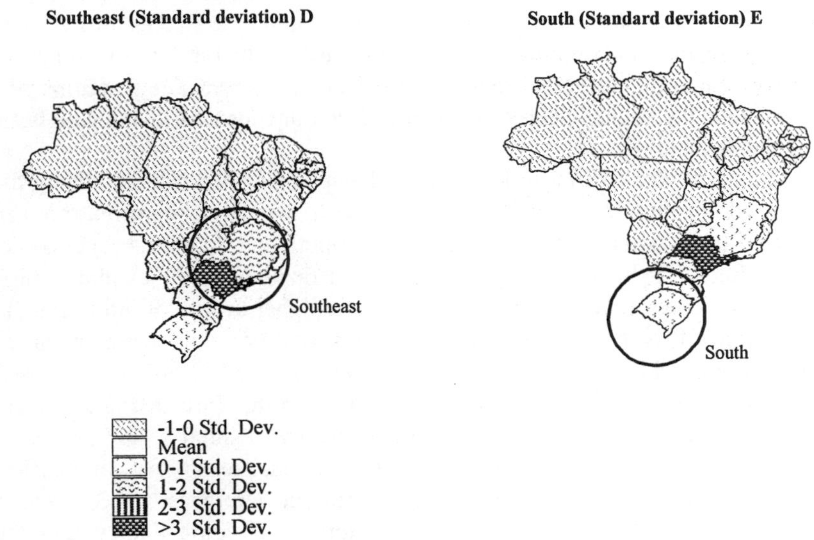

Figure 4.4 *Forward effects: standard deviations*

4.5 CONCLUSIONS

The motivation of this chapter has been to explore the relationships between the Brazilian regions. As we have seen, there are a great number of methodologies that can be used to analyse the interdependencies between sectors and regions. In this chapter, the analysis was carried out by means of the hypothetical extraction method. The results of the methodology applied to the 1996 Brazilian interregional input-output table enable us to conclude that the economic growth of the North, Northeast and Centre-west macro-regions, in terms of backward effects, is more dependent on the performance of the rest of the national economy, and primarily on the Southeast and South, than on their own economies.

Based on the analysis of the backward and forward effects we can observe the importance of São Paulo state in the national context, in that the majority of Brazilian states have a strong relationship with São Paulo state. In other words, the growth of the other Brazilian states is influenced to a great extent by the growth of São Paulo.

The methodology enables us to construct a hierarchy, in terms of backward and forward dependence, of the Brazilian states. As we can see the states with the higher degree of independence are located in the Southeast and the South of Brazil.

The result also enables us to compare the degree of dependence among the states within the macro region. In this respect, we can observe that both in terms of backward and forward linkages the South and Southeast regions (i.e. the regions that have the highest share of Brazilian GDP) exhibit a high degree of dependence within the region (i.e. a higher degree of intra-regional interaction). It is interesting to recall that this kind of result corroborates the ideas developed by Thompson (1965) and Hewings *et al.* (1998).

On the other hand, the states located in the North, Northeast and Centre-west (i.e. the regions with lower levels of income) exhibit a low degree of dependence within the macro region. Based on these results we can confirm that an increase in final demand in the North and Northeast regions would induce effects to a higher degree in the Southeast region than within the region. This kind of result is very important for the policymakers if they want to implement policies which are designed to reduce disparities across regions. In particular, regional policies should be implemented in such a way as to take into account as much as possible the existing structure of economic interactions.

A further step in the study of interactions among the Brazilian states can also be realised through the implementation of this methodology at the sectoral level. Hence, in future we will also measure the linkages among the states and sectors.

NOTES

1. The authors gratefully acknowledge Prof. Geoffrey J. D. Hewings and Prof. Yasuhide Okuyama for the many useful and constructive comments. Perobelli would like to thanks the financial support of Capes and CNPq. Haddad acknowledges financial support of Capes and CNPq.
2. For further example of the application of the methodology of linkages, Key sector and Fields of Influence for the Brazilian economy, see Sonis, *et al.* (1995) and Haddad (1999).
3. For more details see Cella (1984).
4. For more details see Clements (1990).
5. The method is based on Dietzenbacher *et al.* (1993).
6. The backward dependence of a buying region (or sector) with respect to a selling region (sector).
7. The forward dependence of a selling region (or sector) with respect to a buying region (sector).
8. For further applications of this method see Van Der Linden (1998), Dietzenbacher and Van Der Linden (1997) and Sonis, *et al.* (2000).

9. This section is based on Dietzenbacher, *et al.* (1993).
10. The regions will be represented by superscripts $I,J= 1,\ldots,N$ and the products by subscripts $i, j= 1,\ldots, n$.
11. The vector f can be partitioned in the same way.
12. In order to represent these regions we will use the superscript R.
13. For a detailed discussion about the portioning structure see Miller and Lahr (2001).
14. As far as we understand the regional development may be correlated also with the interaction between the Brazilian states and other countries. But, the main aim of this chapter is the internal interaction.
15. The complete list of Brazilian states and regions is presented in the Appendix.
16. For more details about the matrix see Haddad et al. (2002).

APPENDIX

Table 4.A1 Brazilian macro regions and states

Macro region	State	Macro region	State
North (N)	Acre (AC)	Southeast (SE)	Espírito Santo (ES)
	Amazonas (AM)		Minas Gerais (MG)
	Amapá (AP)		Rio de Janeiro (RJ)
	Para (PA)		São Paulo (SP)
	Rondônia (RO)		
	Roraima (RR)		
	Tocantins (TO)		
Northeast (NE)	Alagoas (AL)	South (S)	Paraná (PR)
	Bahia (BA)		Santa Catarina (SC)
	Ceará (CE)		Rio Grande do Sul (RS)
	Maranhão (MA)		
	Paraíba (PB)	Centre-west (CW)	Distrito Federal (DF)
	Pernambuco (PE)		Goiás (GO)
	Piauí (PI)		Mato Grosso (MT)
	Rio Grande do Norte (RN)		Mato Grosso do Sul (MS)
	Sergipe (SE)		

REFERENCES

Cella, G. (1984), 'The input-output measurement of interindustry linkages', *Oxford Bulletin of Economics and Statistics*, **46** (1), 73–84.

Chenery, H. B. and Watanabe, T. (1958), 'International comparisons of the structure of production', *Econometrica*, (26), 487–521.

Clements, B. J. (1990), 'On the decomposition and normalization of interindustry linkages', *Economic Letters*, **33** (4), 337–40.

Dietzenbacher, E. and Van der Linden, J. A. (1997), 'Sectoral and spatial linkages in the EC production structure', *Journal of Regional Science*, **37** (2), 235–47.

Dietzenbacher, E., Van der Linden, J. A. and Steenge, A. E. (1993), 'The regional extraction method: EC Input-output comparisons', *Economic Systems Research*, **5** (2), 185–207.

Diniz, C. C. (1993), 'Desenvolvimento poligonal no Brasil: nem desconcentração nem contínua polarização', *Revista Nova Economia*, **3** (1), Belo Horizonte.

Diniz, C. C. and Crocco, M. A. (1996), 'Reestruturação econômica e impacto regional: o novo mapa da indústria brasileira', *Revista Nova Economia*, **6** (1), Belo Horizonte.

Guilhoto, J. J. M., Sonis, M., Hewings, G. J. D. and Martins, E. B. (1994), 'Índices de ligações e setores-chve na economia brasileira: 1959/90', *Pesquisa e Planejamento Econômico*, **24** (2).

Haddad, E. A. (1999), *Regional inequality and structural changes. Lessons from the Brazilian experience*, Ashgate.

Haddad, E. A., Azzoni, C. R., Domingues, E. P. and Perobelli, F. S. (2002a), 'Macroeconomia dos estados e matriz interestadual de insumo-produto', *Revista Economia Aplicada*, **6** (4), out/dez.

Haddad, E. A., Domingues, E. P. and Perobelli, F. S. (2002b), 'Regional effects of economic integration: the case of Brazil', *Journal of Policy Modeling*, **24**, 453–82.

Helpman, E. and Krugman, P. R. (1985), *Market Structure and Foreign Trade: Increasing Returns, Imperfect Competition, and the International Economy*, Cambridge, MA: MIT Press.

Hewings, G. J. D., Sonis, M., Guo, J., Israilevich, P. R. and Schindler, G. R. (1998), 'The hollowing out process in the Chicago economy, 1975-2011', Federal Reserve Bank of Chicago.

Hirschman, A. O. (1958), *The Strategy of Economic Development*, New Haven, Conn: Yale University Press.

IBGE (1999), 'Contas regionais do Brasil. 1985-1997', Rio de Janeiro: Fundação IBGE.

Magalhães, A., Sonis, M. and Hewings, G. J. D. (2000), 'Regional competition and complementarity reflected in relative regional dynamics and growth of GSP: a comparative analysis of the Northeast of Brazil and Midwest States of the US', Mimeo.

Miller, R. E. and Blair, P. D. (1985), *Input-output Analysis: Foundations and Extensions*, Englewood Cliffs: Prentice Hall.

Miller, R. E. and Lahr, M. L. (2001), 'A Taxonomy of Extractions', in M. L. Lahr and R. E. Miller (eds), *Regional Science Perspectives in Economic Analysis*, Elsevier Science.

North, D. C. (1959), 'Agriculture and regional economic', *Journal Farm Economics*, **41**, 943–51.

Okazaki (1989), 'The hollowing out phenomenon in economic development', Paper presented at the Pacific Regional Science Conference, Singapore.

Rasmussen, P. N. (1958), '*Studies in Intersectoral Relations*', Amsterdam: North Holland.

Sonis, M., Hewings, G. J. D. and Haddad, E. A. (2000), 'The Region versus the Rest of the Economy: The Extraction Method', in H. Kohno, P. Nijkamp and J. Poot (eds), *Regional Cohesion and Competition in the Age of Globalization*, Cheltenham, UK and Northampton, MA, USA: Edward Elgar.

Sonis, M., Guilhoto, J. J. M., Hewings, G. J. D. and Martins, E. B. (1995), 'Linkages, key sectors, and structural change: some new perspectives', *The Developing Economics*, **33** (3), 233–70.

Sonis, M. and Hewings, G. J. D. (1994), 'Fields of influence in input-output systems', Discussion paper, Regional Economics Applications Laboratory, University of Illinois.

Strassert, G. (1968), 'Zur Bestimmung strategischer Sektoren mit Hilfe von Input–Output-Modellen', *Jahrbücher für Nationalökonomie und Statistik*, **182** (3), 211–15.

Thompson, W. R. (1965), *A Preface to Urban Economics*, Baltimore: John Hopkins Press.

Van der Linden, J. A. (1998), 'Interdependence and specialization in the European Union: intercountry input-output analysis and economic integration', PhD dissertation, University of Groningen, Netherlands.

5. The Changing Structure of Trade and Interdependence in a Mature Economy: The US Midwest[1]

Geoffrey J. D. Hewings and John B. Parr

5.1 INTRODUCTION

Four questions are addressed in this chapter. First, as regions mature, what happens to their overall economic trade relationships, both internally and with other regions? Second, is the process similar at all spatial scales, i.e. for metropolitan, regional, macro-regional and national economies? Third, are regions becoming more alike in terms of economic structure? Fourth, what are the implications for regional competition and policy intervention, as regions become more similar and interdependent?

For the past three decades, changes in the spatial interconnections in the production process have been one of the more visible characteristics defining the reorganisation of trade, both international and interregional. In international trade the process of globalisation finds firms specialising production in establishments, often in different nations, in order to exploit such locational advantages as proximity to markets and easier access to relatively advantageous production inputs (Hummels *et al.*, 2001). Within nations interregional trade appears to be motivated less by advantages of cheaper input costs, labour mobility, etc., than by centripetal forces generated by cumulative causation, such as the interaction of scale economies, transportation costs and specialised labour pools (Fujita and Thisse, 1996; Krugman, 1998; Ciccone and Hall, 1996; Wheeler, 2001). This self-reinforcing process may thus result in greater spatial dispersion of production across regions, suggesting that vertical relations within the productive process may now be more spatially complex than in past decades (Parr *et al.*, 2002). For example, as the data in Table 5.1 reveal, interstate flows have grown more rapidly than the US gross domestic product over the period 1993–2002. What of changes in the spatial organisation of production and interdependence within metropolitan areas? A recent analysis (Hewings and Parr, 2007) finds that the nature of interdependence is different from that characterising regions or larger areas.

Table 5.1 *Growth rates of GDP and interstate commodity flows in the US, 1993–2002*

	Percentage growth rates		
	Gross domestic product	Interstate flows of commodities by value	Difference in percentage points
1993-1997	15.54	18.78	3.24
1997-2002	15.85	22.20	6.35
1993-2002	33.86	45.10	11.24

In order to show the changing significance of internal trade within the US, Table 5.1 contrasts the national growth rates of gross domestic product with the growth rates in the interstate flows of commodities for three recent periods. Our focus, however, is on the Midwest states of Wisconsin, Illinois, Indiana, Ohio and Michigan, and their interactions with the rest of the US (defined as a *de facto* sixth region). Throughout the chapter the term 'region' refers to each of the five constituent states of the Midwest and to the 'Rest of the US region' (RUS), while the Midwest is regarded as a 'multi-state area', as is the RUS when it is being compared to the Midwest. In 1996 the gross area product for the Midwest reached $1.1 trillion, representing about 16% of the GNP of the US, as indicated in Table 5.2, which also shows the contribution of the Midwest to the US economy in other categories such as total output, employment and income. Furthermore, the Midwest exhibited a higher level of labour productivity (output per employee) than the nation as a whole. The Midwest provides an appropriate case study for examining linkages among regions and sectors, since it can still claim to be enjoying certain spatially-specific advantages, especially with respect to intrasectoral linkages in durable manufacturing, involving the exploitation of economies of scale, scope and complexity (Parr *et al.*, 2002; Sonis *et al.*, 2002; Swonk, 1996). The Midwest has also been chosen as the main focus of analysis primarily because it typifies a mature economy that has undergone dramatic structural changes throughout the period since 1980. The availability of a set of models and related databases further reinforces the relevance of this case study.

The chapter is organised as follows. Section 5.2 discusses certain background issues, while Section 5.3 examines the spatial structure of trade flows over the period 1980–2000. A method of measuring vertical linkages is outlined in Section 5.4, and Section 5.5 then provides the results of an application of this measure. In Section 5.6 the focus of attention shifts to the metropolitan level, where the examination of interdependence is expanded to include more than simply intersectoral flows. A final section offers an interpretive commentary on the empirical analysis, and considers some of the policy implications.

Table 5.2 Basic statistics for the Midwest and the RUS, 1996

Multi-state area	Gross area product	Output	Employment	Income	Flows
Midwest	1 119 158	1 965 876	22 610	682 445	(16.7)
	(16.0)	(18.5)	(15.1)	(16.8)	
RUS	5 875 182	8 646 034	126 986	3 384 501	67.2

Note: Monetary values are measured in 1992 $m, and employment is expressed in thousands of full-time and part-time employees. An entry in parentheses is the Midwest value as a percentage of the corresponding US value. The column marked 'Flows' refers to the value of intra-area commodity flows as a percentage of the total US flows. The remaining 16.1% (=100-16.7-67.2) represents the flows between the two multi-state areas as a percentage of total US flows.

5.2 RESEARCH DESIGN AND DATA SOURCES

To capture industrial interdependencies in sequential production processes, this study adapts the method proposed by Maddigan (1981) for measuring the vertical linkages of firms, and applies this to a multiregional general-equilibrium model. The approach is based on an input–output matrix, in order to identify sectoral interdependencies in the sequential production process, from which the forward and backward linkages are derived. These linkages also reflect a complex sequence of signals from various sources of demand: regional (within state), interregional (within the Midwest) and external (the RUS and international). The focus on linkages is in contrast to a recent series of geographic concentration measures (Maurel and Sedillot, 1999; Ellison and Glaeser, 1997). These are essentially modifications of the Herfindahl index, and thus only provide a limited perspective on economic structure. More importantly, such measures are unable to utilise the abundant information on linkages embedded in an input–output table, and may generate misleading interpretations about industrial structural change. As an alternative, feedback-loop analysis, elaborated by Hewings *et al.* (1998) and Sonis *et al.* (1993; 1995), offers another way of interpreting the sequential production process, when an input–output table or a trade flow matrix is available. A complementary approach, structural-path analysis, has the great advantage of being able to identify the spatial structure of trade flows, and thus examine the nature and strengths of interregional connectivity. However, although these methods highlight the value of trade flows within interregional and/or intersectoral feedback loops, there are difficulties of measuring in detail the degree of sequential-production linkage at the sector level.

In order to achieve a comprehensive coverage of internal and external effects among sectors and regions, the Midwest Regional Input–Output Econometric Model (MW-REIM) is utilised, details of which are provided in Israilevich *et al.* (1997). The model is based on 1992 and 1997 multiregional input–output tables (compiled using state input–output tables and interstate trade derived from the Commodity Flow Survey for 1993 and 1997), as well as annual regional data that were available from 1969 to 2000. The primary

Table 5.3 Percentage of sector output and value of total output for each region

	Region					
Sector	IL	IN	MI	OH	WI	RUS
1	1.7	1.8	0.7	1.1	1.3	2.7
2	0.4	0.4	0.3	0.5	0.2	1.3
3	**5.4**	5.5	4.9	5.1	5.7	**6.0**
4	5.1	3.7	2.6	3.9	**8.3**	3.7
5	3.3	4.3	2.4	3.3	1.9	3.2
6	1.8	**6.8**	1.9	3.9	1.8	1.2
7	2.8	3.9	4.3	4.4	3.4	1.3
8	**5.7**	5.3	**5.5**	**5.6**	**8.9**	2.5
9	3.9	4.2	1.1	3.7	5.2	2.5
10	2.9	**11.7**	**22.1**	**11.9**	4.8	2.5
11	**7.2**	**7.2**	**5.5**	**6.2**	**9.9**	**7.8**
12	2.5	4.5	3.8	3.6	4.9	**3.7**
13	57.3	40.7	44.8	46.6	43.6	61.6
Total output	562 573	264 978	425 662	485 934	226 729	8 646 034

Note: The projected output levels are obtained from the MW-REIM. Value of total output is in 1992 $m. The three most important sectors, apart from sector 13, are shown as bold for each region (IL = Illinois; IN = Indiana; MI = Michigan; WI = Wisconsin; RUS = Rest of the US region).

focus of the model is on the manufacturing sectors. The MW-REIM is a multiregional, dynamic general-equilibrium model, which covers the five Midwest regions and the RUS, and includes 13 sectors for each region. This model is able to develop quantitative economic linkages among the system variables via its two major components, an input–output module and a time-series module, and can also identify comprehensively the sectoral and regional linkages. In addition, it is possible to extract forward and backward linkages by generating annual input–output tables and forecasts for the period 2001–2030 for the 13 sectors specified in the model. The procedure for extracting the annual derived input–output tables is explained in Israilevich *et al.* (1997). Table 5.3 illustrates the output levels by sector for the six regions, and the sectors are described in Table 5.1 of the Appendix.

5.3 THE SPATIAL STRUCTURE OF TRADE FLOWS

Using this 13-sector, six-region framework, a flow analysis was utilised to identify major linkages within the system, the method for estimating the interregional-trade flows being identical to that used by Sonis *et al.* (2002). These flows are presented in Table 5.4, and the trade-flow matrix is summarised in Table 5.5. The upper part of the table shows that the intraregional trade flows for the six regions (i.e. the diagonal sum of regional flow matrices) reached $4.8 trillion in 2000, and accounted for more than 80% of total trade flow in the US. Within the diagonal summation, intrasectoral flows (flows between establishments within the same sector) sharply increased from 37.3% (= 31.0/83.23) of intraregional flows in 1980 to 46.4% (=37.5/80.8) in 2000, while the intersectoral component in intraregional flows decreased from 62.7% to 53.6% over the same period. These findings are consistent with current trade theory which would suggest that trade between regions with high levels of per capita income, similar endowments and good inter-connectivity moves from domination by intersectoral trade to domination by intrasectoral trade, as firms exploit scale economies at the establishment level and produce multiple products to meet consumer demands for greater choice in goods that are close substitutes. Using a more disaggregated set of trade flows for interaction among the Midwest regions only, Munroe *et al.* (2007) found a similar domination by intrasectoral trade.

Interregional flows are also decomposed into their intrasectoral and intersectoral components, from the off-diagonals of the flow matrix, and the results are shown in the middle part of the Table 5.5. The total trade flows were identified through five feedback loops, hierarchically ordered according to the intensity (sum of flows) of trade through the loops. Sonis *et al.* (2002) identified five interregional trade loops in the six regions for year 1992, and showed that the Midwest economy is well developed and bilaterally balanced: for each flow, there exists a corresponding counter-flow of comparable size. This part of the table indicates that although interregional trade only gradually increased, the intrasectoral component increased from 7.5% in 1980 to 10.0% in 2000, while the intersectoral component remained about the same (at around 9.2%) over the same period.

Table 5.4 Interregional flows, 1980, 1990 and 2000

	Region					
	IL	IN	MI	OH	WI	RUS
1980						
IL	166 547	5 265	6 454	3 008	7 360	108 959
IN	5 556	67 589	7 834	4 283	1 639	42 983
MI	5 629	5 377	111 447	8 650	4 050	85 895
OH	3 187	3 646	16 729	141 792	1 671	71 706
WI	12 525	1 715	7 725	2 631	38 944	74 011
RUS	66 399	28 981	79 177	55 792	57 524	3 375 637
1990						
IL	194 114	5 129	7 503	3 374	8 407	114 071
IN	5 888	80 577	9 057	5 149	2 009	45 116
MI	6 445	6 244	129 992	10 899	4 097	95 311
OH	3 739	4 632	21 846	166 148	1 980	78 525
WI	14 248	2 345	8 534	3 430	39 437	85 095
RUS	75 465	33 485	84 414	61 209	65 737	3 480 674
2000						
IL	250 092	6 575	11 074	4 831	12 750	138 268
IN	7 667	117 757	15 072	7 331	3 341	55 170
MI	8 830	9 980	182 003	15 648	7 130	119 586
OH	4 940	6 813	32 691	225 379	3 135	94 225
WI	17 354	3 680	14 351	5 216	54 958	114 564
RUS	94 582	44 694	108 467	77 607	91 839	3 965 839

Note: Values are expressed in 1992 $m.

The lower part of Table 5.5 summarises the trade flows at the aggregated level of the two multi-state areas: the Midwest (MW) and the RUS. In the absence of detailed analyses of these tables (i.e. without more sectoral disaggregation), there is no clear evidence of the time trends of this trade. A case might be made for a tendency toward consolidation within the overall Midwest economy. Thus, the sum of Midwest trade flows (the Midwest-to-

Midwest matrix) has increased in a logarithmic manner with increases in both the diagonal and off-diagonal sums. Flows between the Midwest and the RUS also increased in both directions over the period. Meanwhile, the percentage of trade within the RUS decreased, mainly due to its larger geographical scale. This trend partially reflects the increased level of international trade over these decades and the more rapid increase in population and employment in the RUS relative to the Midwest.

Table 5.5 Summary of trade flows, 1980, 1990 and 2000

	1980	1990	2000
Total US flows	4 688 314	4 964 328	5 933 438
Intraregional flows	3 901 955	4 090 943	4 796 029
	(83.2)	(82.4)	(80.8)
Intrasectoral	(31.0)	(35.5)	(37.5)
Intersectoral	(52.2)	(46.9)	(43.3)
Interregional flows	786 359	873 385	1 137 409
	(16.8)	(17.6)	(19.2)
Intrasectoral	(7.5)	(8.5)	(10.0)
Intersectoral	(9.3)	(9.1)	(9.2)
MW and RUS flows			
MW to MW	(13.7)	(15.0)	(17.3)
MW to RUS	(8.2)	(8.4)	(8.8)
RUS to MW	(6.1)	(6.5)	(7.0)
RUS to RUS	(72.0)	(70.1)	(66.8)

Note: Values are expressed in 1992 $m. An entry in parentheses indicates percentage of total US flows.

The flow analysis of this section provides a clear indication that over past decades intrasectoral trade has expanded significantly, with respect to intraregional as well as interregional trade. In addition, the Midwest-to-Midwest flow increased over time, providing a possible interpretation that while hollowing out may be occurring at the regional- and metropolitan-area levels, the process does not appear to be evident at the Midwest level. The increased density of the freeway network and the associated reductions in real transportation costs (Parr *et al.*, 2002) have increased the geographical range over which it is economically feasible to ship products, even for just-in-time production regimes. However, the flow analysis needs to be extended in sectoral detail, since an aggregated sectoral analysis is only of limited use in measuring the strength of sequential production linkages.

5.4 MEASURING VERTICAL LINKAGES

The Maddigan (1981) approach to measuring sectoral interdependencies is based on an input–output matrix, and involves a method of linking sectors through production functions via forward/backward linkages in this matrix. Production relationships in the measure thereby capture simultaneous network characteristics, e.g. two sectors may serve as input suppliers to each other. Traditional methods, such as analysing relationships among outputs or the use of value-added-to-sales ratios, understate a firm's or a sector's participation in a linear succession of production stages. Simply put, it is too restrictive to define a sector's production function in a single processing chain.

The core of the Maddigan index is its utilisation of the forward- and backward-matrix derived from the Leontief framework. Her method is applied in a multiregional framework in the following way. Two matrices, A and B, represent relative net inputs and relative net outputs, respectively, and capture net production relationships for each sector's upstream and downstream linkage. Thus:

$$A = I - [x_{ij} / (z_j - x_{jj})] + [y_{ij}]$$

$$B = [x_{ij} / (z_i - x_{ii})] - [y_{ij}] - I$$

where

I = identity matrix (m x m);
x_{ij} = the value of the output of sector i used as an input by sector j;
z_j = the total value of the output of sector j (where $j = 1, \dots, m$);
y_{ij} = $\{ x_{ii} / (z_i - x_{ii})$ if $i=j$; 0 if $i \neq j$; $i, j = 1, \dots, m\}$.

The dimension of m in the model is 78 (i.e. 13 sectors in each of the six regions), and each i and/or j represents a particular sector in a given region. The elements in matrix A are negative, since vertically-linked input values reduce the value of net output consumed internally, while the elements of the matrix B are positive values, and represent the allocation of sector outputs.

An index synthesising these two matrices now can be formulated for an explicit expression of the production-line linkage of this sector. Using the rows and columns of matrices A and B, respectively, a sector is characterised by two matrices, C and D, where the column vector of $c_{ij} \in C$ is the column i of the input matrix of sector j, and the row vector of $d_{ij} \in D$ is the row i of the output matrix of sector j. Formally, for $i, j=1, \ldots, n$ $(n \leq m)$ each element is defined as follows:

$$c_{ij} = a_{s(i)s(j)}$$

$$d_{ij} = b_{s(i)s(j)}$$

where

$s(i)$ = region s in which sector i is located;
c_{ij} = percentage value of the net output of regional sector $s(j)$ contributed by regional sector $s(i)$;
d_{ij} = percentage value of the net output of regional sector $s(i)$ used as an input to regional sector $s(j)$.

Finally, in the multiregional framework, what we term the vertical connection index VC for sector k can be defined as follows:

$$VC_k = 1 - \left[1 \middle/ \prod_{r=1}^{6} \left(\sum_{R=1}^{6} \sum_{j=1}^{13} c_{kr,jR}^2 \right) \left(\sum_{R=1}^{6} \sum_{j=1}^{13} d_{jR,kr}^2 \right) \right]$$

The diagonal elements of C and D are unit values, since $j=k$ and $R=r$. The VC index lies between 0 and 1, and retains three of the properties noted by Maddigan (1981).[2]

By its construction the VC index does not include the intrasectoral component of intraregional trade. However, in our multiregional setting, it is a comprehensive measure, covering a sector's forward and backward linkages to different sectors (the intersectoral components in intraregional and interregional trade) and to the same sector (the intrasectoral components of interregional trade). Hence, the VC indices for Midwest trade are likely to be higher than those for individual regions (compare the entries in Tables 5.6 and 5.7). The values of the VC index for each sector in the five Midwest regions and the RUS are shown in Table 5.6. For a flow analysis of the Midwest alone (i.e. excluding the RUS), the value of R and r is reduced to 5 (from 6), and the two matrices are reduced to a dimension of 65x65 (from 78x78). This partitioning can be applied to a single region, or even to a single sector, even though the index will become significantly smaller with reductions in the total value of in-flows and out-flows for smaller C and D matrices.

Table 5.6 Value of VC index in each sector by region, 1980, 1990 and 2000

	Sector												
	1	2	3	4	5	6	7	8	9	10	11	12	13
1980													
IL	.56	.08	.12	.09	.08	.28	.16	.06	.09	.06	.10	.11	.010
IN	.14	.03	.12	.08	.05	.07	.11	.05	.08	.05	.07	.06	.012
MI	.05	.04	.09	.05	.09	.34	.15	.08	.09	.03	.11	.05	.010
OH	.07	.06	.07	.06	.10	.27	.12	.08	.12	.08	.08	.10	.009
WI	.06	.01	.07	.08	.02	.09	.02	.01	.01	.01	.01	.01	.004
RUS	.27	.23	.09	.19	.15	.15	.12	.05	.11	.14	.14	.11	.015
2000													
IL	.18	.12	.13	.08	.10	.25	.13	.05	.07	.07	.11	.09	.008
IN	.12	.04	.12	.10	.07	.09	.14	.04	.05	.05	.08	.05	.012
MI	.08	.06	.07	.05	.11	.30	.14	.08	.06	.04	.12	.05	.012
OH	.10	.09	.09	.06	.10	.23	.13	.08	.09	.07	.09	.08	.010
WI	.13	.02	.08	.02	.02	.11	.02	.01	.01	.02	.02	.01	.004
RUS	.20	.27	.10	.17	.12	.11	.09	.03	.13	.15	.16	.11	.017
Trend													
IL	-	+	+	-	+	-	-	-	-	+	+	-	-
IN	-	+		+	+	+	+	-	-		+	-	+
MI	+	+	-		+	-	-		-	+	+		+
OH	+	+	+		+	-	+		-	-	+	-	+
WI	-	+	+	-		+				+	+		-
RUS	-	+	+	-	-	-	-	-	+	+	+		+

Note: The notation +/- indicates the increase/decrease in the *VC* index for each sector in each region from 1980 to 2000.

5.5 RESULTS AND INTERPRETATION OF THE *VC* INDEX

In this section the analysis is concerned solely with the two multi-state areas (the RUS and the Midwest), and the results are displayed in Table 5.7. In the upper part of this table the estimated values of the *VC* index are presented for the RUS and the Midwest in 1980, 1990 and 2000. The *VC* index for each sector of the Midwest economy is necessarily less than that for the RUS, since the total values of the input coefficients are lower. The *VC* index for a given sector in an area measures the strength of sequential production linkage. For both areas the *VC* index was high in resources (sectors 1, 2, and 3) and low in services (sector 13), and varied widely within manufacturing.

From 1980 to 2000 most sectors exhibited smooth transitions. However, for the transportation-equipment sector (sector 10) the Midwest showed a major, temporary retreat in 1990, towards less interdependence. This reflected a downturn of production by domestic automobile producers in the 1990s, and also a more dispersed pattern of direct investment by foreign-owned automobile producers, involving locations other than Southeast Michigan. Nevertheless, along with the two metal sectors (sectors 6 and 7), sector 10 has remained one of the most vertically-connected sectors, both in the Midwest and the RUS (Klier, 1998). Regarding the downward trend for resources (sector 1), this might be explained by increased exports of agricultural output, with the agriculture part of the sector becoming more dependent on international trade and more internally interdependent (Frank and Henderson, 1992).[3] This upper part of Table 5.7 provides an overall indication that the increase in Midwest-to-Midwest trade, which has contributed to the hollowing-out process in individual regions, resulted from: i) a significant increase in intra-area, intrasectoral trade in services (sector 13); and ii) steady increases in such interaction in most manufacturing sectors; and also iii) increased values for the *VC* index in sectors 2, 3 and 5. Perhaps the time interval of 20 years (from 1980 to 2000) is not long enough to judge structural changes for all sectors, but their overall trend is consistent with intuition and the results from previous studies.

The middle part of Table 5.7 indicates the rank order of the *VC* index for the 13 sectors in each of the two areas. The heavily-shaded areas denote the four highest-ranked sectors for 1980 and 2000, while the lightly-shaded areas represent the four lowest-ranked sectors. Within this hierarchy of values for the *VC* index, sectors 1, 2, 6 and 10 are the highest-ranked sectors in year 2000 both in the RUS and the Midwest. In the same year, the lowest-ranked sectors were 4, 8, 12 and 13 in the RUS, and 8, 9, 12 and 13 in the Midwest. The low rank of the service sector is mainly due to the fact that most of its production is for final consumption, so that its output linkage becomes less

Table 5.7 *VC index and internal consumption in RUS and Midwest, 1980, 1990 and 2000*

	Sectors												
VC Indices	1	2	3	4	5	6	7	8	9	10	11	12	13
RUS 1980	.92	.64	.62	.55	.54	.90	.67	.48	.58	.74	.60	.49	.40
RUS 1990	.84	.69	.66	.51	.56	.88	.65	.40	.57	.71	.60	.50	.45
RUS 2000	.77	.71	.67	.51	.58	.87	.66	.37	.56	.73	.62	.48	.48
RUS *VC* trend	-	+	+	-	+	-	-	-	-	-	+	-	+
MW 1980	.83	.45	.49	.37	.33	.74	.51	.41	.38	.67	.43	.34	.23
MW 1990	.72	.53	.53	.33	.35	.69	.47	.35	.33	.64	.41	.33	.27
MW 2000	.62	.56	.54	.33	.37	.69	.49	.32	.29	.66	.43	.30	.29
MW *VC* trend	-	+	+	-	+	-	-	-	-	-		-	+
Ranking	Numbered sectors arranged in rank order of *VC* value												
RUS 1980	1	6	10	7	2	3	11	9	4	5	12	8	13
RUS 2000	6	1	10	2	3	7	11	5	9	4	13	12	8
MW 1980	1	6	10	7	3	2	11	8	9	4	12	5	13
MW 2000	6	10	1	2	3	7	11	5	4	8	12	13	9
	Sectors												
Flows	1	2	3	4	5	6	7	8	9	10	11	12	13
RUS 1980	39	9	3	19	22	28	6	18	15	15	30	17	48
RUS 2000	35	4	3	14	23	20	5	14	12	14	21	13	58
RUS trend	-	-		-	+	-	-	-	-	-	-	-	+
MW 1980	26	6	2	11	19	27	5	6	10	12	13	6	30
MW 2000	15	1	2	12	21	21	5	9	11	14	14	7	34
MW trend	-	-		+	+	-		+	+	+	+	+	+

Notes: The shading is explained in the text. The notation +/- indicates an increase/decrease in the value of *VC* for each sector over the period 1980–2000. The lower part of the table marked 'Flows' refers to intra-area, intrasectoral flows as a percentage of total US flows, together with the trends. The RUS total internal consumption (which is intra-area, intrasectoral trade as a percentage of total RUS flows) amounts to 17.8, 18.6, 19.1% in 1980, 1990, 2000, respectively, and for the Midwest the corresponding figures are 15.1, 15.4, 15.7%. The trend of sectoral internal consumption for the US as a whole is approximately the same as that for the RUS.

important, accounting for a smaller percentage of the total value of output. The rank is improving significantly in both areas, however, implying that its outflows and inflows are becoming more linked to the production of other sectors. In both areas this upward trend is also observed in sectors 2, 5 and 6, while there is a downward trend in sectors 1, 7 and 8.

The lower part of Table 5.7 shows the intra-area, intrasectoral flows as a percentage of total US flows. As can be observed, it is only sectors 5 and 13 that contributed to such internal trade expansion within the RUS. In other words, intrasectoral trade actually decreased in percentage terms in the other 11 sectors of the RUS trade flows. For the Midwest, by contrast, it can be seen that such trade has increased not only in sectors 5 and 13 but also in most of the manufacturing sectors. Note that sector 13 covers all private and government services, as well as transportation, communication and utilities. The sector accounts for about 60% of the total output, and given its upward growth trend, this sector must have been the major force boosting the total value of intrasectoral trade within each area.

An alternative way of looking at the *VC* index for each sector in the Midwest is to take its value, and divide this by the corresponding value for the RUS. The sectors are then ordered as 10, 1, 8, 6, 3, 7, 11, 2, 12, 4, 9, 5, 13 in 1980, and 10, 8, 3, 1, 6, 2, 7, 11, 4, 5, 12, 13, 9 in 2000. The ratios (not shown here in tabular form) indicate the relative levels of interdependence, as well as their trends, e.g. compare sectors 1 and 10 for both years. It transpires that along with its high level of production (shown in Table 5.3), heavy manufacturing (sectors 8 and 10) provides the underpinning for sectoral interdependence within the Midwest. This modified *VC* index generally registers low values in sectors having high levels of product diversification (sector 9), leading to increased monopolistic competition, or high sunk costs (sector 5), or final-stage consumption (sector 13). The main contribution of the electronics industry (sector 9) is not generated by its backward dependence on intermediate inputs (since many of the components are imported from overseas) but through the provision of a variety of products.

Finally, the original *A* and *B* matrices were reduced, in order to examine intrasectoral flows between the two areas (the RUS and the Midwest), and to derive an inter-area, intrasectoral measure of linkage for the 13 sectors. This modification allowed us to identify the major sectors that contributed to increases in the intrasectoral components of inter-area trade in the feedback-loop analysis. Table 5.8 shows that in 2000, sector 10 in the RUS recorded the highest *VC* value, followed by sector 13, and the remainder in the order 1, 8, 11, 6, 9, 4, 5, 7, 2, 12, 3. In 1980 the order was approximately the same, but only the service sector showed a significantly increasing trend. In the

Table 5.8 *Inter-area, intrasectoral VC values for RUS and Midwest, 1980 and 2000*

	Sector												
	1	2	3	4	5	6	7	8	9	10	11	12	13
RUS 1980	.402	.026	7E-6	.041	.018	.140	.018	.170	.050	.550	.133	.008	.334
RUS 2000	.143	.004	4E-6	.033	.015	.059	.009	.083	.039	.521	.072	.004	.434
RUS trend	-	-	-	-	-	-	-	-	-	-	-	-	+
MW 1980	.073	.009	7E-9	.005	.002	.054	.005	.008	.001	.475	.019	.001	.001
MW 2000	.019	1E-4	5E-9	.009	.004	.032	.004	.012	.001	.485	.023	.001	.001
MW trend	-	-	+	+	+	-	-	+	+	+	+	+	+

Note: The expression 7E-6, for example, is equal to .000007. The notation +/- indicates an increase/decrease in the value of the VC index over the period 1980–2000.

Midwest, the order was also approximately the same for both years (though somewhat different from that of the RUS in each year), but displayed an increasing trend in the majority of sectors.

5.6 THE INTRAMETROPOLITAN SCALE

The nature of interdependence within metropolitan areas could be expected to follow the structures revealed among nations and among regions within nations. In both of the latter cases, greater trade among nations (regions) has come at the expense of intranational (intraregional) flows. In addition, an examination of the Chicago metropolitan economy revealed that a similar process had also occurred when the metropolitan economy was viewed as a single area. Essentially, a hollowing-out process has occurred with greater dependence on external markets for inputs and outputs (Hewings *et al.*, 1998). Hence, there would be an expectation that a similar process might have occurred *within* the metropolitan area, thus providing a consistent pattern at the national, regional and metropolitan scales.

Table 5.9 Interzonal interdependence within the Chicago metropolitan area derived from intersectoral trade flows only

	1	2	3	4
1	**89.96**	2.40	2.17	2.21
2	2.97	**90.30**	2.77	2.83
3	1.44	1.49	**89.81**	1.38
4	5.63	5.81	5.25	**93.58**
Total	100.00	100.00	100.00	100.00

Notes: Entries show the percentage (of the total) of each column entry from the Interzonal Leontief inverse matrix based on trade flows (the bold diagonal entries are the intrazonal values).

In a recent study by Hewings and Parr (2007), the Chicago metropolitan area was divided into four zones (central area or CBD, rest of the City of Chicago, inner suburbs, and outer suburbs). An examination was made of the interzonal impacts in a series of layers, beginning with only intersectoral trade flows, then adding the impacts associated with income flows (from journeys to work) and finally the impacts generated by the spatial distribution of consumption expenditures. The summary results, shown in Tables 5.9 and 5.10, provide a different outcome from the interregional results. Table 5.9 shows the entries from the interzonal–intersectoral flows analysis and the associated Leontief matrix reveals a structure that is dominated by intrazonal flows. The degree of interzonal interdependence so revealed is very small. However, when the effects of income generation and flows (from workplace to home), together with the spatial distribution of consumption expenditures, are considered, a very different picture of interdependence within the metropolitan area is revealed (Table 5.10). In contrast to Table 5.9, there is much more interzonal dependence, reflecting the fact that the nature of the interdependencies within a metropolitan area are derived from commuting flows and consumption–expenditure flows rather than intersectoral flows.

Table 5.10 Interzonal interdependence within the Chicago metropolitan area derived from trade, income and consumption flows

	1	2	3	4
1	**48.90**	11.29	17.48	13.82
2	5.97	**47.47**	5.69	6.60
3	18.98	11.56	**49.87**	14.69
4	26.15	29.67	26.96	**64.89**
Total	100.00	100.00	100.00	100.00

Note: Entries show the percentage (of the total) of each column entry from the Interzonal Total Impact inverse matrix based on trade, income and consumption flows (the bold diagonal entries are the intrazonal values).

5.7 CLOSING COMMENTS

This chapter has introduced and compared various complementary methods for measuring inter-connection in the production processes within the Midwest economy. Employing a flow analysis, attention was first given to the values of intraregional and interregional trade flows over the past two decades. Being consistent with previous literature, this method provided clear evidence that intrasectoral trade had grown noticeably in intraregional trade and even in interregional trade. Furthermore, the Midwest-to-Midwest flows increased during the period of estimation (1980–2000). It remains to be seen whether the hollowing out process observed for the Chicago metropolitan area represents a spatially-hierarchical process, evident first at a small geographic scale. It would seem reasonable to expect that continued improvements in logistics and communications would effectively undermine any (short-term) advantages offered by the Midwest. Between 1990 and 2004, the economy of the Midwest lost 12% of its manufacturing jobs (for Illinois the loss was much higher, almost 24%, while for the nation the loss was 19%). As has been noted, however, job losses may not be entirely reflected in losses of production, since productivity gains in the Midwest have been dramatic.[4]

It has been argued that traditional notions of agglomeration economies as a dominant factor in location may need to be revised by consideration of economies that are less spatially constrained (Parr *et al.*, 2002). With lower transaction costs, firms can optimise production at establishments located in different states through specialisation, in order to take advantage of economies of scale. The production of intermediate goods in an establishment located in one region that previously may have undergone two or more transformations have now been redistributed across several regions. Thus, linkages in the commodity chain are now more spatially dispersed. The evidence provided in this chapter would suggest that there still exist some spatially-specific economies, but these are realised at the scale of the Midwest rather than an individual region or metropolitan area. The evidence for this is derived from the observed substitution of interregional for intraregional trade in the regions of the Midwest.

Examining the structure of trade at the national level, we find that there has been significant growth in the intrasectoral component of trade, although it is the service sector (covering about 60% of total output) that has heavily contributed to such an expansion. In percentage terms intrasectoral trade has actually diminished in most sectors of RUS trade flows. Compared with the RUS, vertical linkages in the Midwest are the most significant in heavy manufacturing sectors such as transportation equipment (particularly automobile production) and industrial machinery. Apart from the transportation–equipment sector, the intrasectoral component in interregional trade has not been prominent in many manufacturing sectors in the RUS, nor in the Midwest. As with other methodologies used to measure economic activity, this index approach has certain limitations, including a loss of sectoral detail within intersectoral relationships, as well as the difficulty of measuring the statistical significance level of the *VC* index, and incorporating differences due to geographic scale. Yet, the index approach, combined with the information on internal flows, provides a useful tool for analysis at the sectoral level and some important insights on structural change in the Midwest economy.

Given the nature of this volume, it is appropriate to reflect briefly on the policy implications of the foregoing analysis. To do so, we first consider the overall nature of the Midwest economy. This has been established for well over 100 years, during which period it has undergone numerous structural transformations, the one described above being the most recent. From virtually any perspective the Midwest economy must be considered a successful one, which continues to be a major contributor to the US economy. One of the important axioms of sub-national development is that successful economies are ones that are able to adjust to changing external (and sometimes internal) conditions, thus continuing to be competitive. Generally speaking, this adjustment process involves the re-marshalling of resources, broadly defined. It includes the redeployment of factors of production across sectors and over space, the re-utilisation and rationalisation of social and economic overhead capital, the encouragement of inflows of capital and specialised labour, etc. Such a process of adjustment is driven primarily by the market, through a variety of price signals, although the market may not always operate efficiently nor sufficiently promptly. Moreover, the process of adjustment may be seriously impaired by such influences as resource depletion, factor immobility and factor specificity, inappropriate spatial structures, and isolation from national cores. The presence of these influences (individually or in groups) typically causes the emergence of a problem region. The Midwest has generally not suffered from difficulties of this kind, except for localised or temporary instances. It has thus avoided overall economic decline, and maintained a degree of competitiveness. For this reason alone, policy intervention would seem to be largely unnecessary or inappropriate, particularly in view of the highly complex nature of technical and economic change that has occurred over recent decades. But denying a role for public policy in the development process is probably too extreme a stance to adopt.

In considering the role of policy, we begin at the level of the individual state (treated above as a region) within the Midwest. Clearly, each state has a perception (not always clearly defined) of its own economic and social circumstances. It also has a set of concerns with respect to current needs, the internal distribution of state expenditures, and the overall impact of these. Furthermore, each state has a set of policy instruments, although these tend not to differ substantially among states. Bearing in mind what was said above regarding the ability to adapt to change, there appears to be several emphases that the individual state may usefully pursue. These include: i) a realistic assessment of the factors likely to influence development over the medium and long term, along with the specification of a set of attainable objectives (political systems are usually averse to such approaches); ii) the efficient delivery of services for which the state has a primary responsibility or over which it exerts a strong influence, particularly in the areas of public health, education and certain aspects of infrastructure provision, including the support of public transport; iii) the identification and elimination of physical bottlenecks and related frictions (both spatial and temporal), an emphasis which may take various forms, ranging from the upgrading of skills in the workforce to the strategic programming of major transportation projects, notably highway development. One difficulty of policy intervention at the state level is that it is usually unrelated to comparable interventions taking place in neighbouring states of the Midwest. At this level, there is still scope within public policy intervention for avoiding bottlenecks, especially in relation to larger infrastructure projects. While no multi-state tier of government for the Midwest exists (nor is this likely to emerge), serious attention might be given to the possibility of interstate compacts, either on a bi-lateral or multi-lateral basis, by which states are able to cooperate, in order to avoid incompatibility of investment decisions or the wasteful duplication of facilities.

Historically, the fostering of interstate relations has been the responsibility of the federal government or its agencies. One possibility here is for the federal government to avoid these problems of incompatibility and duplication by allocating funds among states in such a manner that difficulties of this kind are minimised. There is also the potential for federal government to play a pivotal role in the interstate regulation of transportation and telecommunications. These sectors lie at the heart of the information economy, upon which so many of the developments described in this chapter depend. In the past the federal government has also made various forays into regional policy with the creation of multi-state entities such as the Tennessee Valley Authority, the Bonneville Power Administration, and the Appalachia Regional Commission, each representing a significant intervention. However, political realities, together with the present levels of development and prosperity within the Midwest, would appear to rule out intervention along these lines. We conclude this chapter by asserting that just as the regional economy has become more structurally sophisticated in recent decades, so must the nature and scope of policy intervention, if this is to be effective. The days of 'smokestack chasing' and 'industry retention' or ill-judged spending on infrastructure are long gone (or certainly should be), and in settings such as the Midwest many of the traditional approaches to regional policy are likely to be blunt and ineffectual. Successful regional policy at any level must involve an appreciation of existing economic and technological complexities, as well as a carefully-designed response to these.

NOTES

1. The authors wish to thank the Carnegie Trust for the Universities of Scotland for its support in the preparation of this chapter. The computational assistance of John Seo is gratefully acknowledged.
2. The three properties of the Maddigan index are as follows: i) the index increases (decreases) when an input sector becomes relatively more (less) important by accounting for a larger (smaller) percentage of the total value of output of the sector being supplied; ii) the index increases (decreases) when relatively more (less) of the output of a supplying sector is used as an input to another sector; iii) the index increases if there is any increase in vertical connections (other than as a result of conglomerate merger activity) between a new sector and the sector's established product line.

3. Over the past decades vertical connections have increased significantly in US agriculture in order to exploit economies of scale and monopolistic market power in the supply of more versatile and differentiated products.
4. In the Chicago metropolitan area, for example, manufacturing production fluctuated over the period 1970–1990. However, the 1990 level was greater (in real terms) than the 1970 level, although there were 44% fewer employees.

APPENDIX

Table A1 Sector mnemonics in the Midwest-REIM

Mnemonic	Sector title	SIC
1	Agriculture, forestry and fisheries	01,02,07,08,09
2	Mining	10,12,13,14
3	Construction	15,16,17
4	Food and kindred products	20
5	Chemicals and allied products	28
6	Primary metal industries	33
7	Fabricated metal industries	34
8	Industrial machinery and equipment	35
9	Electronic and other electric equipment	36
10	Transportation equipment	37
11	Other non-durable manufacturing products	21-23,26,27,29-31
12	Other durable manufacturing products	24,25,32,38,39
13	TCU, service, and government enterprises	40-42,44-65, 67, 70,72,73,75,76

Aggregation Scheme	
1,2,3	Primary
4,5,11	Non-durable manufacturing
6,7,8,9,10,12	Durable manufacturing
13	Services

REFERENCES

Ciccone, A. and Hall, R. (1996), 'Productivity and the density of economic activity', *American Economic Review*, **86**, 54–70.

Ellison, G. and Glaeser, E. (1997), 'Geographic concentration in US manufacturing industries: A dartboard approach', *Journal of Political Economy*, **105**, 889–927.

Frank, S. D. and Henderson, D. R. (1992), 'Transaction costs as determinants of vertical coordination in the US food industries', *American Journal of Agricultural Economics*, **74**, 941–50.

Fujita, M. and Thisse, J.-F. (1996), 'Economics of agglomeration', *Journal of Japanese and International Economics*, **10**, 339–78.

Hewings, G. J. D. and Parr, J. B. (2007), 'Spatial Interdependence in a metropolitan setting', *Spatial Economic Analysis*, **2**, 7–22.

Hewings, G. J. D., Sonis, M., Guo, J., Israilevich, P. and Schindler, G. R. (1998), 'The hollowing-out process in the Chicago economy, 1975-2011', *Geographical Analysis*, **30**, 217–34.

Hummels, D., Ishii, J. and Yi, K.-M. (2001), 'The nature and growth of vertical specialization in world trade', *Journal of International Economics*, **54**, 75–96.

Israilevich, P. R., Hewings, G. J. D., Sonis, M. and Schindler, G. R. (1997), 'Forecasting structural change with a regional econometric input–output model', *Journal of Regional Science*, **37**, 565–90.

Klier, T. H. (1998), 'Geographic concentration in the US manufacturing: evidence from the U.S. auto supplier industry', Discussion Paper, Research Department, Federal Reserve Bank of Chicago, Chicago.

Krugman, P. (1998), 'Space: The final frontier', *Journal of Economic Perspectives*, **12**, 161–74.

Maddigan, R. J. (1981), 'The measurement of vertical integration', *Review of Economics and Statistics*, **63**, 328–35.

Maurel, F. and Sedillot B. (1999), 'A measure of geographic concentration in French manufacturing industries', *Regional Science and Urban Economics*, **29**, 575–604.

Munroe, D. K., Hewings, G. J. D. and Guo, D. (2007), 'The role of intraindustry trade in interregional trade in the Midwest of the US', in R. J. Cooper, K. P. Donaghy and G. J. D. Hewings (eds), *Globalization and Regional Economic Modeling*, Heidelberg, Springer-Verlag.

Parr, J. B., Hewings, G. J. D., Sohn, J. and Nazara, S. (2002), 'Agglomeration and trade: Some additional perspectives', *Regional Studies*, **36**, 675–84.

Sonis, M., Oosterhaven, J. and Hewings, G. J. D. (1993), 'Spatial economic structure and structural changes in the European common market: Feedback loop input–output analysis', *Economic Systems Research*, **5**, 173–84.

Sonis, M., Guilhoto, J. J. M. and Hewings, G. J. D. (1995), 'The Asian economy: trade structure interpreted by feedback loop analysis', *Journal of Applied Input–Output Analysis*, **2**, 24–40.

Sonis, M., Hewings, G. J. D. and Okuyama, Y. (2002), 'Vertical specialization and spatial production cycles in interregional trade: Feedback loops analysis of the Midwest economy', in G. J. D. Hewings, M. Sonis and D. Boyce (eds), *Trade, Networks and Hierarchies*, Heidelberg, Germany: Springer Verlag.

Swonk, D. C. (1996), 'The Great Lakes economy revisited', a series paper for the 1996 Workshop, The Midwest Economy: Structure and Performance, Chicago, IL: Federal Reserve Bank of Chicago.

Wheeler, C. H. (2001), 'A note on the spatial correlation structure of county-level growth in the US', *Journal of Regional Science*, **42**, 433–49.

6. Mature Industries and Declining Regions: An Analysis of the Spanish Case[1]

Josep-Maria Arauzo-Carod and Elisabet Viladecans-Marsal

6.1 INTRODUCTION

During the 1970s and 1980s most advanced industrial economies suffered major regional decline, reflecting the consequences of the crisis in the early to mid-1970s. The economic downturn was felt most severely in the more mature industrial sectors of coal, steel, iron, shipbuilding and textiles (Rodwin and Sazanami, 1991; Heim, 1997) and the regional impact was most marked in areas specialising in these activities, e.g. several British, Belgian, Spanish and French regions. The decline was characterised by high rates of regional unemployment with the loss of many jobs in large manufacturing corporations, a fall in economic growth and a drop in GDP per capita. On the basis of these characteristics, the areas are referred to as either Old Industrial Regions or Mature Industrial Regions.[2] Today, it is particularly important to analyse the restructuring processes experienced in these regions and to examine how these areas have gone about transforming the skills of their labour force and entered into a new phase in which the market for their previous specialisation has declined. More specifically, it is interesting to study the decline in manufacturing activity, the change in specialisation patterns and the potential increase in private and public services.

The restructuring processes mentioned above have operated over different time scales according to the specific activity. Thus, the readjustments in the iron and steel industries had largely been concluded by the end of the 1980s, while in the textile industries this process has yet to be finished. In fact, the European textile firms of the former EU15 countries are now having to face an increase in international competition due to the enlargement of the EU in 2004 and the 2005 WTO agreement which has liberalised the international commerce of textile products. Thus, the sector faces further cutbacks in employment levels[3] and the closure of the firms affected which, in some cases, have transferred their production to other countries. This increase in international competition and the pressure on large firms to shift their plant localisation is beginning to be felt in other industrial sectors, such as electric components and transport materials, in particular the production of vehicles.

Here, we wish to examine in detail the evolution in the geographic performance of a specific set of industrial sectors, namely those of a mature technological level (mature or traditional sectors). These sectors have experienced a constant decline within the manufacturing sector as a whole, with the process unfolding in the context of the modernisation of manufacturing industries in general. If we examine the Spanish industrial mix, it becomes apparent that the mature manufacturing sectors – those most affected by the restructuring process, still weigh more heavily in Spain than they do in other OECD countries. This is somewhat surprising, particularly if we consider the great difficulties these sectors face in generating added value and competing internationally and, moreover, if we take into account the restructuring process described above.[4] Nevertheless, in spite of this process and because of the rigidities of Spain's industrial structure, the Spanish pattern of specialisation has become even more focused around these manufacturing activities in recent decades. In fact, in 1966 the mature industrial sectors accounted for 70.3% of all industrial activities and by 2002 they still represented around 60%. This share is considerably higher than the average figure for the EU, which in 2002 stood at 48.7% (Myro and Gandoy, 2005).

Many studies have emphasised the great geographic inequality in the spatial distribution of manufacturing, both in terms of the economy as a whole and of individual industrial activities (see Viladecans, 2004). Thus, manufacturing has become concentrated in the more populated areas. At the same time, given that the characteristics of the Spanish regions are far from being homogeneous, the marked regional specialisation levels reflect comparative advantages and an area's manufacturing tradition. The concentration of manufacturing activities in certain Spanish regions seems to suggest that the restructuring process has been more successful in these regions and so the problems that this process brought with it have been highly localised in space.

We have chosen to organise the rest of this chapter as follows. In the second section, we outline the characteristics of the Spanish regions so as to illustrate the heterogeneity of the Spanish regional map in terms of its surface area, population, employment in the industrial and service sectors, unemployment rates and GDP per capita. Following this description, we analyse the evolution in these regional economies from the early 1970s to the present day and highlight the differences in regional performance. In the third section, we focus our study on the mature industrial sectors and examine the Spanish geography of these activities, their evolution and regional specialisations. In the fourth section, after defining our criteria, we seek to identify the regions of Spain that are in decline, in other words, the heavily industrialised regions, specialised above all in the mature sectors, which have had to face major restructuring processes since the 1970s.

Finally, in the fifth section we characterise some of the elements that might determine the success or otherwise of the restructuring processes in the declining regions. We finish the chapter by summarising the conclusions that can be drawn from the preceding discussion.

6.2 SOME INSIGHTS INTO THE SPANISH REGIONAL ECONOMY

6.2.1 The Territorial Units of Spain

Before analysing Spain's regional economy, it is necessary to understand how the country is organised territorially. Spain is organised for administrative purposes into municipalities, provinces and regions, the so-called Autonomous Communities (AC). More specifically, the 8108 municipalities lie within the 50 provinces and 17 AC (plus the autonomous cities of Ceuta and Melilla in North Africa). The Autonomous Communities, which occupy Level II of the Eurostat's Nomenclature of Territorial Units for Statistics (NUTS), can be considered the equivalent regional division.

Spain's regions are far from being homogeneous and present significant differences in terms of their surface area, population, employment structure, unemployment and GDP per capita (see Table 6.1). More specifically, the Spanish regions vary greatly in size with the regions of Castile-La Mancha, Andalusia, and Castile and Leon accounting for more than half the territory of Spain. By contrast, the area of three other regions – Balearic Islands, Cantabria and La Rioja – represents just 3% of the whole territory. However, the regions with a relatively small surface area tend to concentrate a greater proportion of the whole population. This is the case of Catalonia, the Valencian Community and the Madrid Community which account for 15.8, 13.5 and 10.6% of Spain's population, respectively.

Similarly, there is a marked territorial concentration of economic activities within just a few of the more developed regions, in particular Catalonia, the Madrid Community and the Basque Country. It is here that the most advanced and specialised services and the most innovative manufacturing

Table 6.1 Some insights into the Spanish regions – 2005

Regions	Surface area (%)	Population (%)	Jobs (%)	Manufacturing jobs (%)	Service-sector jobs (%)	Unemployment rate	GDPpc (Spain=100)*
Andalusia	17.36	17.80	15.66	10.44	15.43	13.83	75.56
Aragon	9.45	2.88	2.94	3.81	2.67	5.65	108.48
Asturias	2.10	2.44	2.14	2.54	2.07	9.56	85.68
Balearic Islands	0.99	2.23	2.37	1.22	2.66	7.48	118.78
Canary Islands	1.48	4.46	4.47	1.76	5.15	10.67	90.75
Cantabria	1.04	1.27	1.27	1.22	1.27	8.10	94.61
Castile-La Mancha	18.59	4.30	4.04	4.59	3.50	9.39	83.66
Castile and Leon	15.74	5.69	5.34	5.56	4.98	8.54	98.72
Catalonia	6.36	15.86	17.39	24.07	16.92	6.64	118.21
Valencia	4.61	10.64	10.90	13.15	10.43	8.01	99.40
Extremadura	8.25	2.46	1.99	1.23	1.87	15.31	73.52
Galicia	5.86	6.26	5.89	6.41	5.44	9.13	85.42
Madrid	1.59	13.52	15.02	10.46	17.59	5.92	123.94
Murcia	2.24	3.03	3.03	2.84	2.58	7.42	79.75
Navarre	1.94	1.35	1.46	2.08	1.35	5.95	122.93
Basque Country	1.40	4.82	5.08	7.36	5.13	6.40	121.81
Rioja, La	1.00	0.68	0.74	1.20	15.43	6.53	113.93
SPAIN	100.00	100.00	100.00	100.00	100.00	8.70	100.00

Note: * 2003 data.
Source: Authors' own based on data from INE (Spanish Statistical Institute).

activities have located and where the leading firms have chosen to site their headquarters. The concentration of manufacturing activity is particularly high, with Catalonia alone accounting for 24.1% of all Spanish industrial employment. This region, together with the Madrid Community and the Basque Country, accounts for 42.5% of this employment in this sector. In the case of service activities, the spatial distribution is a little more homogeneous as public services by their very nature tend to be more equally distributed throughout a territory. However, the uneven distribution of economic and industrial activities results in markedly different economic performances across the regions. The unemployment rate varies greatly between regions and is above the Spanish average in the regions of Extremadura, Andalusia, the Canary Islands and Asturias. At the same time, all these regions have a level of GDP per capita below that of the Spanish average.

6.2.2 The Evolution in the Regional Economies Over Recent Decades

If we examine the spatial distribution of economic activity in the Spanish regions during the second half of the 20th century some clear patterns emerge. The Spanish economy underwent rapid growth during the 1960s and the first half of the 1970s, largely as a result of economic liberalisation and the opening up of the Spanish economy to international markets. This was followed by a period of intense economic crisis during the second half of the 1970s and the early 1980s. Following Spain's entry to the European Community in 1986, the country enjoyed a new period of economic growth, but this was cut short by the economic crisis of the early 1990s. Finally, from the mid-1990s to the present day, the Spanish economy has enjoyed considerable economic growth. In fact, and with the exception of the most recent stage during which the economy has grown at a faster rate than that recorded in most other European countries, the Spanish economy has stayed largely in line with average European rates of growth. The changes in the Spanish economy during these years can best be described by examining four factors: population, industrial activity and specialisation indexes, unemployment and GDP per capita.

The demographic data reveal the migration of population from the less developed (agrarian-based) to the more developed (industrialised and/or services-based) regions. For the period 1950–2005, the change in population in the less developed regions (Andalusia, Aragon, Asturias, Cantabria, Castile and Leon, Castile-La Mancha, Extremadura, Galicia, Navarre and La Rioja) was below the Spanish mean, while the opposite was true for the more developed regions (Balearic Islands, Basque Country, Canary Islands, Catalonia, Madrid Community, Murcia and Valencian Community), as can be seen in Figure 6.1.

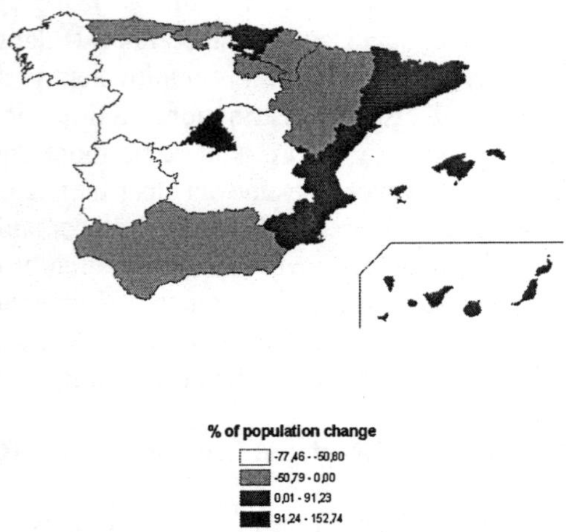

Notes: Positive values: population change above Spanish mean.
Negative value: population change below Spanish mean.

Figure 6.1 Population change between 1950 and 2005

Nevertheless, there were some exceptions to this pattern, namely Asturias and Cantabria, for instance, whose share of the Spanish population began to fall after the economic crisis of the early 1970s.

The industrial mix in Spain experienced an important shift towards manufacturing activities during the 1950s and the 1960s (a process that had come to an end by the mid-1970s) and towards service activities throughout the period analysed. This shift saw the loss of 1 292 000 jobs in the agricultural sector between 1955 and 1995, most of which were located in Andalusia (442 000), Castile-La Mancha (207 000), the Valencian Community (140 000), Extremadura (135 000) and Castile and Leon (117 000). The majority of these regions were heavily dependent on agriculture and their income level was well below the Spanish mean (their GDP per capita in 1955 was only 74.3% of the Spanish mean). Similarly, between 1955 and 1995 employment in manufacturing activities as a share of total employment fell from 29.3 to 21.2%, while employment in service activities rose from 45.3% in 1955 to 63.8% in 1995.[5]

As can be seen in Figure 6.3 (a-d), there has been a marked process of de-industrialisation since the 1970s, during which time manufacturing activities have shed about 587 000 jobs in Spain. Those job losses have been mainly in the mature industries, including textiles (200 000), metallic minerals and steel (101 000), wood and furniture (81 000) and minerals and non-metallic mineral products (70 000) and were due largely to closures and bankruptcies of big established firms.

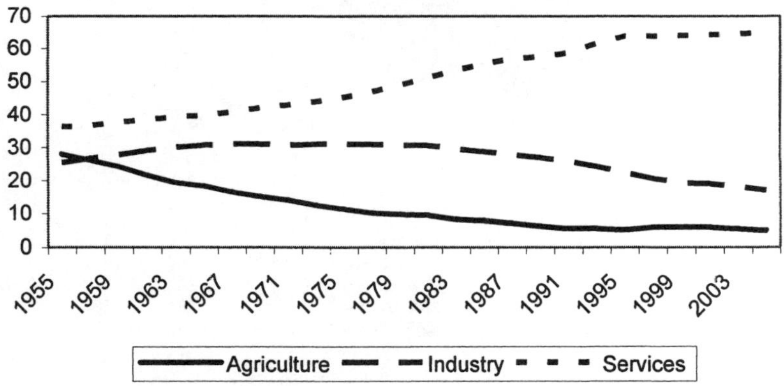

Source: Authors' own using data from BBVA Foundation and Funcas.

Figure 6.2 *The evolution in Spain's industrial mix (1955–2005) (% of employment in each sector)*

The unemployment rate shows a similar picture.[6] With the exception of the specific shocks caused by manufacturing crises (Catalonia and the Basque Country) and by the adjustment process in the mature regions (Asturias and Cantabria), the agrarian regions typically have higher unemployment rates than the Spanish mean.

6.3 MATURE MANUFACTURING ACTIVITIES

6.3.1 Data

Our main data describing the industrial mix in the Spanish regions are taken from the BBVA Foundation and range from 1955 to 1995. As well as providing data for a considerable period of time (almost half a century), the

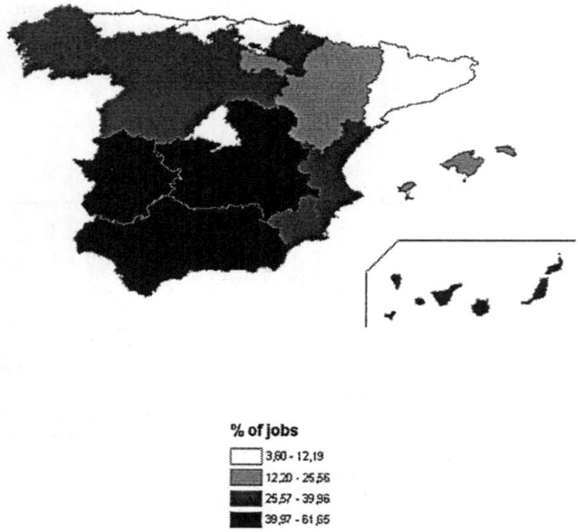

Figure 6.3a Industrial mix between 1955 and 2005: percentage of job in agriculture in 1955

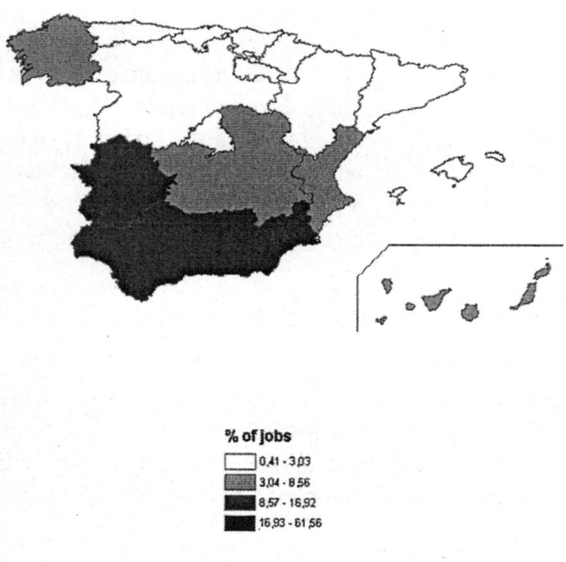

Figure 6.3b Industrial mix between 1955 and 2005: percentage of jobs in agriculture in 2005

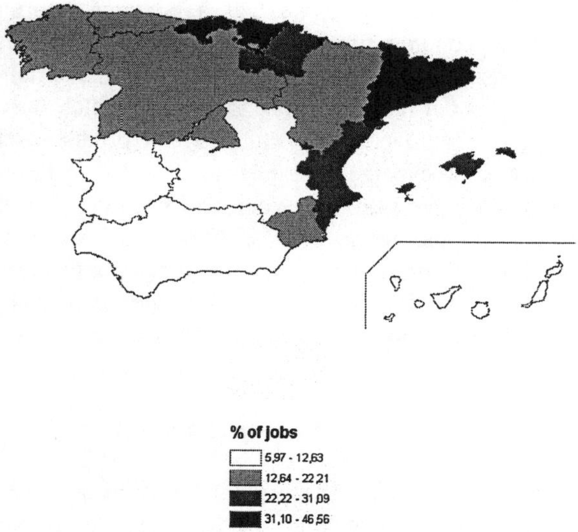

Figure 6.3c *Industrial mix between 1955 and 2005: percentage of jobs in manufacturing in 1955*

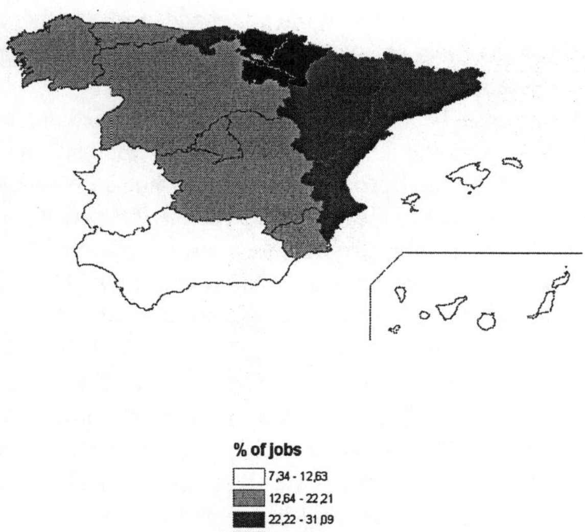

Figure 6.3d *Industrial mix between 1955 and 2005: percentage of jobs in manufacturing in 2005*

database has additional advantages in that it contains information about several fundamental economic transformations and also allows us to work at a provincial level. Its main drawback is the size of the industrial aggregation in that it includes just ten manufacturing sectors, of which only six can be identified as mature manufacturing sectors (Metal products; Non-metallic mineral products; Food, beverages and tobacco; Textiles, leather and clothes; Pulp, paper, publishing and printing; Wood and furniture).[7] In addition, we have access to data for the period 1995 to 2004. These are drawn from the *Encuesta Industrial* (Industrial Survey EI)[8] and include a greater aggregation by sector (up to 13 manufacturing sectors). All other statistical information comes from the Spanish National Statistical Institute.

6.3.2 The Evolution in Mature Industrial Activities

Traditional industrial activities still represent a significant share of Spain's manufacturing activities, although they have suffered a continuous fall in employment over recent years, both in relative and absolute terms. This process has unfolded within the general context of the modernisation of manufacturing activities which has seen sectors with a greater capacity to generate added value being promoted. However, despite this modernisation and because of the rigidities of Spain's industrial structure, traditional activities still account for 59.6% of the value added of Spanish manufactures (2002), a lower value than in 1966 (70.3%) but undoubtedly higher than that recorded elsewhere in western Europe.

Manufacturing can be divided into mature, intermediate and advanced activities, according to the level of technological intensity and the market growth experienced by their products. Table 6.2 shows the industries that are included here within each of these groups. The territorial distribution of these activities in Spain is far from homogeneous, because of the high concentration of mature activities in just a few regions. Nevertheless, these territorial differences have been reduced in recent years. Thus, while in 1955 employment in the mature manufacturing activities was concentrated within five provinces[9] (Barcelona, Madrid, Vizcaya, Valencia and Guipúzcoa), accounting for 52% of all employment in these activities,[10] by 1995, this figure had fallen to 44% (mainly because Barcelona's share fell to just 17.6%). It appears to be the case that this reduction in traditional manufacturing activities shows a different intensity at the territorial level.[11] We should also add that these changes in industrial mix have been accompanied by a de-concentration of manufacturing activities (similar to that experienced by most European countries; see, for example, Chapman and Walker, 1991; Delisle and Lainé, 1998; Scott, 1982) towards regions of lower density (Costa and Viladecans, 1999). However, the extent of this phenomenon appears to have varied according to the sector.

Table 6.2 *Manufacturing industries grouped according to their technological intensity*

Advanced industries
Manufacturing of office machinery and computers
Manufacturing of radio, TV, communication equipment, medical, precision and optical instruments, watches and clocks
Intermediate industries
Manufacturing of chemicals and chemical products
Manufacturing of rubber and plastic products
Manufacturing of machinery and equipment
Manufacturing of other transport equipment
Mature industries
Manufacturing of basic metals and fabricated metal products
Manufacturing of other non-metallic mineral products
Manufacturing of food products, beverages and tobacco
Manufacturing of textiles, leather and clothes
Manufacturing of pulp, paper and paper products
Manufacturing of wood and furniture

Source: Adapted from Myro and Gandoy (2005).

Alañón (2004) has demonstrated a positive spatial autocorrelation across mature industries at the provincial level in 1955. In other words, the spatial values of the variable are not random but rather depend on the distances and economic relations between neighbouring provinces.[12] This result implies that at this time mature industries were geographically concentred (at the industry level, though not for all mature activities) in selected neighbouring provinces. However, in 1995 only half the mature industries showed a positive spatial autocorrelation, indicating that these activities have become more dispersed.

If we examine the territorial distribution of these mature activities in Spain, there is no obvious pattern of developed regions in contrast with less developed regions (in which these activities should supposedly be located). In fact, the distribution pattern is one of industrialised regions versus less industrialised regions (with the mature activities being located in the former). The evolution of employment in these industries shows how mature industries have suffered considerable losses, while intermediate industries have undergone rapid growth and advanced industries have increased their share slightly.

Taking the description of Spain's industrial map one step further, we next analyse the specialisation of the regions in the various industrial activities. To do this we calculate a Location Quotient index of geographical specialisation, which compares the regional concentration of each industrial activity with the country's mean concentration. An index with a value greater than one indicates that the sector has a greater representation in a particular region than in Spain as a whole. By contrast, an index with a value less than one indicates that the sector does not have such a strong presence in the region as in the country as a whole.

Table 6.3 Specialisation indices

1975		1985		1995	
Metallic minerals, iron and steel:		Metallic minerals, iron and steel:		Metallic minerals, iron and steel:	
Asturias	(7.0)	Asturias	(8.4)	Asturias	(11.4)
Basque Country	(3.0)	Basque Country	(3.2)	Cantabria	(4.1)
Cantabria	(2.8)	Cantabria	(2.9)	Basque Country	(3.0)
				Navarre	(2.8)
Food and beverages:		Food and beverages:		Food and beverages:	
Canary Islands	(3.3)	Canary Islands	(3.1)	Canary Islands	(2.8)
Extremadura	(3.0)	Extremadura	(3.0)	Extremadura	(2.8)
Murcia	(2.7)	Murcia	(2.4)	Murcia	(2.2)
Andalusia	(2.2)				
Transport materials:		Transport materials:		Paper and printing:	
Galicia	(3.2)	Galicia	(2.1)	Madrid Community	(2.2)

Source: Authors' own based on data drawn from the BBVA Foundation.

We calculate the index of geographical specialisation for all the regions and each industrial activity from 1975 to the present day. Our results show that, in most cases, the industries that were most highly specialised at the regional level were the mature industries such as Metallic minerals, iron and steel, and Food and beverages. Moreover, this specialisation was found to be particularly marked in two types of region: those dependent primarily on agriculture (Extremadura, Galicia and the Canary Islands) and those dependent primarily on mature manufacturing activities (Asturias, Cantabria and the Basque Country). In line with other similar analyses, our results also show that this level of specialisation is decreasing over time. The index values are smaller today than those recorded thirty years ago, while the number of specialised regions is also decreasing over time. The results for 2004 also corroborate these findings. Table 6.3 shows the regions and sectors with a specialisation index greater than two for the years 1975, 1985 and 1995 (years in which the database is homogeneous). We choose to record the regions and sectors with a specialisation greater than two as we wished to highlight those exhibiting the most intense patterns of specialisation.

As can be seen, regional specialisation occurs each year in just four sectors – namely: Metallic minerals, iron and steel; Food and beverages; Transport materials (in just one region); and Paper and printing (also in just one region). Among the more surprising results recorded is the specialisation index for Metallic minerals, iron and steel in Asturias. Not only did this region specialise in this mature industry at the beginning of the period under analysis (specialisation index = 7.0), but it increased its level of specialisation: 8.4 (1985) and 11.4 (1995).

6.4 DECLINING MANUFACTURING REGIONS (DMR) IN SPAIN

While it is relatively simple to identify the existence of regional decline in the 1970s and 1980s, it is not so easy to identify which regions were involved or to accurately measure the rate of that decline.[13] The European Commission has proposed its own definition of declining industrial regions. To be classified as such it is necessary, first, to be industrial (that is, to be specialised in manufacturing activities) and, second, to present a certain specific criterion of decline. While it is not difficult to apply these two broad criteria, the definition lacks accuracy. To counter this, the European Commission provides greater details as to what an industrial region is (see, for example, the *Fifth Periodic Report* and the *Third Periodic Report*). But determining the set of industrial regions is only the first step. It then has to be decided which of them is in decline. Here, the European Commission relies on rates of unemployment and job loss. However, the use of both variables is not without their problems. Unemployment rates are not always the most appropriate measure as business cycles do not occur simultaneously, while there are certain specific institutional factors within a country that might lead to more regions suffering problems of unemployment than in other countries. Similarly, job loss figures have certain drawbacks as, at times, this variable can reflect the consequences of investment and technological change and increases in labour productivity.

Table 6.4 *Declining manufacturing regions in Spain*

	A	B	C	D	TOTAL
Andalusia	--	--	--	--	0
Aragon	X	X	--	--	2
Asturias	X	X	X	X	4
Balearic Islands	--	X	X	X	3
Canary Islands	--	--	--	--	0
Cantabria	X	X	X	X	4
Castile-La Mancha	--	--	--	--	0
Castile and Leon	--	X	--	X	2
Catalonia	X	X	X	X	4
Valencia	X	X	--	X	3
Extremadura	--	X	--	--	1
Galicia	--	--	--	--	0
Madrid Community	--	--	X	X	2
Murcia	--	--	--	X	1
Navarre	X	--	--	--	1
Basque Country	X	X	X	X	4
Rioja, La	X	X	--	--	2

Notes: A: Percentage engaged in manufacturing sector greater than the Spanish average in 1975.
B: Specialisation index in declining manufacturing sectors (*Metallic minerals, iron and steel* and *Textile, leather and footwear*) greater than 1 in 1975.
C: Decrease in manufacturing employment greater than the Spanish average (1975-2005).
D: GDP growth rate below that of the Spanish average (1975–2003).

Source: Authors' own.

Taking all these factors into consideration, we may assume that a region can be considered to be a *Declining Manufacturing Region* (DMR) if it presents the following characteristics:

a) The region was specialised in manufacturing activities at the beginning of the period.

b) The region was specialised in those manufacturing sectors most severely affected by the industrial decline following the crisis of the early 1970s.
c) The percentage of the workforce engaged in manufacturing in the region fell at a faster rate than that for the country as a whole during the period under analysis.
d) The GDP per capita growth is below that of the Spanish mean during the period under analysis.

Table 6.4 shows the results of applying these criteria to the Spanish regions for the period 1975–2005. The first two criteria refer to the situation in the regions at the beginning of the period, i.e., in 1975.

In the case of the first criterion, which considers the relative weight of industrial employment compared to employment in all sectors in each region in 1975, industrial employment was higher than the Spanish average in eight regions (at a time when 27.4% of all employment was concentrated in the industrial sectors). Most notable among these regions are the Basque Country (with 45.6% of its workforce employed in the industrial sector), Catalonia (with 39.8%), La Rioja (36.8%) and Navarre (35.9%) (Figure 6.4a). The second criterion identifies those regions that had a high level of specialisation in declining manufacturing activities. Here, the industries that have been considered as such are *Metallic minerals, iron and steel* and *Textile, leather and footwear,* on the basis that both activities suffered the greatest decreases in employment following the crisis in the early 1970s. After calculating the specialisation index for both sectors and each region, ten regions out of the 17 were found to have been specialised in one of these activities. These findings demonstrate the considerable presence of mature industries in Spain and their spatial concentration in many regions of the country (Figure 6.4b).

In the case of the *Metallic minerals, iron and steel* sector, the degree of specialisation was particularly high in Asturias, the Basque Country and Cantabria (with an index of specialisation[14] of 6.0, 3.0 and 2.8, respectively). In the case of the *Textile, leather and footwear* sector, the specialisation index values were not so great, while the regions with the highest degree of specialisation in the mid-1970s were the Balearic Islands (1.9 specialisation index), Valencia (1.7), La Rioja (1.6) and Catalonia (1.5).

The third and fourth criteria consider the evolution in the industrial/regional economies from the mid-1970s to the present day. The third criterion, which identifies the regions in which the decrease in industrial activity has been greater than the Spanish average over the period, is met by six of the 17 Spanish regions. The fall in activity has been particularly marked in Asturias, the Basque Country and Cantabria (Figure 6.4c).

Figure 6.4a *Declining manufacturing regions: percentage of manufacturing employment higher than the Spanish average in 1975*

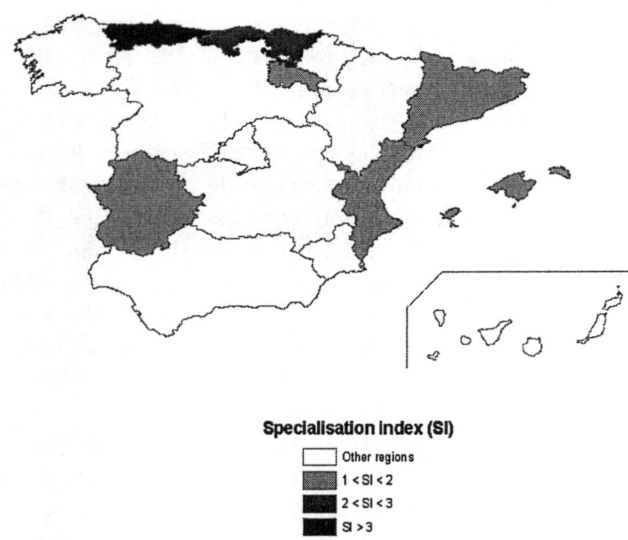

Figure 6.4b *Declining manufacturing regions: specialisation index (SI) in declining manufacturing sectors (Metallic minerals, iron and steel and Textile, leather and footwear) in 1975*

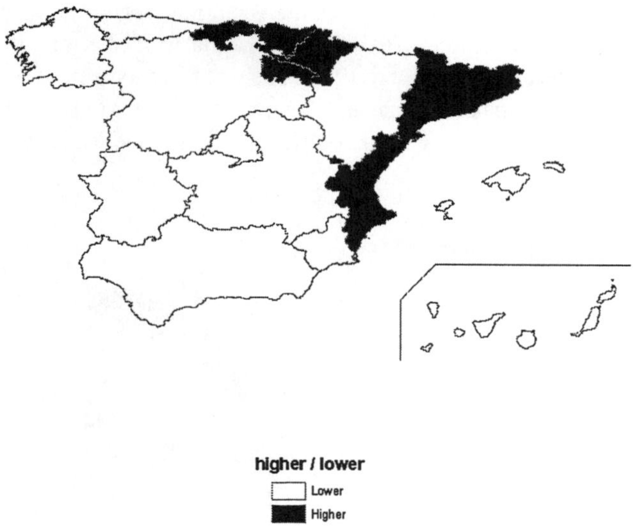

Figure 6.4c *Declining manufacturing regions: decrease in manufacturing employment greater than the Spanish average (1975–2005)*

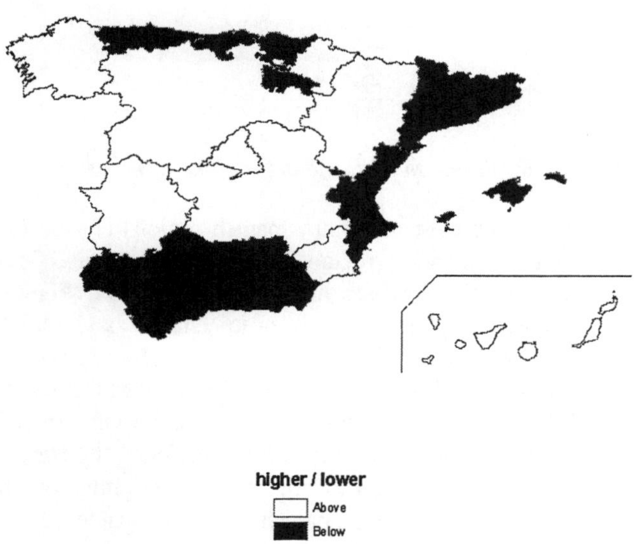

Figure 6.4d *Declining manufacturing regions: GDP growth below that of the Spanish average*

Finally, the last criterion describes the economic performance of a region by analysing the evolution in its GDP per capita from the mid-1970s to the present day. Here, eight of the 17 Spanish regions present a growth rate below that of the Spanish mean for the period. The regions with the worst performances were Asturias, the Balearic Islands and the Madrid Community (Figure 6.4d).

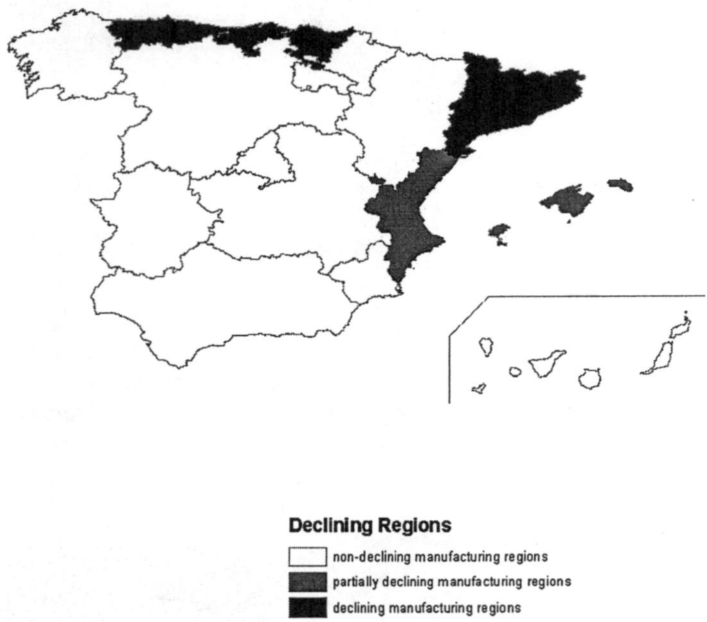

Figure 6.4e Declining manufacturing regions: synthesis

On the basis of this evidence, four Spanish regions can be identified as being declining manufacturing regions: Asturias, Cantabria, Catalonia and the Basque Country, all of which meet all four criteria (Figure 6.4e). In addition, two other regions (the Balearic Islands and the Valencian Community) may be considered declining manufacturing regions, albeit to a lesser degree, as they fulfil three of the four criteria. Our results match fairly well with the Spanish regions designated Objective 2 areas[15] in EU Regional Policy (for the period 2000–2006), given that Catalonia, the Basque Country and the Balearic Islands are considered as being eligible for Objective 2 funding. Cantabria is the only Spanish region considered eligible for Objective 1 provisional support (phasing-out region).

6.5 THE CHARACTERISATION OF SPAIN'S MATURE INDUSTRIAL REGIONS

The economic performance of the declining manufacturing regions identified above varies considerably. Table 6.5 shows that the GDP per capita in these regions was well above the Spanish average at the beginning of the period analysed (1955), i.e. just before Spain entered its main period of industrialisation (1959–1975). However, by 1975 the relative GDP per capita in these regions had fallen, leaving them only slightly above the Spanish average.

Table 6.5 Relative situation of Spain's declining manufacturing regions (average Spanish GDP pc=100 and growth rates of this average)

Regions	1955	1975	1996	2004	1955 –75	1975 –96	1996 –04	1955 –04	1975 –04
Asturias	119.8	105.0	85.9	87.1	-14.8	-19.1	1.2	-32.7	-17.9
Basque Country	190.4	135.7	112.3	124.5	-54.7	-23.4	12.2	-65.9	-11.2
Cantabria	123.5	103.7	90.6	94.3	-19.8	-13.1	3.7	-29.2	-9.4
Catalonia	158.5	127.7	122.5	115.5	-30.8	-5.2	-7.0	-43.0	-12.2

Source: Authors' own based on data drawn from Parellada (2005).

After the onset of decline beginning in the mid-1970s, and following the restructuring processes in the manufacturing sector, only two of these regions continue to exhibit a higher GDP per capita than the Spanish average (the Basque Country and Catalonia: 124.5 and 115.5, respectively, where the average Spanish figure is equal to 100). Although the mature manufacturing activities have declined, they continue to retain a certain importance in these regions, while the industrial mix has become more diversified and export oriented.

The four declining manufacturing regions can be divided in two groups: one comprising the neighbouring regions of Asturias, Cantabria and the Basque Country – all of them located in the North of Spain, and the other comprising Catalonia.[16] The first of these two groups is characterised by a high degree of specialisation in heavy industries such as the Metallic minerals, iron and steel, while the second group is characterised by its specialisation in light industry, such as Textiles, leather and footwear. The specialisation indexes for the heavy industry sector in 1975 were 6.95 (Asturias), 2.78 (Cantabria) and 3.21 (the Basque Country), while the specialisation indices for textiles, also in 1975, were 1.70 for Catalonia. If we examine the same specialisation indices 20 years later (in 1995), we find an increase in specialisation in Metallic minerals, iron and steel products in Asturias and Cantabria and a decrease in specialisation in the same sector in the Basque Country and in Textiles in Catalonia. Thus, two of these regions have become more specialised in mature activities (Asturias and Cantabria), while the other two have become less specialised (the Basque Country and Catalonia).

If we examine the general economic performance of these regions (Table 6.6), the two regions that have increased their specialisation in these mature sectors have seen their GDP per capita fall below the Spanish average, while the two regions in which this specialisation has decreased have maintained their GDP per capita above the Spanish average (albeit that it has fallen in relative terms). Table 6.6 shows the extent of the shift in specialisation in these traditional activities to a new scenario in which the numbers engaged in these sectors have fallen considerably (above all in the Basque Country).

Table 6.6 Weight of mature manufacturing activities as a percentage of all manufactures (jobs)

	1975		1995		1995-75	
Regions	Jobs	%	Jobs	%	Jobs	%
Asturias	61 904	80.2	35 905	69.3	-25 999	-42.0
Basque Country	129 804	44.2	57 739	33.8	-72 065	-55.5
Cantabria	23 791	53.3	14 928	49.0	-8 863	-37.3
Catalonia	424 545	56.3	269 126	48.9	-155 419	-36.6

Source: Authors' own based on data drawn from Parellada (2005).

The fall in relative GDP per capita in the Basque Country and Catalonia should be seen within the context of an intense income convergence process at the Spanish regional level. Furthermore, since the early 1960s a process of rapid industrialisation has permitted less developed regions to grow quickly. Thus, the changes in the relative wealth of regions such as the Basque Country and Catalonia should be attributed to the rapid growth of the poorer regions rather than to a deterioration in their economic conditions.

Nevertheless, the growth of each region is highly dependent on its development strategies. Thus, the stagnation suffered by the economy in Asturias can be largely attributed to the failure of its industrial and development policies for the restructuring of its manufacturing sector. In Asturias the industrial restructuring process has coincided with a similar process in the agricultural sector and, until recently, the region has suffered a deficient provision of infrastructure, given the difficult relief features that have to be overcome. Yet, the main factors contributing to the failure of this restructuring process have been the cost of public policies of support, above all for the mining industry,[17] and the strategic decisions taken by local firms (with a marked preference for low added value industries, and the virtual absence of investment and R&D activities, among others). Since the end of the 1990s, the economic performance of this region seems to have picked up: job opportunities are being created, the service sector has increased its share of the economy, while the industrial sector, including some heavy industrial activities such as iron and steel, has introduced more sophisticated production processes and is gradually increasing productivity.

Cantabria, by contrast, has had to face a slightly different situation. Here, the lack of growth can be attributed to the region's industrial mix, which has tended to concentrate around sectors of weak demand, and to the virtual absence of entrepreneurial activity. In common with Asturias, however, the regional economy has begun to show signs of growth in recent years.

By contrast with the experiences of these two regions, the economies of the Basque Country and Catalonia illustrate how regions specialising in mature activities have, nevertheless, managed to transform their manufacturing sectors. In the Basque Country, the shift towards more knowledge intensive activities has proved to have been a good decision. In particular, the restructuring of the region's iron and steel firms has been highly successful. There are also a number of non-economic factors that explain the success of the Basque Country. These include its fiscal system – the same as that employed in Navarre but different to that operated in the rest of Spain – which ensures the regional government has access to more public funds for financing its policies and, in particular, for restructuring its industrial sector. Catalonia is another success story. In the 20 years between 1975 and 1995, its once great textile industry had been reduced to half its size (in terms of labour force). The restructuring saw the closure of the less efficient firms and the delocalisation of assembly processes to the developing countries, but the company headquarters, R&D laboratories and other leading departments have remained in Catalonia. Today, the restructuring of the region's textile activities continues apace but the industry still accounts for 11.6% of all manufacturing jobs in Catalonia (2004). If we examine the shift in the Catalan economy, in which the share of manufacturing jobs has fallen while that of the services sector has risen, more than 66% of job losses have occurred in the textile sector.

6.6 CONCLUSIONS

The international crisis of the early 1970s set in motion a process of manufacturing restructuring in areas specialising in heavy industrial activities. This was the case of many European regions and, as we have seen, was the case with most of Spain's industrialised regions. These regions, which included some of the wealthiest in Spain, suffered massive losses of employment, a sharp rise in their unemployment rates, and a reduction in the growth of their GDP per capita.

In this chapter we have presented an overview of recent trends in the manufacturing industries of the Spanish regions and, more specifically, of the restructuring process that was set in motion in the second part of the 1970s. We have focused this analysis on both mature industries and mature regions. In line with the recommendations of the European Commission, we have designed a four-item indicator of mature manufacturing regions which has enabled us to identify four regions that fully meet all these characteristics: Asturias, the Basque Country, Cantabria and Catalonia, as well as two regions that fulfil them partially (the Balearic Islands and the Valencian Community).

This analysis must be set within the context of a structural change that was underway by the 1950s, as a result of which most of Spain's regions shifted from an agrarian-based economy to one that was based on manufacturing and services. As a consequence of this rapid growth (resulting from the substitution of activities by others that were considerably more productive), the less developed regions began a catching-up process (one that has yet to be fully completed) with the more developed regions, which at that time were highly specialised in mature manufacturing activities.

Subsequent changes illustrate how some of these mature regions (above all, the Basque Country and Catalonia) succeeded in transforming their economic structures into more modern ones, while others have yet to record similar success stories (as is the case of Asturias).

The success or otherwise of the restructuring process of the mature manufacturing regions can be attributed to a variety of reasons. These include a change in the industrial mix to incorporate a better and more productive manufacturing structure that was less dependent on traditional manufactures, and, in some cases, the implementation of public policies aimed at improving the competitiveness of certain specific manufacturing activities (such as the mining industry in Asturias).

In this chapter we have limited our discussion to a certain number of specific sectors and regions. A broader analysis of Spain's manufacturing system would show that there are several other restructuring processes in motion, some of which are of considerable importance, such as those being implemented in the manufacture of electrical machinery and apparatus and the manufacture of other transport materials, and in which more than one Spanish region is heavily engaged (including Aragon, Castile and Leon, Galicia and Navarre). In short, Spain's manufacturing system has undergone a major transformation since the mid-1970s. However, this process is far from finished if we take into consideration the growing competition within Europe caused, in part, by the enlargement of the EU.

NOTES

1. The authors are grateful to the CICYT (SEJ2007-65086/ECON and SEJ2007-64605/ECON) for their financial support and to Daniel Miravet for the preparation of the maps.
2. See Danson (2005), Brown et al. (1996), Singh and Borzutzky (1988) and Massey and Meegan (1982), among others.
3. Job losses in the textile industry are currently running at 20 000 per year.
4. See Myro (1992) for an analysis of the restrictions inherent to the restructuring process of the 1980s and the need for a more intense transformation of the Spanish manufacturing sector.
5. See Suárez-Villa and Cuadrado-Roura (1993) for an analysis of the dynamics and sectoral transformation of the Spanish regions.
6. Unfortunately, our data set for the unemployment rate does not cover all the period and ranges only between 1975 and 2005.
7. It is important to bear in mind that this classification also has the disadvantage of grouping together activities of very different characteristics. However, it is adequate for identifying sectorial specialisation in mature industries.
8. The EI data set comes from the Spanish National Institute of Statistics (INE).
9. The province occupies Level III of the Eurostat's Nomenclature of Territorial Units for Statistics (NUTS).
10. The province of Barcelona accounted for more than 25% of the employment figures.
11. See Arauzo and Viladecans (2004) for a detailed analysis of mature manufacturing activities in Spain during the second half of the 20th century.
12. This implies that the values of a variable in a territory are not only explained by internal conditions, but also by the values of the same variable in neighbouring or nearby provinces.
13. See Heim (1997) for an extensive discussion of this question.
14. An index of specialisation measures the difference in the concentration of a specific industrial activity in a given region compared to its presence in the country as a whole. An index with a value equal to 1 indicates that the presence of this industrial activity in the region is the same as in the whole of the country. An index with a value less than 1 indicates that the sector does not have such a strong presence in the region as in the country as a whole. By contrast an index with a value greater than 1 indicates that the sector has a greater representation in a particular region than in Spain as a whole. The higher the index recorded, the more specialised the region can be considered.
15. Objective 2 of EU Structural Funds recognises those areas (normally lying close to the EU average level of development) that are affected, among others, by structural difficulties in their manufacturing industries.
16. Note, were we to include the partially declining manufacturing regions (Valencian Community and Balearic Islands), then they would be grouped with Catalonia on the basis of their industrial mix. Thus, the second group would also comprise neighbouring regions.
17. Asturias has traditionally been associated with coal mining. The restructuring of this sector started later than similar processes implemented in the rest of Europe and centred above all on the provision of public subsidies to the more uncompetitive firms and on the maintenance of inefficient production methods. In fact, it was not until the end of the 1990s, and following

pressure from the European Union, that this policy changed in an attempt at making the sector more productive and competitive (del Rosal, 2004).

REFERENCES

Alañón, A. (2004), 'Evolución espacial de las manufacturas tradicionales en España. 1955–1995', *Economía Industrial*, **355-56**, 145–54.

Arauzo, J. M. and Viladecans, E. (2004), 'El impacto regional de la renovación de la industria tradicional', *Economía Industrial*, **355-56**, 133–44.

Brown, L. A., Lobao, L. M. and Verheyen, A. L. (1996), 'Continuity and change in an old industrial region', *Growth and Change*, **27**, 175–205.

Chapman, K. and Walker, D. F. (1991), *Industrial Location. Principles and Policies*, Oxford: Basil Blackwell.

Costa, M. T. and Viladecans, E. (1999), 'Concentración geográfica de la industria e integración económica en España', *Economía Industrial*, **329**, 19–28.

Danson, M. (2005), 'Old industrial regions and employability', *Urban Studies*, **42** (2), 285–300.

del Rosal Fernández, I. (2004), 'La reconversión del carbón, una dependencia plena de la decisión pública', *Economía Industrial*, **355–356**, 155–66.

Delisle, J. P. and Lainé, F. (1998), 'Les transferts d'établissements contribuent au desserrement urbain', *Économie et Statistique*, **311**, 91–106.

Heim, C. E. (1997), 'Dimensions of decline: industrial regions in the United States and Europe, 1970–1990', *International Regional Science Review*, **20** (3), 211–38.

Massey, D. and Meegan, R. (1982), *The Anatomy of Job Loss: The How, Why, and Where of Employment Decline*, London: Methuen.

Myro, R. and Gandoy, R. (2005), 'Sector Industrial', in J. L. García Delgado and R. Myro (eds), *Lecciones de Economía Española*, 7th edn,, Thomson – Civitas: Madrid.

Myro, R. (1992), 'Segunda reconversión y política industrial', *Papeles de Economía Española*, **50**, 186–91.

Parellada, M. (2005), 'Distribución Territorial de la Renta', in J. L. García Delgado and R. Myro (eds), *Lecciones de Economía Española*, 7th edn., Thomson – Civitas: Madrid.

Rodwin, L. and Sazanami, H. (eds) (1991), *Industrial Change and Regional Economic Transformation: The Experience of Western Europe*, London: Harper Collins Academic.

Scott, A. J. (1982), 'Locational patterns and dynamics of industrial activity in the modern metropolis', *Urban Studies*, **19**, 111–42.

Singh, V. P. and Borzutzky, S. (1988), 'The state of the mature industrial regions in Western Europe and North America', *Urban Studies*, **25**, 212–27.

Suárez-Villa, L. and Cuadrado-Roura, J. R. (1993), 'Thirty years of Spanish regional change: Interregional dynamics and sectoral transformation', *International Regional Science Review*, **15**, 121–56.

Viladecans, E. (2004), 'Agglomeration economies and industrial location: city-level evidence', *Journal of Economic Geography*, **4**, 565–82.

PART II

MATURE REGION–INDUSTRY CASE STUDIES

7. Birmingham's Marshallian Knowledge: A Constraining Geo-Historical Context for Domestic Saloon Manufacturers?

Peter Clark

7.1 INTRODUCTION

The closure of the Birmingham Longbridge plant of MG Rover in 2005 signified the final moment in the long story of domestic manufacturers trying to make the home base of the Birmingham Region internationally competitive in saloon car production and distribution. Manufacturers in the region survived the American invasion by Ford in 1912, then General Motors at Luton a decade later, but came under pressure from German and Swedish competitors by the early 1960s (Whipp and Clark, 1986). State-led intervention sought to impose rationalisation (1967–1972) and then to provide large investments to fund the next design cycle. Eventually the state incentivised the sale of the remaining rump onto British Aerospace who then sold it on to BMW. In 2005 a flurry of newspaper reflections and academic studies mused over the missed opportunities to save the sector at various key moments in the past. The period 1967–1975 is widely cited in the sketchy counterfactuals. My chapter contends that these optimistic reflections are a-historical musings which lack understanding of the geo-historical competitive constraints inhering in the British situation, and especially in the role of the Birmingham region. The argument is that by 1958–1964 the British owned car industry constantly faced a problem of making sufficient surplus profits to sustain ever more costly design cycles. Therefore, this chapter examines how the Marshallian origins of the regional social structure and the national cultural conventions of coordination (Storper and Salais, 1997) explain the evolution of the British saloon car sector.

Typically the analysis of national social structures is too abstracted from regional specificities (Gregory, 1982; Whipp and Clark, 1986; Urry, 1995; Clark, 2000, 2003). For example, the issue of sectors and regions was central to the lively debate at the Academy of Management in Boston 1997 between Dunning and Porter regarding the analytic usefulness of the diamond framework (Dunning, 1997; Porter, 1997). However, neither the eclectic theory nor the diamond framework is sufficient (Clark and Mueller, 1996; Clark, 2000, 2003). The Dunning-Porter debate omits three important elements. First, an analysis of the geo-historical context of firms in the automobile sector and their long-term possibilities of engaging in cycles of innovation-design cycles which were sequentially radical and incremental (Abernathy, 1978). Second, Storper and Salais's (1997) theory of cultural conventions suggests that the conventions of coordination of the Birmingham car firms were mismatched for the longer term production of cars in the British market place and for exporting overseas. Third, there were the unintended consequences of the division of knowledge between firms along the value chain (Jacobides and Winter, 2005). Value chains vary consequentially in their division of knowledge between different stages. If, as in the British case, focal firms in the value chain possess low modularity in their inter-firm relationships then there are problems of control over the surrounding national networks which lead to problems of moving abroad. In the Birmingham region the key saloon car producers were heavily dependent on the specialised knowledge of external parties such as Lucas for new knowledge about electrics and then electronics. This division of knowledge was shaped by the social composition of the sectoral institutions, especially the roles of management, the lack of electrical and automobile engineering graduates in the technostructure and by the role of craft workers in relation to the product and the production process.

From the early 20th century into the 1970s the Birmingham Region was the main home base for the British saloon car industry and its suppliers. British-owned firms were losing their share of the domestic market from the mid-1950s onward. The remaining firms were encouraged by state policies to merge into a single large company. This un-rationalised conglomerate lacked the relevant cultural conventions of coordination appropriate to the emerging context of mass customisation. By 1975 the decline was being transformed into a hollowing out. Yet in the period 1975–2005 saloon car production and distribution in the UK was successfully managed by new continental European entrants (e.g. BMW) and by the Japanese firms: Honda, Nissan (Renault) and Toyota. American entrants were already established.

To refine the analysis I compare the founding of the saloon car sector in the Birmingham Region, England and in the Detroit Region, USA. The regional space of the Detroit and of the American infrastructure for a manufacturing-distribution complex was quite different to the Birmingham Region. The differences inform why the Birmingham Region became a constraining home base for the British saloon car industry in the 20th century. The aim is to explain why a particularly historical trajectory was highly unlikely. The specific argument is that, if Henry Ford had been transported to Birmingham in say 1904 and he had started to envisage the Ford T, then even his remarkable capacities for agency and control (Emirbayer and Mische, 1998) would most probably have failed. This chapter focuses on analysing and explaining the probability of that failure.

National contexts possess distinctive elective affinities (Clark, 2000; Sorge, 2005) which provide the zones of manoeuvre and organisational spaces for firms. Only exceptional agency and strategic fortune can reconfigure these national zones of manoeuvre. Strategic choice is constrained (Penrose, 1959; Child, 1972, 1997; Clark, 2000). My claim is that this geo-historical account of the global and local aspects of international business systems requires a sociological visage to confront and complement the dominance of economists and geographers (Clark, 2000; Sorge, 2005). The challenge is to ground the normative fantasies of policy economists in the critical realism of explanations which respect geo-history (Callinicos, 2006). With regard to the Birmingham Region I shall deploy critical realism to construct a configuration of features in that founding period at the opening of the 20th century and seek to show that this ensemble unintentionally shaped the eventual demise of the sector in the 1970s and its closure in 2005 (Clark, 2000; Calincos, 2006). These features include the:

- Social composition of the institutions of the sector (Clark, 1987, 2000; Church, 1994);
- The elective affinity between the Marshallian inheritance and the culture of coordination requisite for long term survival (Storper and Salais, 1997);
- Sectoral innovation-design cycles (Abernathy, 1978) and the extent of design modularity amongst firms constituting the whole value chain (Jacobides and Winter, 2005).

The analytically structured narrative combines these three strands to argue that the historical inheritance of the Birmingham region did initially enable the founding of automobile firms able to supply British consumers in effective competition with overseas entrants. However, the long-term positive feedback mechanisms obstructed the emergence of those features necessary to compete when the automobile sector evolved into mass customisation and faster design cycles typified by Japanese firms such as Toyota and Honda and by several European firms (e.g. BMW).

7.2 WHY HENRY FORD WOULD HAVE FAILED IF STARTING FROM THE BIRMINGHAM REGION, ENGLAND[1]

7.2.1 1911: American Standardised Kits and British Consumers of the Ford T

In 1911 Henry Ford did enter the British automobile market from Detroit, USA by exporting knock-down kits to Trafford Park, Manchester, England where they were assembled. Ford gained one-quarter of the British market in 1912, whilst Rover, at that time the world's largest producer of cars, bicycles and saloon cars was third with sales of 1200 cars. The largest British supplier was Wolsey with a similar share of the market to Ford. Two questions arise. First, could Henry Ford have started out from the Birmingham–Coventry corridor in which much of the British owned saloon car industry surfaced? Unlikely. Why? Second, why didn't any British car firms supply the small car segment earlier and why did British owned firms continue to produce saloon cars in relatively small volume until the closure of the domestic industry in 1980s? We shall compare the contexts of Rover and of Ford.

The case of Henry Ford's launching of the Ford T from Detroit, Michigan, USA in 1908 provides a useful instance for exploring, testing and clarifying the debate over competition between contexts (Clark, 2000). This section contends that the contextual intra-dependencies in the British and American contexts provided zones of manoeuvre to the domestic car firms which were very different. The specifics of those zones of manoeuvre were consequential. There was competition between contexts. Detroit, USA was the more enabling context for low cost, standardised products like the Ford T.

Could Ford have successfully launched the 'T' in Birmingham and gained such a world position? We shall argue that Ford would have found the Birmingham region and the British context contained restricted zones of manoeuvre. Ford would not have been able to dominate the context. However, if in 1906, the year Ford advertised the forthcoming Model T, we had applied the analytic frameworks from contemporary resource based strategic analysis to the British situation and to the historic role of the industrial district, then we might well have concluded that there were very favourable circumstances. It might have been difficult to imagine impediments. In contrast the theory of competition between contexts proposed by Clark (2000) emphasises both the hidden configurations of elements and the specificities of regional contexts.

7.2.2 Henry Ford and the Detroit Region, 1890s–1920s

In Detroit and in the USA it was possible for an entrepreneur to create high volume capabilities and to draw upon societal capacities as well as regional opportunities which were distinctive. Much attention has been given to Ford's design of the Ford T and to the production process within which it was assembled, but this is only one part of the story. Throughout the early period a very high proportion – sometimes 90 percent – had been sold before the Ford T had been built. Henry Ford had established this practice of advertising and forward selling in 1906, two years before the launch of the Ford T. The American market place for all kinds of goods was, by 1900, serviced by highly competitive firms in both distribution and manufacture. Distribution, marketing and advertising were exceptional American developments in the 19th century through both direct selling in urban centres (e.g. Chicago) and indirect selling through mail order. These methods of distribution were accompanied by well developed systems of financial repayment through standardised means, including advanced and delayed payment. Ford was therefore able to activate a capacity absent in Europe to skilfully advertise his forthcoming car in 1906 and commenced receiving payment before delivery. These inflows of revenue came directly to Ford and meant that a major source of uncertainty faced by European manufacturers was removed and that significant onward planning was possible. In contrast British manufacturers tended to wait for orders, or, to stock a small number of chassis-wheel assemblies (e.g. Austin). The British manufacturers were always in a dependent relationship to the powerful distributors who made quite limited feedbacks of information.

The American market differed from Europe in that there was a large potential segment beyond the middle class urban population. In the US the low cost of housing and of primary consumption products increased the disposable income available for major products like sewing machines, household furniture and later, automobiles. The market was prepared for standardised, function goods which could be customised in the home. Consumers were experienced in learning about future purchases through catalogues, exhibitions, public trials and newspapers.

It may be argued that America was the first nation to take industrial knowledge seriously. The position of supporting and related sectors in the American context which formed the pre-existing potentials for Ford was highly favourable. Henry Ford was able to draw upon the world's must highly developed blend of capacities in functional design and in the standardisation of components in the contemporary world. This was a system of knowledge, forms of organisation and artefacts which had already delivered standardised military equipment, sewing machines, bicycles, wood frame houses and so on (Hounshell, 1984). The production process had mastered the notion of flow (Chandler, 1977) at high speed (Rosenberg, 1976) and created systems for handling agriculture (e.g. meat) and fluids (e.g. oil) by 1900. The career of Duke and American Tobacco in the cigarette industry provided the template of world markets through rationalisation and vertical integration (Chandler, 1977).

Henry Ford was able to engage with suppliers and able to receive functional standardised components at an early stage and know that these would fit into sub-assemblies which could then be assembled into the Ford T. Detroit contained many of the necessary ingredients. The machine tool industries occupied a virtuous relationship in respect of their capacities to supply and to diffuse new learning (Rosenberg, 1982).

Ford created one of the most frequently cited templates of corporate organisation, yet at the launch of the Ford T there was considerable rivalry amongst the scores of small and the few medium sized firms. That soon changed and by the 1920s the US became a three-firm nation. Henry Ford was an excellent example of entrepreneurship and the building of an actor network with extensive pillars of power and cognition. Ford's Model T, when introduced in 1908 for less than $1 000, was a spectacular success. The Model T became the world's best selling model and totally dominated North American markets, before being replaced in 1926. The Ford T sold in cities, but more importantly it sold to the majority of Americans in rural and small towns where the roads were often rudimentary. Prior to the appearance of the Model T most American automobiles were open top, mainly wooden, very shaky for the passengers, unreliable and hard to service.

The Model T established the essentials of the future design hierarchy around the chassis to include the frame, engine, transmission, brakes, wheels, radiator and all other mechanical components except the outer body (Abernathy, 1978:13; Abernathy and Clark, 1985). Ford's entrepreneurship provided a synthesis of the existing practices and introduced key innovations such as three-point suspension that separated the engine and the frame, thereby reducing shake and potential disassembly. The Model T incorporated features appropriate to the rough roads and rural market (e.g. high ground clearance) of the USA and not found in the contemporary European markets. In the years before the launch the competition between alternative power sources had been decided in favour of the internal-combustion gasoline engine. Before 1908 Ford had redesigned the engine and the power train for greater reliability and ease of maintenance. The design hierarchy embodied in the Model T of 1908 remained stable with incremental extensions for the next 18 years. The car was light, open, single colour, with wooden body work and suspended on a chassis.

The Model T was designed to be manufactured from standardised components insourced from suppliers and then assembled by skilled workers building up sub-assemblies into the final product. At the stage of connecting innovation to design, Ford and his inner network of collaborators articulated the process of assembly with the product. It was some years later after the successful launch that innovations in the assembly were introduced. Most famously these introduced the template of the modern assembly line, initially for magneto assembly (1913) and then for the final assembly (Abernathy, 1978). There was a dramatic transformation in the organisation of work by reducing the work cycle from more than nine minutes to less than two minutes. Ford managed to apply the principles of manufacturing imagined for gun making in the late eighteenth century, but not achieved until the mid-nineteenth century, to a product with very much more complexity. The simplification of work reduced the training time, removed the more powerful craft groups, and in the American context enabled the increase of managerial influence on the shopfloor.

The template of continuous production had emerged earlier and become established in several industries – for example, the process industries (e.g. sugar, petroleum), the meat industry and cigarette making. After 1885 the bicycle industry increasingly utilised flow line methods. The basic principles of continuous flow had fascinated and preoccupied American business men from the late eighteenth century onward.

Prior to the launching of the Model T the servicing of the car was done through the local garage and by the driver undertaking irregular, running repairs. The Model T was designed to replace the continuous and irregular servicing by regular preventative servicing by trained agents working to the formalised instructions from Ford. Consequently, Ford set out to control the distributors and to have them inside his network.

Ford announced the car in 1906 through an extensive advertising campaign which invited future owners to start making down payments. Advertising and financial services were already established in a standardised format. Part of the inflow of prepayment was utilised to fund some activities.

The Model T differed from automobiles regarded as the leading edge of European designs in several key respects. First, the engine was technically coarser, but was more robust and therefore did not require such fine machine tolerances. This led Europeans to look down on the engine.

7.2.3 Ford Overseas in Britain

Ford was able to appropriate existing templates of corporate growth by developing large scale assembly of the American domestic product overseas (Hughes, 1990). At an early stage, 1911, Ford planned entry into Britain and became the leading distributor in 1912. However, the conditions necessary for Ford to develop his network resided in the USA generally and in Detroit specifically. These essential features were absent in England for each of the clusters of features and for the overall configuration. Consequently when British manufacturers like Lanchester sought to engage in standardised production in Coventry, this was undermined by the poor quality of components and the requirement for skilled workers to finish their preparation. It was more than a decade before the British automobile firms developed a competitive small car to Ford and it was the McKenna Duty in the 1930s that provided an umbrella for the cumulative pathway through the zones of manoeuvre for British firms.

7.2.4 England: Birmingham Region, 1900–1912

Birmingham was known in the late nineteenth century and early twentieth century as an *industrial district* and cited by Marshall as an excellent example. The economist Marshall claimed that Birmingham was an industrial district characterised by remarkable flexibility. In the Marshallian market model the specialised firms serve varied demands. These firms develop routines for searching for new clients and other routines to maintain flexibility. The Birmingham Region might therefore seem to be the ideal place from which to establish a saloon car industry. Birmingham was a core region in relation to the metropolitan core of London in the development of Britain and its Empire. The region had once been the centre of the arms industry for the British Empire and was close to, and part of, the industrial revolution (Rosenberg, 1969). Gun making was an industry of great innovation, especially in the reduction in the weight of guns that had enabled their transfer into shipping and then their mobility on land. As an industrial district there were both large firms and many small firms based in streets and even homes, which combined and recombined as the product itself altered and as demand fluctuated. Metal work related to domestic and mercantile trade had blossomed, creating many small firms run by entrepreneurs. Firms were flexible in their responsive to demand because the craft skills were located with the workforce rather than owned by the employer. The putting system and tendencies to quasi-autonomous group working were endemic. Within the industrial district complexity was handled between firms which were proximate and could establish working relationships. The canal system, said to be more extensive than Venice, linked the industrial district and connected it to the export ports of Bristol, London and Liverpool. So extensive was the generation of local wealth that banks (e.g. Midland) later became national major players. Birmingham was also a centre for chartists and for strong associations amongst the workforce. It has been argued that these labour associations were so embedded and extensive that the days of the week possessed their local distinctiveness. St. Monday was for recovering from the weekend exertions and ebullience.

The seeming flexibility of firms in the Birmingham industrial district was deceptive. Although enabling the early phase of the automobile industry, the industrial district lacked the capacity to follow the evolution of the sector (as Abernathy, 1978). Birmingham was slow to transform its sectoral-clustering in the crucial decades between 1890 and 1920 when the bicycle sector was displaced by the incoming automobile sector. Birmingham possessed a distinctive, pre-existing structure of social and economic relationships which contained specific powers and liabilities. The pre-existing filiere and structure enabled small firms to start up and small entrepreneurially-led firms (e.g. Jaguar) to enter and survive. Also, car firms that combined the price-range with medium volume cheap cars (e.g. Austin) were able to survive in the face of Ford and GM until the 1960s, because the advantages to American incomers were slight as long as they also faced the British market.

Demand conditions in the British market were quite different to those in the American market inherited and created by Henry Ford after 1908. British demand was heavily located in the urban environment; London dominated middle class areas, at the moment when the residual financial benefits of British mercantile capitalism based on world shipping were being heavily eroded. The variegated, stratified, middle class competed and displayed themselves through small distinctions in taste. Also, net disposable income of British consumers was distributed so that the middle classes could afford expensive cars, but the large working class segments were less well off than their American cousins. The consumers demanded expensive, customised automobiles supplied by niche-oriented car firms. These niche firms created a segmented market articulating distinctions of taste, and their large number cannot be equated with the notion of high rivalry (as Porter, 1997). The barriers to entry were quite low. The British automobile industry was typified by many entrants at its founding, by a higher rate of survival than the USA, and by more new entrants in the late 1920s (e.g. Jaguar, Hillman, Jowett). There was high market segmentation and a degree of rivalry.

The consumer–producer nexus was controlled by the increasingly powerful distributors who remained independent from the production firms, and tended to retain their detailed knowledge of the consumer. The niche car firms focused upon small volume production runs. The car firms were often founded by entrepreneurs who dominated the board level. The owners were driven by pecuniary rationality rather than the technical rationality more characteristic of their French and German rivals. There was often low representation of engineering competencies at the top levels. The techno-structure was weakly developed relative to other leading international nations (e.g. Germany and France). At the 'frontier of control' between employers and employees in the industrial district the employers formed defensive alliances in order to promote their interests and to co-align the actions of their rivals.

The owners and managers of car firms tended to rely on market-like mechanisms outside and within the firm. There was no equivalent habit to the American system of manufacture, and consequently the potentials for standardisation established in the different sectoral cluster of food, drink and tobacco was absent (Clark, 1987). Yet the car was an enormously complex configuration of thousands of components. There was low standardisation coupled to the irregular use of external consultants to supply designs for the cars and to itemise the components to be sought from external suppliers. Consequently, there was a low accumulation of knowledge within the firm and the associations of the work community. Within the firms and within the industry there was only a slight development of advanced knowledge. Universities were similar to the American pattern whereby General Motors sponsored Flint University in Detroit. In England the major civic universities in industrial cities were wary of corporations. The history of Lucas cites the post-war unwillingness of the prestigious University of Birmingham to become connected in the development of industrial engineering. Yet Lucas was a knowledge community with considerable significance.

Low standardisation slowed the development of costing procedures and reduced the opportunities for reflective monitoring by management. Management played only a partial role in assembling practices (Clark, 1972). Management, as an occupational strata, were poorly developed in business knowledge or engineering knowledge. There was little systematic recording and compilation of data to provide standardised costs (c.f. the cigarette industry). The incipient techno-structure was slow to develop. Hence reflective learning was inhibited. A large part of shop floor management was effectively delegated to the relatively autonomous skilled craft groups. Low standardisation also impacted design of the car and of the production process. This was reinforced by the tendency to coordinate the firm by buying in expertise for car design in the founding period. The specifications to suppliers and the monitoring of contracts were considerably less developed than in the American, French and German competitors.

The dominant work practices of the industrial district provided a pre-existing structure in which skilled craft groups offered their expertise in a market relationship with employers that was mediated by the role of Birmingham associations of trade and labour. The shop floor was populated by craft groups whose expertise was learnt through participation in the everyday experience controlled by the trade association. Consequently, skilled machinists were needed for the constant adjustment of both outsourced and internally manufactured components. The high levels of uncertainty were handled by recruiting a skilled workforce which was heavily involved in self-organised training and cognitive apprenticeship. The expertise of these workers was required to correct the low standardisation of the components and to respond to the seasonal variability of demand (Turner *et al.*, 1967; Clark, 2000). There was an ample supply of skilled and semi-skilled workers willing to engage in the apprenticeship system, but this was a largely tacit rather than explicit form of learning. Because the workforce from the Birmingham and Coventry area were required to undertake such complex work on the production process, the firms were unable to contemplate the struggle over the appropriation of expertise and learning. This contrasts with British firms in food, drink and tobacco who successfully appropriated major elements in the Taylorist approaches (Clark, 1987: 282–86). Indeed standard work study in automobile industry was still in its infancy even in the 1950s and 1960s (Clark, 2000).

The supporting sectors did develop. In electronics Lucas became a designer–supplier to the British-owned firms, developing from headlights into the full panoply of electrical components. But Lucas did not acquire an international status and some of their systems became known in America as 'the prince of darkness'.

7.3 CULTURAL FOUNDATIONS OF SECTORS WITHIN CAPITALISM

Storper and Salais (1997) contend that understanding the trajectories of sectors and their firms requires a thorough analysis of the cultural foundations of capitalism. Why? Because the successful coordination of complex, heterogeneous ensembles involves interpretation, precedent and convention to collectively discover and constitute coherent patterns. To understand economic actors requires examining both the conventions governing pragmatic situations and frameworks depicting collective action in economic activity. Because all economic activity involves coherence between the constraining structures and the conventions that enable action there is a pragmatic problem of establishing and maintaining a common existence for the multiple actors involved. Conventions for coordination represent a framework of foreseeable action for economic actors. Conventions define the common existence for all actors in a particular world and they face four external requirements: profitability is an essential measure of performance; the size of the international market for saloon cars; the establishment of an identity which refers to the taken-for-granted network of roles, and this rests on deeply sedimented competencies and characteristics of groups; conventions of participation in specific places like the Birmingham region.

Coordination in the automobile sector connects the consumers who utilise the saloon cars to those who supply the cars. Their relationship is mediated by innovation-design as indicated by Abernathy (1978). To be efficient the coordination of the heterogeneous collection of actors must be aligned with the coordination of expectations. It is the conventions in the ensemble that give or exclude the possibility of long-term survival.

Conventions are systems of expectation with respect to the competencies and behaviours of others. Actors will generate conventions around the enacted and pragmatic situations in which they find themselves. Therefore within the sectoral ensemble collective actions become coupled and they collectively evolve in a (subsequently) trajectory that is specific, local and possibly durable. Learning co-evolves and because anticipations are so central to the ensembles there is a subtle experience of recurrence and temporality that is specific to particular places. So, as situations recur actors bring to bear their enacted repertoires.

Sectors will vary in their conventions of coordination depending on three key dimensions. First, the market for cars can vary from measurable uncertainty to uncertainty where the risks cannot be calculated. Therefore, sectors will vary between the production of generic or customised saloon cars. Second, the resource inputs to the sector will vary from standardised to specialised. Therefore the conventions and anticipations governing resource inputs vary. Specialists occupy a key position in coordination and the non-routine technology tends to be strongly equated with the specialist. A convention of standardisation enables interchangeable and reproducible resources. These two different forms of coordination affect supplier–consumer relationships. Third, the economies of sector will vary from scale to scope economies. There are different conventions for high volume and high variety. These dimensions provide a typology of four ensembles which can be used to trace the historical trajectories of sectors and to simulate policy implications.

The notion of ensemble is deployed to characterise four major forms of coordination between consumption and production. The four ensembles are:

a) In the *Interpersonal Ensemble*, specialised and customised products reflect the perceived desires of the consumers and the conventions as in the fashion, design and craft intensive work situations. The conventions rest on reputation, confidence and specificity of image. The conventions for labour deliberately group individuals into networks but each person is responsible for the adjustment to unknown factors. These are the Marshallian features found in early 20th century Birmingham. These conventions shape certain economic activities although they are probably less prevalent than in the pre-Fordist period. The conventions in the interpersonal ensemble of specialised, dedicated outputs with high uncertainty and the economies of variety locate the basis of competition in quality and the evaluation of quality is in terms of price. There is the extreme state of uncertainty where conventions indicate few points of reference to start the process between supplier and consumer. These designed-based outputs include such diverse items as the barrister in a complex commercial dispute defending property rights. Great attention is focused upon meanings, experience and a common language.

b) In the *Market Ensemble*, the conventions for the standardised mass customisation of products and services can be expressed in formalised, codified norms. Competition is based on price and rapidity of supply. These are not universal conventions but are found in particular locales. The convention for labour is coordination through small groups who have their responsibility for developing knowledge. In the Market Ensemble the conventions about outputs are of standardisation even when a run of production is dedicated to specific clients. The convention of standardisation is located in the language.

c) The *Industrial Ensemble* contains generic standardised products whose coordination conventions are mediated by notions of efficiency. This was the world which Ford exemplified, yet the possibilities for this world gradually receded. The convention for labour is the coordination of individuals who experience variations in the demand for their labour. In contrast the conventions in the industrial ensemble of standardised, generic outputs are of predictable risk with economies of scale. Uncertainty is around the business cycle and fluctuations in demand and the response to those uncertainties will be in terms of forecasting the medium term. The basis of competition is price and the evaluation of quality is against generic industrial standards. The conventions about the constitution of outputs construct uncertainty as predictable risk. Norms of quality are codified and the constitution of outputs is mediated by the material context of buildings, equipment and raw materials.

d) *Intellectual Resources Ensemble*. The conventions for the coordination of creation have to cope with a diverse, multiplicity of production methods and types of output. Conventions are embodied in professional and scientific rules. The firms possess internal labour markets and labour adjusts to the unknown through unemployment. In the ensemble of intellectual resources the conventions guide the processes whereby the qualities of existing objects are altered. Action also proceeds in a context of uncertainty. Workers in this context survive longer when they develop knowledge that has applicability across a number of consumers. Their conventions involve local codification.

Each of these ensembles is distinguished by fundamentally different routines for enacting and absorbing uncertainty and they are coordinated around different principles for connecting the market to its suppliers. The four ensembles are mutually exclusive.

The effects of the national state on the capacities of actors are crucial. The issue of policy choice hinges around the relationship between the state and international competitiveness. The administrative rationality by the state is much more limited than current policy theories allow. These theories are confined to an idealist notion of the institutional architecture and omit attention to the processes of coordination and the frameworks of action. In effect they use institutions as vehicles of policy to replace culture. Their separation between institutions and culture elides the features of action and coordination. It is practically impossible to simplistically adopt exogenously determined best practices. Conventions and frameworks are frequently problematic when existing routines do not match the tests of linking the external with the internal.

In the Marshallian model the specialised firms serve varied demands and tend to develop different sets of routines for searching for new clients and other routines to maintain flexibility. Although there are multiple pathways from the intellectual Marshallian ensemble, there is a constant threat of coordination failures because the surrounding conventions and institutions take shape to protect their actors from threat. British car firms varied in the understanding of their zones of manoeuvre around and then away from the conventions of coordination in the Marshallian ensemble. For example, Rover remained very much within its world view both before the onset of bankruptcy (1928–32) and well into the 1960s when the firm sold out to Leyland (Whipp and Clark, 1986). In the 1950s, when Rover developed the 4WD Land Rover, they created a robust design regime which could be stretched into other major architectural variations (e.g. Range Rover). Moreover, the 'vehicle' could be customised to meet the market requirements of Spanish farmers in the Picos, strategic forces in flexible assault groups, gentrified city dwellers and expeditions everywhere. The conventions of coordination from the Marshallian ensemble provided a viable and profitable framework. In contrast the Rover saloons, re-established in 1932 for the slowest moving market segment, encountered severe competition in the late 1950s from the upper market, medium volume speciality manufacturers such as BMW, Mercedes, Volvo and Saab. These firms had successfully established the appropriate conventions of coordination, but Rover saloon production had not.

7.4 INNOVATION DESIGN CYCLES AND VALUE CHAIN MODULARITY

The British automobile sector largely failed to generate the profits necessary to undertake the radical innovation-design cycles necessary to generate new automobiles, new modes of production and new modes of distribution. The problems became more evident after the mid-1960s. Abernathy (1978) demonstrated that this problem also arose in the American auto industry in the 1960s. The American industry did develop more effective control over the modularising knowledge down the value chain.

In the 1970s Utterback and Abernathy (1975) examined longitudinal data sets from many different sectors (e.g. lamps) to construct a template of the sector life cycle from founding to maturity. Their template stated that when a sector was founded there were high rates of innovation in the product but these soon declined into incremental innovation. A similar pattern was found for the production process but with a lagging of a decade or so. These two curves indicated that a mature sector was typified by incremental innovation – therefore the strategic future should be high volume and low costs. This future was extrapolated from the past. Very quickly this template of the sector life cycle became a cognitive pillar within the North American business community, but much less so in Europe. Then Abernathy (1978) led a project to test the template of the sectoral life cycle on the very extensive data archives available from Ford. This compact and truly remarkable synthesis revealed problems with the sector life cycle by demonstrating crucial dislocations to the notion of maturity. Abernathy realised that the elegant and beautiful model of the sector as neo-biological life cycle was flawed. Two streams of new analysis stimulated his thinking. One stream was the comparison between the costs of car production, especially inventory, in the USA and Japan. The other stream consisted of a series of small-scale studies of innovation in the car and in its process production. Analysis of innovation in the car revealed that new knowledge in the 1960s had increased the rate of innovation above an incremental level in most facets of car design (Abernathy, 1978). Equally, developments in administrative technologies provided the opportunity – if grasped – to alter modes of production to combine volume with customisation. Then Abernathy, in close collaboration with Kim Clark developed a completely novel view of the sectoral life course, which sought to anticipate emergent tendencies and disruption to existing relations between the product/producer and between them and the consumer. The revised model of the sectoral life course pre-figured a wholly different kind of knowing. It is therefore curious that the analysts involved with the British saloon car sector have made so little use of this perspective (e.g. Church, 1994; Holweg and Oliver, 2005).

Second, Jacobides and Winter (2005) contend that the vertical structure and division of specialised knowledge between firms along the value chain is a feature which can enable or constrain the capacity of firms to move their products abroad. If the value chain possesses dense linkages to the rest of the sector and to the domestic context then the value chain is institutionally non-modular. This was the case with the Birmingham region for the automobile sector until the late 20th century. Non-modularity means that critical specialised knowledge (e.g. Lucas and electrics) cannot be substituted for. Therefore British automobile firms were disabled when they attempted to move production abroad in order to achieve higher volumes, because they could not replicate their value chain. On the other hand if within the value chain all critical linkages can be substituted, then there is institutional modularity. In this situation moving aboard requires focal firms to impose scope. This is more like the situation created by Ford in the post-1918 period.

7.5 CONCLUSION

In the period 1890s–1920s the Birmingham industrial district possessed a form of flexibility in which specialised firms were able to serve varied demands. However, this much vaunted Marshallian notion of flexibility was deceptively promising. The firms did install many routines to maintain flexibility, but developed few routines to carry themselves beyond the founding period or beyond the unique British consumer. For critical knowledge they relied initially on specialist suppliers and later relied upon the supportive role of the state, especially after the 1970s. The Birmingham Region and the British market place were only partly the ideal location from which to establish an international saloon car industry. There were many inherent problems, especially with conventions of coordination. Consequently, domestic car firms were faced with producing in a space in which the complex coordination and knowledge chaining necessary for the successful selling of domestic products in sufficient volumes to fund the next design cycle was chronically problematic. This interpretation does much to challenge those whose ever bright optimism continuously evoked a promising future for British car firms. This chapter agrees with Storper and Salais (1997) who propose four elements that should be found in the practically useful analytically structured narrative. These are: realism and indeterminacy; attention to prior collective experiences and emergent forms; historicity and endogeneity; diachronic analysis to explain path dependent phenomena. These four requirements for serious diagnostics have shaped the analytically structured narrative presented in this chapter.

NOTES

1. This section is based on Clark (2000).

REFERENCES

Abernathy, W. J. (1978), *The Productivity Dilemma. Roadblock to Innovation in the Automobile Industry*, Baltimore: John Hopkins.

Abernathy, W. J. and Clark, K. B. (1985), 'Innovation: mapping the winds of creative destruction', *Research Policy*, **14**, 3–22.

Callinicos, A. (2006), *The Resources of Critique*, Oxford: Polity

Chandler, A. D. (1977), *The Visible Hand*, Cambridge M. : Belknap.

Child, J. (1972), 'Organizational structure, environment and performance: the role of strategic choice', *Sociology*, **6**, 1–22.

Child, J. (1997), 'Strategic choice in the analysis of action, structure, organizations and environment: retrospect and prospect', *Organisation Studies*, **18** (1), 43–76.

Church, R. (1994), *The Rise and Decline of the British Motor Industry*, Cambridge: Cambridge University Press.

Clark, P. A. (1972), *Organizational Design: Theory and Practice*, London: Tavistock.

Clark, P. A. (1987), *Anglo-American Innovation*, Berlin: de Gruyter.

Clark, P. A. (2000), *Organisations in Action. Competition Between Contexts*, London: Routledge.

Clark, P. A. (2003), *Organizational Innovations*, London: Sage.

Clark P. A. and Mueller, F. (1996), 'Organisations and nations: from universalism to institutionalism', *British Journal of Management*, **7** (2), 125–40.

Dunning, J. H. (1997), 'Knowledge Capitalism: Competitiveness Evaluated. A Macro-Organisational Viewpoint', in M. Das (ed.), *Knowledge Capitalism*, Academy of Management.

Emirbayer, M. and Mische A. (1998), 'What is agency?', *American Journal of Sociology*, **103** (4), 962–1023.

Gregory, D. (1982), *Regional Transformation and the Industrial Revolution. A Geography of the Yorkshire Woollen Industry*, London: Macmillan.

Holweg, M. and Oliver, N. (2005), 'Who Killed MG Rover?', Cambridge: Jude Institute of Management Working Paper.

Hounshell, D. A. (1984), *From the American System to Mass Production, 1800-1932*, Baltimore: John Hopkins University Press.

Hughes, T. C. (1990), *American Genesis*, Baltimore: John Hopkins University Press

Jacobides, M. G. and Winter, S.G. (2005), 'The co-evolution of capabilities and transaction costs', *Strategic Management Journal*, **26** (5), 395–413.

Porter, M. E. (1997), 'Location, knowledge creation and competitiveness', in M. Das (ed.), *Knowledge Capitalism*, Academy of Management.

Rosenberg, N. (1976), *Perspectives on Technology*, Cambridge: Cambridge University Press

Rosenberg, N. (1982), *Inside the Black Box: Technology and Economics*, Cambridge: Cambridge University Press.

Sorge, A. (2005), *The Global and the Local: Understanding the Dialectics of Business Systems*, Oxford: Oxford University Press.

Storper, M. G. and Salais, R. (1997), *Worlds of Production. The Action Framework of the Economy*, Cambridge MA: Harvard University Press.

Turner, H.A., Clack, G. and G. Roberts (1967), *Labour Relations in the Motor Industry. The study of Industrial Unrest and International Comparison*, London: Allen & Unwin.

Urry, J. (1995), *Consuming Places*, London: Routledge.

Utterback, J. M. and Abernathy, W. J. (1975), 'A dynamic model of product and process innovation', *Omega*, **3** (6), 639–56.

Whipp, R. and Clark, P. A. (1986), *Innovation and the Auto Industry: Product, Process and Work Organization*, London: Pinter.

8. Life After Longbridge? Crisis and Restructuring in the West Midlands Auto Cluster

David Bailey and Seiji Kobayashi

8.1 INTRODUCTION

Recent events have highlighted the structural changes and shifts unfolding in the auto industry and their impacts on the West Midlands auto industry. These have included the collapse of MG Rover in 2005, the closure of the Jaguar's Coventry plant, Peugeot's closure of its Ryton plant, and Ford's sale of Land Rover and Jaguar to Tata in 2008. Whilst the latter has been welcomed as securing jobs in the short–term, longer term the question remains as to whether plants and jobs will remain in the UK. Indeed, all of these events have highlighted the difficulties involved in supporting and developing the auto 'cluster' in the region.[1]

This chapter explores the background to, and implications of, the MG Rover (MGR) collapse. Section 8.2 sets the scene by outlining the form of the auto cluster in the West Midlands, putting this into broader context by examining structural changes in the industry. These include: greater pressure on firms to recover costs when technological change has been intensifying; increases in the costs of new model development; increased international sourcing of modular components; and a shift of assembly operations towards lower cost locations. Such issues make maintaining the cluster more difficult for policy makers. In Section 8.3, it then looks at what went wrong at MGR, finding much of value in recent analyses which have stressed the long-term problems and 'cycle of doom' at the firm and its inability to recover costs for new model development. Given this, BMW's withdrawal from the firm in 2000 left MGR virtually dead on its feet, with a very limited time horizon in which to find a partner. By 2002/3 it was clear to many that the firm was running out of time. The chapter goes beyond this in Section 8.4 and looks at other contributing factors to the demise of MGR, in particular government policy mistakes over the years, including a misguided 'national champions' approach involving forced mergers in the 1950s and 1960s, a failure to integrate activities under nationalisation in the 1970s, a mistaken privatisation to British Aerospace in the 1980s, and the failure of competition policy in allowing the sale to an inappropriate owner in BMW in the 1990s. Adding in the considerable volatility of sterling and the scene was set for the firm's demise.

In Section 8.5, the chapter then looks at policy issues and the impact of the collapse on the local economy; whilst this has been a substantial shock to the West Midlands economy, the impact would have been much greater if the firm had collapsed in 2000. The five years of continued production over 2000–2005 allowed many suppliers to adjust and diversify away from MGR, aided by the work of the first Rover Task Force (RTF1), which may have 'saved' as many as 10 000–12 000 jobs. Key policy issues such as retraining needs are discussed. More broadly, the work of the local RDA in its clusters policy is examined. The MGR case raises the critical issue of the coordination of regionally based 'cluster' policies where the 'cluster' in question crosses administrative regional boundaries, calling for a more coordinated industrial policy to join up regionally based policies. Finally, Section 8.6 concludes the paper.

8.2 THE WEST MIDLANDS AUTO 'CLUSTER' IN A RESTRUCTURING GLOBAL INDUSTRY

8.2.1 Putting the WM Auto 'Cluster' into Context

The West Midlands is still seen as the core of the British automotive industry and until 2001 accounted for just over 20% of UK jobs in manufacturing (i.e. assembling) cars – or 16 300 jobs in 2001 (DTI, 2001). The concentration in the components sector has been even more marked, with the region accounting for 35 800 jobs in 2001, or one in three British jobs in that sector (ibid.). Furthermore, the industry includes electrical equipment for engines and retail sales and distribution of spares and parts. In related industries, a half of UK tyre production took place in the region in 2001, and 20% of jobs in the processing and shaping of glass (e.g. windscreens) (ibid.). The plastics and metal products clusters also support the industry. Whilst Bentley (2000) put the number working in the three automotive industries in the region in 1997 as 76 000, taking a broader definition to include automotive-related industries gives a figure of around 120 000 people in 2001, 17% higher than a decade before (DTI, 2001), and higher than the 100 000 figure given in RTF (2000).[2] The latter figure accounts for around 5% of regional employment and 5% of regional GDP (RTF, 2000). However, with recent assembly plant closures and job losses in the components industry, Donnelly et al. (2005) put the numbers of workers in the broadly-defined auto industry in the region as low as 65 000. The industry has also been a major focus for inward investment in the region over the last decade, with nearly 40% of all jobs created by FDI being in auto or auto-related industries (RTF, 2000).

Whilst centred in the West Midlands, the cluster clearly spreads into adjacent regions, with 41 000 employees in the neighbouring East Midlands region, another 19 700 in the South East region (plus 1 600 in the motor sport cluster), 12 800 in the South West, 12 000 in Wales (no assemblers but suppliers for Honda and engine manufacturing for Ford and Toyota), and some 24 000 in the North West. Whilst the latter was seen by the DTI (2001) as a distinct cluster, parts and accessories manufacturing is under-represented in the region, indicating that suppliers from elsewhere (such as the West Midlands) are used, and that these are in fact interrelated clusters or even part of a national cluster. Thus if a West Midlands 'cluster' can be identified, in reality it forms part of an inter-regional or national auto network extending well beyond the West Midland region into several other regions. Indeed, the transfer of some Land Rover production by Ford from Birmingham to its Halewood plant on Merseyside, and the sourcing of engines by BMW from Birmingham for second-generation MINI production in Oxford, are both indicative of the deep inter-linkages across administrative regions.

8.2.2 Restructuring and Global Shifts in the Industry

A key feature of the industry is too much assembly capacity, with 20% over-capacity in assembly in the European market. The weakest firms are under intense pressure, with MG Rover the first to go under, but Fiat was also in such intense difficulties that GM paid $1.5 billion in a divorce settlement in 2005 (although new model launches have since helped Fiat). Such pressures are intensified by the rising costs of new model development. In contrast with what is expected under the 'life-cycle' model of industry development, the 'crisis of cost recovery' facing firms has intensified over time, as from the 1960s onwards the industry entered a period of renewed technological development, driving up model development costs (see the chapter by Clarke in this volume). In today's terms, the cost of getting a genuinely new car to market lies in the range of £400 million to £1 billion. As a result, large-scale production over different brands using a platform-sharing approach is vital to generate the cash for future model development.

Meanwhile, the major manufacturers are shifting operations to new, low cost locations in emerging markets such as Central and Eastern Europe. For example, a key factor in the recent decision by Peugeot to close its Ryton plant near Coventry with 2 300 job losses was the opening of a €1 billion

Peugeot plant in Slovakia to make the 207 model – the replacement for the 206 made at Ryton. Labour costs in Slovakia are one quarter of those in Britain, hence the decision by the Peugeot strategic decision-makers to cut capacity in Western Europe. In this regard, Britain's much-vaunted flexible labour markets make it as easy to destroy jobs as it does to create them, evidenced by the fact that cutting of each job in France would have been up to three times as expensive for Peugeot as doing it in the UK. This, plus the lack of much in the way of domestic mass manufacturing, means that production activities in Britain are particularly exposed when transnationals look to cut capacity in the auto industry and in manufacturing more generally (Bailey and Cowling, 2006).

Given such shifts, within the West Midlands vehicle assembly is now concentrated in Birmingham and Solihull, with parts and accessories manufacturing in Sandwell, Walsall, Telford, Warwick and Newcastle-under-Lyme. With the collapse of MG Rover and the Peugeot closure in late 2006, volume production has ended and what is left of assembly is high-value luxury branded niche-production (Land Rover and Jaguar in Birmingham and Solihull) currently in the process of being sold by Ford to Tata. Meanwhile, Nanjing has planned to restart small-scale MG sports car assembly at Longbridge. However, the relaunch has been delayed several times and if it does go ahead will at best be very small scale 'kit car' production (less than 15,000 units) with complete 'knock-down' kits shipped in from China. Longer term there is much scepticism over Nanjing's ability to develop new models and their commitment to production in the West Midlands. Recent delays and the pull-out of a major supplier, StadCo, have heightened scepticism over the project. The takeover in late 2007 of Nanjing by the larger state-owned Chinese firm Shanghai does offer more hope in this regard, though, given Shanghai's much larger size and its commitment to R&D activities in the region through its joint venture with Ricardo. Finally, it should be noted that the region also contains various lesser-known but significant producers such as LDV Vans (commercial vehicles), Morgan (sports cars), Massey Ferguson (tractors) and Carbodies (taxis).

Amongst first tier suppliers there has also been a process of concentration and specialisation around global players, a process accelerated by recent takeover activity. In the West Midlands this leaves first tier firms such as GKN (drivelines), Dana (axles), Bosch (lighting), Delphi (engine management), Johnson Controls (air conditioning and heating), Faurecia (seating), Lear (seating and interiors), Denso (starters and alternators), TRW (steering, and safety systems), Rockwell (chassis) and Siemens-VDO (instrumentation). Significant second tier suppliers include Sarginsons Precision Components and Zeus (aluminium castings), Brandaur (pressings), Radshape (sheet metal forming) and Premier Stampings (die forgings) amongst many others. The cluster is underpinned by research, consultancy and support organisations such as Ricardo (engine and drivetrain research), MIRA (an automotive research and development technology centre), the Warwick Manufacturing Group, and the Society of Motor Manufacturers and Traders (SMMT) Industry Forum. The region also benefited from the decision by Ford in 2006 to invest £1 billion in research and development into cleaner technology and hybrid engines in the UK, with significant funding coming to Ford's development centres at Gaydon and Whitley. With these now passing to Tata as part of the latter's takeover of Jaguar and Land Rover, there are hopes that Tata will bring a significant R&D presence to the region. Strengths also remain in engine production and research, as evidenced by the number of patents in this area and BMW's investment in its Hams Hall engine plant near Birmingham, which supplies engines for MINI production at Oxford.

8.2.3 Diversity and Heterogeneity in the WM Auto 'Cluster'

As Tilson (1997) noted, the diversity of component manufacturing in the region is a strength that enables it to supply a wide range of products. Many of these firms have been dependent on local assemblers, though, with 70% selling their products locally (ibid.). Many are small, family owned firms with fewer than 250 employees (Lees and Smith, 1993). Tilson found many were experiencing squeezed profits through downward pressure on costs from OEMs (Original Equipment Manufacturers). This is not surprising in what is basically an oligopsonistic global industry with that part located in the West Midlands cluster being dominated by a few giant firms at the top of the hierarchy. Whilst most firms surveyed were optimistic about prospects, a 'hard core (23.5%) anticipated increasing difficulties through consolidation and decimation of the supply chain' (Tilson, 1997). Tilson also suggests that a fifth of the region's component suppliers could be forced out of business by emerging technologies (ibid.), in addition to firm deaths due to other dangers such as the pressure to meet quality standards, the reduction in supplier numbers, global competitive pressures, and investment constraints. These new and emerging technologies include fuel cells, electronic braking and steering, and space-frame technology involving extruded metal parts that are riveted or glued together with plastic or lightweight panels attached (*The Economist*, 23/02/02).

The wide differences across the cluster also extend to which firms are networked and engage in collaborative activities, which markets they serve, their openness to outside the cluster, and the trajectories of individual plants: expansion at Honda (Swindon) and BMW MINI (Oxford), as against cuts in shifts at GM Vauxhall and imminent closure for Peugeot at Coventry. Tully and Berkeley (2004) draw on Gordon and McCann's work to identify three different groups of firms in the West Midlands Auto 'cluster'. The first group (30% of the sample) had no cooperation or long-term interaction with competitors or customers, reflecting a 'pure agglomeration model'. Firms in this group were atomistic and urban based, and their co-location was in line with Marshall's original three key rationales of accessing labour, infrastructure and a flow of ideas and information. The second group (45%) of the sample, engaged in supply chain cooperation. This involved more sophisticated relationships with customers and suppliers which were stable, long term and static. Such relationships are in fact often underpinned by OEM schemes to drive up supply quality and to reduce costs. A final group (25% of the sample) of firms engaged in collaboration with competitors and agencies in a 'social network model' à la Granovetter. These relationships are characterised by trust and a lack of opportunism, and encompass a range of more complex interpersonal relationships outside of the core business remit, reflecting recognition of the need to work together for common, beneficial goals. As Tully and Berkeley (2004) explain in detail, the more sophisticated a firm's relationships with suppliers, customers and competitors, the more positive was their outlook, the more informed they were about market trends, and the more likely they were to have links with local universities. They also undertook more investment in new technologies, had better links beyond the region, and were more likely to be market leaders and outsourced more. Overall, firms engaged in more sophisticated networking relationships were more likely to have undertaken the risk to shift away from high volume and low-value added standardised production to low-volume, high-value added, and customised design-led niche activities. Given the structural changes taking place in the industry, such strategic moves are vital for firms to survive. Yet as Donnelly et al. (2005) stress, many small and medium-sized firms in the sector 'lack the capacity to upgrade their skills, processes or R&D capacities on their own, and so outside assistance is required. Otherwise many small firms will fail'. This raises a crucial role for policy.

8.2.4 Modularisation and an End to the 'Geography of Proximity'?

Industry-wide trends in the automotive sector threaten established local production systems such as the West Midlands cluster. The internationalisation of component sourcing by assemblers has been accelerating in recent years (Sadler, 1999; RTF, 2000). For example, Rover under BMW control shifted £1 billion of its £4 billion annual spending on components out of Britain (*Financial Times*, 24/06/99; 27/04/02) and before collapsing was planning anyway to source significant amounts of components from China. Manufacturers have also been demanding lower costs and wish to deal with fewer suppliers, so as to ease the management of the supply chain. As noted, the net effect has been to force suppliers to become 'world class' in part through mergers and acquisitions, leading to a wave of rationalisations similar to that for final stage manufacturers, with first tier suppliers being hit the hardest. In the future, there are likely to be far fewer first tier suppliers and those that remain will have closer links with the major manufacturers and will be expected to take on stronger R&D roles (RTF, 2000; Bergner, 2000).[3] They are also likely to exert greater power over remaining lower level suppliers (McIvor et al., 1998).

Whilst suppliers continue to build supplier parks across the world (see Larsson, 2002), some commentators such as Wells and Rawlinson (1994) foresee a possible shift away from a post-Fordist system based around just-in-time, lean production and the 'geography of proximity' (Lagendijk, 1997) to a 'post-Japanisation' phase characterised as 'at supplier cost'. The end result of these processes may be a 'trend away from the agglomeration or clustering of component suppliers around OEMs' (Bentley, 2000). This in turn would threaten the stability of local auto production systems, via increased global sourcing. Equally it should be noted that trends such as modularisation may well involve a greater local presence by first-tier suppliers, as they take on assembly and sub-assembly responsibilities from OEMs. Often these are in close proximity to the final assembly because of transport costs, the need to manage increased variety under 'delayed differentiation', and the desire to develop organisational proximity through interactive and collective learning processes between different firms' employees. However, as Larsson (2002) notes, these first-tier suppliers may have little incentive to source components locally for the modules they prepare, preferring global sourcing, with the result that 'automotive clusters that are associated with modular production could have a limited impact for regional development' (ibid.). Modularisation involves the supply of complete systems or modules which are slotted onto platforms at the assembly line. In turn, these are likely to necessitate more R&D by first-tier suppliers and more generally more intense and interactive relations between OEMs and first-tier suppliers (*The Economist*, 23/02/02). It is clear though that modularisation enables suppliers to integrate several systems, and as a process is leading to increasing specialisation and integration in the supply chain. This in turn further promotes concentration and global sourcing, adding further competitive pressures to local auto production systems.

The key points to emerge from this overview of the industry and cluster are threefold. Firstly, across the industry, even the major firms are under intense pressure, given the rising costs of new model development which necessitate large-scale production, platform-sharing strategies and/or joint ventures across firms in order to survive. Secondly, global assembly and components sourcing is increasingly a key threat to localised 'clusters' such as the West Midlands or UK auto components cluster, thus making a cluster policy in this area both more necessary yet also more much difficult to sustain. The ending of mass production in the West Midlands through the collapse of MGR and the closure of the Peugeot plant, and the shift in the region to smaller-scale higher-value luxury niche production, is itself evidence of this. Thirdly, at the local level, the West Midlands cluster is very diverse, comprising both high and low technology applications, and can be seen as a series of interlinked networks ranging from local firms just supplying locally through to global supply chains dominated by the big players.

8.3 MG ROVER: A BRIEF ANATOMY OF FAILURE

At the time of its collapse in 2005 MGR was producing just over 100 000 units a year when it needed to be in the 2–3 million per annum range to generate anywhere near enough cash for new module development. Not surprisingly, media attention focused on the failure of Phoenix management over the proceeding five years and the tiny size of the firm. However, from a wider perspective the firm should be viewed as the unsustainable rump of a government-created giant which never sufficiently integrated activities and which was never in a position to recover the rising costs of new model development. Its long-term decline and ultimate collapse is tied up in a complex vortex of long-running and interrelated factors, including macro-economic instability, the particular short-termism of British finance–industry relations, fratricidal industrial relations, misguided government policy interventions, and above all the firm's perennial inability to generate the cash needed for new model development (Williams et al., 1994). As noted, this 'crisis of cost recovery' has actually intensified over time.

8.3.1 Long-Running Problems and the Failure of a 'National Champion'

MG Rover itself was the remnant of a government creation of the 1950s and 1960s, the British Leyland Motor Corporation (BLMC, later BL and finally Rover). This was brought together by the government, in effect merging smaller car manufacturers (Austin, Morris, Triumph, Rover and Jaguar) through various stages into a single firm, in probably the most prominent example of the misguided policy of creating so-called 'national champions'. As Owen (1999) commented, the merger 'was a mistake both in concept and in execution, reflecting a naïve belief in the advantages of size and in the ability of charismatic individuals to revive declining companies'. Rather than a 'champion', a mega-merger was forced on reluctant and resistant incumbent managers and not surprisingly, suspicion and rivalry across brands hampered efforts to integrate activities and to strip-out costs leading to continued loss-making. This inability to recover development costs, or the 'cycle of doom' as Holweg and Oliver (2005) term it, went back to the 1950s and 1960s, and plagued the firm across decades. After the oil-price shock of the early 1970s, nationalisation in 1975 cost the government – and British taxpayers – billions of pounds in subsidies.

Under Margaret Thatcher, the company was privatised in 1985 through a sale to British Aerospace (BAe) which was diversifying away from aircraft. After the failure of its regional jet business, BAe sold Rover to BMW for £800 million. As Hutton (1999) noted, the sale of Rover to a foreign firm reflected the twin factors of British short-termism (BAe's desire for cash) and the openness of British industry and government to penetration by foreign investment. Rover therefore became part of BMW, but probably for the 'wrong' reasons. BAe needed cash and BMW wanted the 4WD Land Rover division, as BMW at that time did not have the resources to develop its own model to compete with Mercedes, Ford and GM, all of which had developed such vehicles.

8.3.2 Life under BMW: The 'English Patient'

Whatever the precise reasons for BMW's acquisition, problems were immediately apparent. Rover under government and then BAe ownership had relied heavily on a joint venture (JV) with Honda, through which Honda designs were badged and sold as Rovers to European markets, saving Rover considerable R&D costs and periodically enabling it to make modest profits. Honda executives were unimpressed when Rover was abruptly sold to a competitor, and Honda subsequently sold its 20% stake in Rover. As commentators warned at the time, it was not clear what BMW could actually produce at Rover after the JV with Honda was terminated. Although one joint Honda–Rover model programme, the Honda Civic – Rover 400/45 – continued into production, BMW was now on its own. As Williams et al. (1994a) noted, BMW simply did not have the ability to produce competitive, space-efficient, light cars and Rover jobs were thus at risk. Unfortunately the British government failed to comprehend this despite warnings.

Not surprisingly, through a combination of model obsolescence (as BMW struggled to bring on line new cars), marketing mistakes, and macro-economic factors such as the rise in the value of sterling, Rover sales declined and losses grew to £400 to £500 million per year by the late 1990s.

Whilst BMW invested heavily in the Land Rover plant at Solihull (Birmingham) and at Cowley (Oxford), the position of Longbridge was always vulnerable, with BMW threatening early on to switch production of the new MINI elsewhere if productivity did not improve and deals on working conditions were not agreed (Bailey, 2003). Similarly, in March 1999 BMW threatened to shift production to Hungary if a state aid package was not agreed by the government to build a new model range, the R30. A £152 million subsidy package was agreed with the government in mid-1999 in return for a planned £1.7 billion investment by BMW in Longbridge. After a complaint by Porsche, however, the European Commission decided to investigate the aid package under state-aid rules. Before the prolonged investigation was finished BMW announced in March 2000 that it was pulling out, leading to the buy-out by Phoenix.

It was clear to several commentators (Williams et al., 1994; Bailey et al., 1994) that BMW's purchase of Rover in 1994 was a corporate mistake for BMW and likely to cause severe problems for Rover. As Wolf later commented (*Financial Times*, 22/03/99), 'BMW did not realise how bad a buy Rover would be'. The failure of competition policy contributed to this problem by allowing BMW to buy Rover when it was clear to many that it was an inappropriate owner (Bailey et al., 1994). It is widely accepted that local production systems can be improved by firms – whether domestic or foreign – that bring new technology and investment but can also be damaged through takeovers (Harrison, 1994). Whilst BMW was seen by many as an inappropriate owner at the time of its takeover of Rover, Volkswagen (which had shown interest in 1987) might have been much more suitable in that it could extend its strategy of sharing platforms across brands to MG Rover. It should be stressed that the case here is not anti-inward investment per se, but rather centres on the suitability of acquiring firms (whether domestic or foreign) for the local production system.

Not surprisingly, despite substantial investment in the late 1990s, BMW abandoned Rover in 2000, selling the firm for a symbolic £10 to the Phoenix consortium. Despite the hopes raised in 2000, BMW's withdrawal left MGR virtually dead on its feet. The cancellation of the R30 project, which was the key mid-sized model being developed under BMW as a replacement for the R45, meant that a whole cycle of model development had been missed (Holweg and Oliver, 2005). The firm was now brutally exposed as its aged model line became increasingly unattractive to buyers and it had only a very limited time horizon in which to find a partner. By 2002/3 it was clear to many that the firm was running out of time (Bailey, 2003). The unsustainable position of the firm was evidenced by the way it consumed what assets it had.

8.3.3 Exchange Rate Volatility

Another view is that BMW acquired Rover not only for its 4WD competencies but also to obtain a manufacturing base 'in a country which had lower labour costs than Germany and... a stable labour relations climate' (Owen, 1999). Any such attractiveness was soon wiped out by the marked appreciation in the value of sterling from 1994 onwards. At the time of the takeover in 1994, sterling was valued in the range DM2.40 – 2.50, making car assembly in the UK attractive to transnationals such as BMW. With sustained sterling appreciation during the mid to late 1990s, BMW budgeted for a turnaround plan at Rover with sterling at around DM2.90. Yet by January 2000, sterling had risen to DM3.20, and Church (1999) estimated that by July 1999 the Sterling effective exchange rate was overvalued by around 20%. This over-appreciation exacerbated Rover's problems, making exports extremely difficult at a time when the company was trying to re-orientate its sales towards export markets. Such exchange rate volatility has made conditions for the regions' manufacturers extremely difficult, and has accelerated the shift to sourcing overseas and/or in Euros to pass the exchange rate risk on to suppliers. More broadly, cluster policies and development goals in a manufacturing-orientated region can be undermined by such major exchange rate fluctuations. There may be good reasons for Britain remaining outside of the Eurozone, but the cost of exchange rate volatility has not gone away, and MGR's most recent difficulties are a stark illustration of this – the story could have been quite different in the absence of such sterling over-appreciation.

8.3.4 Phoenix: An Unsustainable Strategy

On taking over in 2000, the Phoenix management set four strategic objectives for the firm: maintaining production at 200 000 cars, bringing a new car to market (the replacement for the medium-sized R45); finding a partner for new model development; and returning to profit. But with a limited and aging product range, and in particular the lack of models in key growth segments (such as compact cars, people carriers and sports utility vehicles), this was always going to be a huge challenge. Whilst sales held up reasonably well in 2000, thereafter they declined rapidly and by 2002/3 it became increasingly clear to commentators that the firm had limited time in which to find a partner to bring new models to market (Bailey, 2003). Re-badging aged Rover designs as MGs bought a little time, but over the next few years MGR sold off its only real assets (land, the profitable parts business and finance arm and later its intellectual property rights) in an increasingly desperate attempt to keep going. By 2004, output had dwindled to around

115 000 units and R&D spending had dried up. No partnership deal had been delivered, despite attempts with China Brilliance and Proton. Increasingly it became clear that the very survival of MGR depended on a deal with Shanghai Automotive to jointly develop cars. However, Shanghai became increasingly concerned about the financial viability of MGR and feared picking up redundancy and pensions liabilities, and talks dragged on for several months before ending in failure in April 2005. At that point the firm was forced into administration, with remaining assets later bought for £60 million by Nanjing, another Chinese firm. All but two production lines were then stripped out and shifted to China.

Nanjing aimed to restart small scale production of MG TF sports cars at Longbridge in 2007, with kits being imported from China. However, to date quality issues have delayed production restarting properly and only a limited number of units have been made. Eventual production volumes are as yet unclear and likely to be far smaller than the 100 000 units a year initially suggested by Nanjing at the time of takeover. The pullout of StadCo as a supplier in 2008 means that if production does go ahead this is likely to be a purely 'screwdriver' operation. Thus whilst of considerable significance in terms of redeveloping the Longbridge site, only 'a few hundred' jobs will be created, with few linkages into the local economy. Other developments look more promising, however. Shanghai Auto (which brought the intellectual property rights to the Rover 25 and 75 and the replacement model for the R45 in development at the time of the MG Rover collapse) has developed a joint venture with Ricardo to develop the new model. In late 2007, Shanghai also acquired the car–making operations of Nanjing with Chinese government encouragement. This is likely to strengthen the likelihood of R&D coming back to Longbridge. Auto assembly (even if on a small scale) does finally seem to be returning to Longbridge, under Chinese state ownership.

8.4 IMPACT AND POLICY RESPONSES

8.4.1 Before the Storm: The Rover Task Force Mark 1 (RTF1)

Caught largely unaware by BMW's plans to sell Rover in 2000, the government hastily set up the RTF Mark 1 under the lead of the RDA, with the support of the Government Office of the West Midlands (GOWM). With the Phoenix purchase, the government saw the Rover issue as closed and felt that no further action was necessary. The GOWM, in private discussions with Stephen Byers (then Secretary of State at the DTI), pressed for a credible package to support and develop the regional economy.[4] It won the argument and the RTF Report of July 2000 focused on the inter-linked themes of modernisation (with funding of £17 million), diversification (£19.7

million) and regeneration (£22 million). The *modernisation* programme recognised that too much of the supply base in the region was in low value-added 'metal bashing' activities, and not enough in higher value-added areas such as electronics, and thus was particularly vulnerable given the global changes taking place. The programme included a number of linked initiatives to improve the competitiveness (in this case productivity) of the cluster. For example, the Accelerate programme was extended from EU Objective 2 areas and funds the creation of supplier networks and contributes to improving activities in firms in order to assist them in becoming more competitive (see Accelerate, 2002).

The *diversification* programme was initially the emergency response to the crisis in 2000 and helped suppliers diversify away from Rover. The diversification programme over 2000–2005 aimed to encourage the application of engineering skills in other industries such as medical and nano technologies. RTF1 assisted with development costs, with the Small Business Service providing a diversification service. Behind this was basically a clusters approach, which evolved over time since the RDA's initial Regional Innovation Strategy (RIS) was the starting point in developing the West Midlands Economics Strategy (WMES). Building on these, the *Agenda for Action* (jointly produced by the RDA and the Regional Chamber) then identified ten targeted clusters comprising: *established* clusters (transport technologies, building technologies, food and drink, tourism and leisure and high value consumer products); *growing* clusters (ICT, specialist business and professional services and environmental technologies); and *embryonic or aspirational* clusters (interactive media and medical technologies).

The RDA was able to use the RTF1 to accelerate and legitimise this emerging clusters policy and to extend it further. Indeed, what was genuinely new in the RTF1 was an attempt to link cluster policies with spatial targeting via corridors of regeneration given continued concern over Rover's prospects. The RTF1 Final Report identified three areas for an in-depth study of their growth potential as 'high-tech corridors': the Central Technology Belt (CTB) running past the University of Birmingham and the University Trust hospitals, past the (down-scaled) Longbridge plant, down to Malvern; the Coventry/Solihull/Warwick (CSW) area; and the Wolverhampton/Telford Technology Corridor (WTTC). For example, the CTB idea specifically related to developments flowing from the University of Birmingham, such as a new medipark (RTF, 2000).

8.4.2 Collapse and Impact

MGR's collapse in 2005 sent shock waves through the region, with the firm's turnover accounting for as much 1% of regional GDP, and with £200 million a year alone in government revenue being lost. The firm ceased operations, owing £1.4 billion to creditors, with £109 million owed to UK-based suppliers (House of Commons, 2006) who eventually saw around one penny in the pound for what they were owed. Severe cash flow problems in the supply chain meant that it was impossible for the administrator to restart production. The pension fund deficit also ran to around £500 million, with the new Pension Protection Fund (backed by private sector money) having to make up the difference. In addition to the 6 000 workers who lost their jobs at Longbridge, the supply chain around MGR in the West Midlands was forecast to be badly hit. The 'jobs multiplier' in the car industry is notoriously difficult to calculate, depending on the degree to which the car is made 'in house' as against the pure assembly of bought-in components, as well as the degree to which components are sourced locally rather than internationally. Given such differences, multipliers have been estimated in the range 1:1 to 1:4. In the case of Longbridge, the combination of in-house engine production plus a shift to overseas sourcing before 2005 meant a relatively low local jobs multiplier nearer to the 1:1 end of the range, implying a maximum of around 12 000–13 000 job losses in the broader economy. This 'back-of-the-envelope' 1:1 calculation in fact matched quite closely the physical 'head-count' figure made by AWM by counting suppliers and the total number of jobs dependent on MGR, which came to around 12 200 as well (*Birmingham Mail*, 2005). Given that a Longbridge shut-down in 2000 would have resulted in an estimated 22 500 job losses as against a maximum of 12 000 forecast in 2005, this suggests that the five years of additional time bought by Phoenix production, combined with the efforts of AWM, enabled suppliers to diversify away from the firm, 'saving' in the region of 10 000 jobs (House of Commons, 2006).

However, the final number of total jobs lost directly was actually estimated to be around 8 500 and not 12 000 (i.e. approximately 6 000 at Longbridge and 2 500 in the wider supply chain), with 'only' 11 major suppliers having ceased operations by March 2006 (ibid.). Whilst still a major shock to the economy, this is significantly lower than initial estimates, owing to a number of factors. Firstly, the Longbridge run-down over 2000–2005 and the diversification efforts of AWM precipitated a greater shift by component suppliers to other activities beyond MGR, in excess of that initially estimated. Indeed, over this time, the number of British firms depending on MGR for over 20% of sales dropped from 161 to 74, with only 57 in the West Midlands region (ibid.). Secondly, the emergency policy response in 2005 involved a package of measures which included a wage-subsidy element for suppliers. This bought some much-needed extra time for firms facing extreme cash-flow problems and saved jobs. It is to this that we turn next.

8.4.3 Picking up the Pieces: The Rover Task Force Mark 2

On alert from the beginning of 2005, AWM and the national government moved quickly when MGR finally collapsed in April 2005. A new Rover Task Force (RTF2) was set up with the aims of: helping suppliers to keep going and to diversify; aiding workers in finding new jobs; and in supporting the broader community. An aid package eventually worth £176 million was made available, including £50 million for retraining (up to level 2 skills), £40 million in redundancy payments, a £24 million loan fund to help otherwise viable businesses, and £41.6 million to support MGR suppliers to sustain trading. Another £7.6 million was announced by AWM in June 2005 to assist with supplier diversification. The tailored support for suppliers proved critical in limiting the damage to the local economy. For example, a £3.4 million Wage Replacement Scheme helped 170 firms and kept around 3 000 workers in place for the critical weeks following the collapse, with 1 329 'confirmed' jobs being saved in this way (RTF2, 2006).

By February 2006, of 6 300 claimants resulting from the MGR collapse, 4 000 were back in work, 2 000 had received some form of training, and 2 300 remained out of work (RTF2, 2006). Whilst portrayed in the media as a great success story in enabling structural change and assisting workers back into employment, the fact that 2 300 remained out of work ten months after the collapse highlights some of the broader structural problems in the local economy. A key issue here is the general and rapid shift away from mass manufacturing towards smaller-scale high value-added work, with manufacturing employment in Birmingham having fallen by 100 000 since the mid 1980s, and forecast to fall by another 10% over the next 10 years (BSEDF, 2005). Whilst services in Birmingham have grown quickly over the last two decades (accounting for 80% of jobs in Birmingham), these have been quite diverse jobs, with some being high-paying professional services, and others being relatively low-paid (often part-time) jobs in retail or wholesaling. Overall employment levels are up compared with 20 years ago, yet unemployment remains stubbornly high relative to the national average, with 7.5% of the workforce in Birmingham unemployed at the time of writing. This is highly concentrated in pockets of deprivation, where people have been unable to cope with the structural change that has unfolded around them. A real legacy effect is the absence of a 'widespread culture of studying for qualifications', with 37% of local 16–64 years old having no recognised qualifications and 31% no NVQ level skills in 2001 (BSEDF, 2005).

This low educational level is a real factor in the MGR example, where most of the growth sectors in the local economy in South West Birmingham are in high-quality public sector activities (e.g. higher education and medical care) requiring level 3 or 4 qualifications. Yet only 10% of Longbridge workers had such level 4 qualifications. More broadly, the new growth sectors (such as aeronautics, where turnover is forecast by some to double by 2020), often require higher skills, which in turn creates a supply–demand mismatch as workers with lower skills leave manufacturing activities such as assembly work but are unable to move into the new growth sectors (*The Birmingham Post*, 20/04/06). In this connection, Cowling and Isles (2005) note that local employers were 'only offering jobs for the most highly skilled workers' which 'leaves a core of around 75% of the Longbridge labour force who are lower skilled operatives'. This raises two key points for the longer-term assimilation of Longbridge workers and indeed for the development of the local economy. Firstly, there is a need for a much greater effort in terms of cross-training initiatives. The aid package outlined above provided assistance for skills upgrading to level 2, but it is clear that a greater effort is required, given the skills needs of growth sectors. Secondly, it suggests a need for geographical mobility on the part of the least skilled workers coming out of Longbridge, yet blue collar mobility is often a real problem in such situations (Cowling and Isles, 2005), especially for workers who had car loans for MGR cars which they were unable to repay after redundancy.

While the 'cost' of closure fell most heavily on those unable to shift into new employment, even those subsequently back in work have experienced substantial losses, as a study for a BBC Radio 4 series has found, with ex-MGR workers back in full-time employment earning around £3 500 a year less and those in part-time £10 500 less, with only those in self-employment now earning more than they did at Longbridge (Armstrong, 2006). Furthermore, nearly a half of ex-MGR workers viewed their 'new' job as worse than their MGR job. With the loss of secure employment, there is also an increased likelihood of suffering unemployment in future years, adding to stress levels. As Armstrong (2006) rightly notes, longer-term policies are required to build skill levels in Birmingham and tackle growing labour market polarisation and inequality.[5]

8.5 SUPPORTING MANUFACTURING CLUSTERS: SOME WIDER POLICY ISSUES

8.5.1 Tensions and Incompatibilities in Cluster Policies

The MGR crisis over 2000 to 2005 prompted an acceleration of regional-level initiatives. The modernisation and diversification agendas recognised the need to shift suppliers away from dependence on MGR; in so doing, they assisted firms looking for new markets and applications and contributed to the 10 000–12 000 jobs 'saved' over 2000 to 2005. More broadly, it recognised the need to move the auto cluster towards high-performance and prestige markets, to diversify the economic base away from a dependence on the auto cluster, as well as the need to overcome a defensive style of capitalism amongst small firms, that mitigates against cooperation and the benefits of innovative activity (see De Propris, 2000).

Secondly, the MGR crises kick-started a regeneration agenda which linked cluster policies with spatial targeting and land-use planning. This has been a particular problem for RDAs more generally, with cluster policies not considering factors such as land-use planning, science and technology and industrial support (Waters and Lawton Smith, 2002). Thirdly, the regeneration approach enabled a sub-regional approach to develop centred on the corridors, which in two cases at least have begun to draw in key local organisations such as universities. As Walters and Lawton Smith (2002) note, universities are assumed to be part of the regional innovation infrastructure and can be a key factor in cluster development, yet they are difficult to coordinate as part of regional strategies because of their autonomous status and more ambivalent relationship with territory than RDAs. Fourthly, the crisis forced AWM to provide a degree of strategic oversight and mobilise a range of other actors to respond to the situation. Again this has been a problem for RDAs more generally; a case in point is collaboration with the Small Business Service where Fuller et al. (2002) note there are no formal links between the two sets of organisations. Yet they note that Rover was 'very important in fostering joint strategy making and action across the region' (ibid.).

Whilst recognising these successes, there are limits to what AWM can achieve in the auto sector for a number of reasons. Firstly, supporting a modernised West Midlands auto cluster based around major assemblers and first-tier suppliers will be difficult to maintain given global shifts in the industry. The last five years have seen the collapse of MG Rover, and closures at Jaguar and Peugeot. Mass production of cars has all but disappeared from the West Midlands, leaving in its wake (for the time being) higher-value niche production at Jaguar, Land Rover and Aston Martin.

Relating back to the earlier discussion of what exactly constitutes a cluster, the auto cluster in question here is in fact a top-down oligopsonistic cluster dominated by global players headquartered elsewhere and which are transnationally footloose, as indicated by the recent closures.

Furthermore, the WM auto cluster at the very least forms part of a national auto network which is being increasingly internationalised through mergers amongst first-tier suppliers to create global firms. This global dimension via the local presence of transnational firms brings challenges as noted, but also opportunities through opening the cluster to new innovation, technology and investment, thus helping to avoid atrophy and decay. More generally, the importance of linking regional clusters with global networks has been emphasised by researchers (Malecki, 1997). One objective of policy thus needs to be to protect and nurture the supply chain and to embed global firms locally so as to avoid them shifting their sourcing elsewhere. A regional development policy is important here in fostering knowledge intensive competencies, and to try to tie transnational firms into these competencies. A regionally based cluster policy in this sense needs to emphasise training, innovation, and the linking of networks of service providers with firms, and could offer a bond between footloose transnationals firms and the locality. In so doing, building regional competencies to tie in remaining operations could avoid the risk of further hollowing out through increasingly global players sourcing overseas. The RTF, building on the work of the Accelerate programme, has begun to do this (Accelerate, 2002) and does genuinely represent a step forward.

In line with Camagni (1991), overcoming competence gaps is closely linked with regional 'collective learning', which is seen as central to the development of a successful milieu or set of territorial relationships 'encompassing in a coherent way a production system, different economic and social actors, a specific culture and a representation system, and generating a dynamic collective learning process' (ibid.). Camagni sees the latter developing tacitly through factors such as the local labour market, customer–supplier technical and organisational interchange, imitation processes, reverse engineering, exhibition of successful 'climatisation', application of general technology to local needs, 'cafeteria' effects and specialised service provision. Similarly, Keeble et al. (1999) see regional collective learning as 'the emergence of basic common knowledge and procedures across a set of geographically proximate firms which facilitates cooperation and solutions to problems' through establishing a common language, trust, shared technological knowledge, as well as tacit codes of conduct. This links in turn with the concepts of the 'learning region' and 'regional innovation systems'. In this sense a regional cluster policy needs to focus on support that takes a holistic approach focused on 'soft infrastructure' and competence building (Lundequist and Power, 2002). Providing the 'soft support' associated with the learning economy is challenging anyway (see Storper, 1997), let alone when the RDA lacks the power and resources over many of the areas in question. Indeed, Fuller et al. (2002) note that RDAs focus primarily on urban regeneration and 'not on building regionally based innovation systems'. For example, in the context of the WM auto cluster, component suppliers have repeatedly indicated shortages in specific skills as a key problem hindering growth (Tilson, 1997). Yet AWM has been handicapped in this respect, with limited resources or powers in this area until very recently. Resources were instead initially under the control of Learning and Skills Councils (LSCs) reporting to the Department for Education, with the RDA having no formal powers, leaving them unable 'to develop fully a skills role within their regions', thereby highlighting the 'nebulous and precarious role of RDAs within the area of learning and skills' (Fuller et al., 2002).

It also needs to be recognised that the cluster already extends well beyond the region and a single regional cluster initiative may be inadequate. Linked initiatives across regions which are coordinated nationally, are required to assist the auto cluster develop. Whilst the regionalisation of industrial policy through decentralisation to RDAs has brought advantages in developing and tailoring policies to suit local needs, it also brings potential dangers in terms of competing and overlapping regional initiatives that are insufficiently coordinated when the cluster in question extends across regional borders. As Peck and McGuiness (2001, quoted in Waters and Lawton-Smith, 2002) note 'the alternative involving greater collaboration between regions in developing cluster strategies has yet to be taken seriously'. Despite the retrenchment of national-level industrial policy and the shift to the regional scale, there is clearly a key role for national-level coordination of RDA strategies, most obviously via the DTI. The administrative (and aspirant political) scale here is the West Midlands yet the relevant economic scale for the auto industry covers at least five English regions (East Midlands, North West, South East, South West and the West Midlands) with each auto cluster forming part of a national or international network.

Paradoxically, simultaneous to this lack of inter-RDA coordination of cluster policies, central government in Britain has been excessively restrictive and interventionist over *which* clusters and technologies to support. Indeed, Charles and Benneworth (2001) argue that it has determined which clusters are important and has refused to delegate science and technology policy to the regional level. This might help to explain why the clusters targeted by different RDAs across the UK are so remarkably similar: many target the same list including automotive, ICT, creative industries and media, food and drink, and biotechnology. The 'astonishing overlap' noted by Andriani et al.

(2005) in cluster policies 'stretches credibility to believe that the UK could be home to half a dozen or more successful biotech clusters. Thus most strategies (if not all) will fail, raising the question of whether this constitutes an appropriate direction of public funds'.

In a similar vein, Waters and Lawton Smith (2002) argue against 'the local application of top-down policy'. This has impacted particularly on the West Midlands where the diversification approach centred on the need to shift into new clusters such as nano-technology. This was initially met with encouragement by central government and was developed as part of the CTB corridor idea. However, central government later ordered AWM to postpone its application for a £90 million nano-technology centre to support this aspirational cluster (*Financial Times*, 02/12/02). This was despite nano-technology having been identified as early in the RIS as a key technology to be developed and diffused (West Midlands First, 1998), and having been stressed as a thematic priority under the EU's Sixth Framework programme. Eventually, the go-ahead for a new centre was given, but only after lengthy delays, much to the frustration of local officials.

The top-down, *dirigiste* approach has been heightened by central government constraints on the limited resources available to the English RDAs, in comparison with devolved agencies in Scotland and Wales. Whilst the total budget of RDAs rose to £1.2 billion in 2003, individual budgets remain dwarfed by their Scottish equivalent. In the West Midlands case, the retention in 2000 of £129 million of RSA assistance within the region that had been earmarked for BMW plus another £176 million in 2005 to pick up the pieces after the collapse, was an implicit recognition that AWM did not have sufficient resources to tackle a problem of this scale. Furthermore, RDAs in general have faced only limited flexibility over the limited spending available to them, being heavily constrained by preordained silos of funding from central government. Although funding is now in the form of a block grant or 'single pot', with the ability to vary across budgets, central government has set very specific targets for RDAs to meet, which considerably limit their room for manoeuvre (Fuller et al., 2002). The paradox thus arises that RDAs are overly constrained in terms of their ability to pursue strategic objectives in terms of regional industrial policy, but that their efforts to foster the development of local clusters or sectors are insufficiently coordinated on a national or interregional basis.

8.5.2 Recognising Policy Shortcomings

Whilst MGR's problems have centred on long-seated organisational and management failures, it needs to be noted that a contributing factor was also a series of policy failures going back to the 1960s, as detailed above. This needs to be recognised for the development of future industrial policy given that the latter is now back on the agenda in some sense. More recently, the central government's response to the MGR affair over the last five years has been typically reactive, with the setting up of the RTF1 representing a knee-jerk response bolted onto other competitiveness programmes (e.g. the Accelerate programme) and RTF2 coming after MGR's final death throes in 2005. The DTI itself only intervened in late 2004 and early 2005, when the firm was rapidly approaching a meltdown.

The 2006 National Audit Office Report into the DTI's handling of the MGR collapse deserves close scrutiny in this respect. In the report, AWM and local actors are given high marks for their response to the crisis. Helping 4 000 workers back into work has indeed been a major achievement, although it should be stressed that there remains much to be done for the remaining unemployed workers. Furthermore, that the knock-on effects of the MGR crash on the supply chain were not as severe as first anticipated was also largely thanks to the work of RTF1 and AWM in working to diversify firms away from MGR over the previous five years. As noted, as many as 10 000–12 000 jobs were saved in this way.

Whilst the media has focused on the NAO's criticism of the government spending £6.5 million to pay the wage bill for five days to see if Shanghai could be tempted back to buy MGR as a going concern, the real criticism of the DTI, only partly recognised by the report, was that it took so long to realise that MGR was struggling, and then rushed into contingency planning that focused too much on how to deal with a collapse of the firm. Less than a year before MGR's collapse, in May 2004 the then DTI secretary Patricia Hewitt stated that she was 'very pleased with Rover's performance' and that the Phoenix four had taken 'very big risks' in starting up Rover again. Yet it was clear to many that MGR was selling off its assets (land, the parts business, its finance arm and later its IPRs) in an increasingly desperate attempt to keep going. This had been clear to several commentators as early as 2002/3 who said that the firm had no long-term future without a partner and that it was running out of time (see Bailey, 2003).

The question arises as to what the DTI was doing at that point; was it helping MGR to do a deal with Shanghai or another firm then? A deal in 2002/3 might well have saved some small-scale production and research and development at Longbridge. That the latter was not saved under Nanjing has been a lost opportunity, even if mass production was always going to go. Yet the DTI was not capable of doing this as the on-going and detailed monitoring of strategic companies is not what it considers a relevant part of modern industrial policy. This lack of information gathering was a key criticism of the DTI in 2000 by the House of Commons Select Committee on Trade and Industry when it looked at the BMW withdrawal (Bailey, 2003a).

By 2004, there was little that the DTI could have done. Overall the NAO Report does not make as comfortable reading for the DTI as some government ministers have claimed. There are important lessons to be learnt for industrial policy design and delivery.

8.6 CONCLUDING COMMENTS

A number of technological and structural changes are unfolding in the auto industry, including more rapid technological change which has driven up the costs of new model development, in turn increasing the pressure on firms to recover costs by more international sourcing of modular components with a greater role for larger first-tier suppliers, and a shift of labour-intensive assembly operations towards lower-cost locations as trade barriers have come down and as globalisation proceeds. All of these trends make maintaining the West Midlands cluster more challenging for firms and policy makers.

In this context, MGR was the unprofitable rump of a former giant which for years had struggled to generate cash for new models, owing in part to a lack of integration across the firm. The firm became reliant on Honda for new models in the 1980s before being acquired by BMW. The latter's withdrawal from the firm in 2000 left MGR virtually dead on its feet, and by 2002/3 it was clear to many that the firm was running out of time. Also significant in the firm's demise, however, were a number of government policy mistakes over the years, including a misguided 'national champions' approach, a failure to integrate activities under nationalisation, a mistaken privatisation, and the failure of competition policy in allowing the sale to an inappropriate owner in BMW in the 1990s. If we add in the considerable volatility of sterling, the scene was set for the firm's demise. A lack of intelligence-gathering on the part of the DTI around 2002/3 did little to help the firm when it was at a critical juncture in the hunt for a partner.

Whilst the MGR collapse was a substantial shock to the West Midlands economy, the impact would have been much greater if the firm had collapsed in 2000. The five years of continued production over 2000–2005 allowed many suppliers to adjust and diversify away from MGR, aided by the work of the first RTF. This was a very real success for local policy makers. The response after the collapse was also a major effort that minimised job losses. Challenges remain, however, especially for the least well-skilled workers as the manufacturing base 'hollows out' locally, with high levels of local unemployment and a low-skills base in much of the Birmingham economy.

More broadly, the chapter examines work of the AWM in its clusters policy. The MGR case raises the critical issue of the coordination of regionally based cluster policies where the cluster in question crosses administrative regional boundaries, calling for a more coordinated industrial policy to join up regionally-based policies. Work by AWM at the regional level would be more effective in building local competencies if the fragmentation of regional level policy delivery was properly addressed; if the RDA was given more resources and flexibility and was freed from the straightjacket of a still top-down government technology policy; if regional industrial policies were coordinated inter-regionally where clusters cut across regions (clearly the case in the auto industry); and if the need for exchange rate stability was recognised nationally in order to enable regionally-based manufacturing strategies to succeed.

A key problem is the relevant regional scale when administrative and economic spaces do not coincide. AWM and the RTF have genuinely attempted to improve regional economic prospects by attempting to diversify, modernise and regenerate those areas most affected by the shake-out in the auto industry and through building regional competencies and social capital There still remains the key question of who is 'joining it all up' properly at the regional, inter-regional and national level.

NOTES

1. The authors wish to acknowledge the support of the ESRC under award number RES-000-22-2478. For more detail on the MG Rover case, see Bailey *et al.* (2008).
2. The automotive industries include motor vehicle production (SIC code 341), bodies (SIC code 342) and parts and accessories (SIC code 343). Related industries include rubbers and plastics, electrical equipment, glass, and so on.
3. This is a global trend: Bergner (2000) notes that between 1988 and 1998 the global number of direct component suppliers to OEMs and the aftermarket shrank from 30 000 to 8 000.
4. Interview with Gerry O'Donnell at Government Office for the West Midlands (GOWM), 09/03/01.
5. International comparisons between policy responses in the MGR case and the Mitsubishi closure in Adelaide are explored by Thomas *et al.* (2008) and Armstrong *et al.* (2008).

REFERENCES

Accelerate (2002), www.accelerate.co.uk.

Andriani, P., Jones, C., Perkmann, M., De Propris, L., Sena, V., Delbridge, R., Möslein, K. and Neely A. (2005), *Challenging Clusters,* London: AIM Research.

Armstrong, K. (2006), *Life after MG Rover: A Report prepared for BBC Radio 4*, London: The Work Foundation.

Armstrong, K., Bailey, D., De Ruyter, A., Mahdon, H., and Thomas, H. (2008). 'The Impact of Plant Closures on Workers: A Comparison of MG Rover in the UK and Mitsubishi in Australia', *Policy Studies*, **29** (3).

Bailey, D. (2003), 'Globalisation, Regions and Cluster Policies: The Case of the Rover Task Force', *Policy Studies*, 2003, **24** (2/3).

Bailey, D. (2003a), 'Does the EU Need a Transnationals Policy? The BMW-Rover

Affair as a Case of Corporate, Strategic and Government Failure', *Birmingham Business School Working Paper*, 2003-11.

Bailey, D. and Cowling, K. (2006), 'The Lesson Peugeot has Taught Britain', *Parliamentary Brief*, June 2006.

Bailey, D., Harte, G. and Sugden, R. (1994), 'British Policy towards Inward Investment', *Journal of World Trade*, **28** (2).

Bailey, D., Kobayashi, S. and MacNeill, S. (2008), 'Rover and Out? Globalisation, the West Midlands Auto Cluster, and the end of MG Rover', *Policy Studies*, **29** (3).

Bentley, G. (2000), 'The Automotive Industry: Change and Challenge for the RDAs', in G. Bentley and J. Gibney (eds), *Regional Development Agencies and Business Change*, Aldershot: Ashgate.

Bergner, R. (2000), *Responding to the Challenges: Demands from Vehicle Manufacturers towards Suppliers are Ever-Increasing*, http://www.autoindustry.co.uk/library.

Birmingham Mail (2005), 'The Real Cost of Rover's Collapse', 21/09/05.

The Birmingham Post (2006), 'Aerospace Creates Three Times as many Jobs as Automotive', 20/04/06.

BSEDF (Birmingham and Solihull Economic Development Forum) (2005), *Economic Review 2005/6*, Birmingham: Birmingham and Solihull Economic Development Forum

Camagni, R. (1991), 'Local "Milieu", Uncertainty and Innovation Networks: Towards a New Dynamic Theory of Economic Space', in R. Camagni (ed.), *Innovation Networks: Spatial Perspectives*, London: Belhaven.

Charles, D. and Benneworth, P. (2001), 'Are we Realizing our Potential? Joining up Science and Technology Policy in the English Regions', *Regional Studies*, **35** (1).

Church, K. B. (1999), 'Properties of the Fundamental Exchange Rate in the Treasury Model', *National Institute Economic Review*, **169**, July 1999.

Cowling, M. and Isles, N. (2005), *Sent to Coventry? The Re-Employment of the Longbridge 5000*, London: The Work Foundation.

De Propris, L. (2000) 'Innovation and Inter-Firm Cooperation: the Case of the West Midlands', *Economics of Innovation and New Technology*, **9**.

Department of Trade and Industry (DTI) (2001), *Business Clusters in the UK – A First Assessment. Volumes 1 and 2*, London: The Stationary Office.

Donnolly, T., Barnes, S. and Morris, D. (2005), 'Restructuring the Automotive Industry in the English West Midlands', *Local Economy*, **20** (3), August 2005.

The Economist (2002), 'Incredible Shrinking Plants', 23/02/02, 99–101.

Financial Times (1999), 'Carmaker's Auction', 22/3/99, 16.

Financial Times (1999), 'Rover Suppliers Warned to be More Competitive', 24/6/99, 9.

Financial Times (2002), 'Defiant Carmaker Foresees a Positive Route to the Future', 27/04/02, 3.

Financial Times (2002), 'Concerns surround Science Centre Proposals', 02/12/02, 4.

Fuller, C., Bennett, R. J. and Ramsden, M. (2002), 'The Economic Development Role of English RDAs: The Need for Greater Discretionary Power', *Regional Studies*, **36** (4).

Harrison, B. (1994), *Lean and Mean. The Changing Landscape of Corporate Power in the Age of Flexibility*, New York: Basic Books.

Holweg, M. and Oliver, N. (2005), *Who Killed MG Rover?*, Cambridge: The Cambridge-MIT Institute.

House of Commons (2006), The Committee of Public Accounts, *The Closure of MG Rover. Report, Together with Formal Minutes, Oral and Written Evidence*, HC10013, London: The Stationery Office Ltd.

Hutton, W. (1999), 'Why Rover was Driven out of UK Hands', in *The Stakeholder Society. Writings in Politics and Economics*, Oxford: Polity Press, pp. 58– 60.

Keeble, D., Lawson, C., Moore, B. and Wilkinson, F. (1999), 'Collective Learning Processes, Networking and "Institutional Thickness" in the Cambridge Region', *Regional Studies*, **33** (4).

Lagendijk, A. (1997), 'Towards an Integrated Automotive Industry in Europe: A "Merging Filiere" Perspective', *European Urban and Regional Studies*, **4** (1).

Larsson, A. (2002), 'The Development and Regional Significance of the Automotive Industry: Supplier Parks in Western Europe', *International Journal of Urban and Regional Research*, **26** (4).

Lees, J. and Smith, P. (1993), 'A Sectoral Study of the Walsall Economy', *Journal of Industrial Affairs*, **2**.

Lundequist, P. and Power, D. (2002), 'Putting Porter into Practice? Practices of Regional Cluster Building: Evidence from Sweden', *European Planning Studies*, **10** (6).

Malecki, E. J. (1997), *Technology and Economic Development: The Dynamics of Local, Regional and National Competitiveness*, London and Boston: Addison Wesley, Longman.

McIvor, R. T., Humphreys, P. K. and McAleer, W. E. (1998), 'European Car Makers and their Suppliers: Changes at the Interface', *European Business Review*, **98** (2).

National Audit Office (2006), *The Collapse of MG Rover*, HC961 Session 2005–6, London: The Stationery Office.

Owen, G. (1999), *From Empire to Europe. The Decline and Revival of British industry Since the Second World War*, London: Harper Collins.

Rover Task Force (RTF1) (2000), *Final Report and Recommendations*, Birmingham: Advantage West Midlands.

Rover Task Force (RTF2) (2006), *Final Update Report: The Work Goes On*, Birmingham: Advantage West Midlands.

Sadler, D. (1999), 'Internationalization and Specialization in the European Automotive Components Sector: Implications for the Hollowing Out Thesis', *Regional Studies*, **33** (2).

Storper, M. (1997), 'The Regional World: Territorial Development in a Global Economy', *New York*: Guilford.

Thomas, H., Beer, A. and Bailey, D. (2008). 'A Tale of Two Regions: Comparative versus Competitive Approaches to Economic Restructuring', *Policy Studies*, **29** (3).

Tilson, B. (1997), *Survey of Firms in the Automotive Components Sector in the West Midlands Region*, Birmingham: Centre for Urban and Regional Studies, The University of Birmingham.

Tully, J. and Berkeley, N. (2004), 'Visualising the Operating Behaviour of SMEs in Sector and Cluster: Evidence form the West Midlands', *Local Economy*, **19** (1).

Waters, R. and Lawton Smith, H. (2002), 'Regional Development Agencies and Local Economic Development: Scale and Competitiveness in High-technology Oxfordshire and Cambridgeshire', *European Planning Studies*, **10** (5).

Wells, P. and Rawlinson, M. (1994), *The New European Automobile Industry*, New York: St Martin's Press.

West Midlands First (1998), *Shaping our Future: Regional Innovation Strategy for the West Midlands. Consultation Document*, Birmingham: University of Birmingham / West Midlands RIS Steering Group.

Williams, K., Haslam, C., Johal. S. and Williams, J. (1994), *Cars. Analysis, History, Cases*, Providence: Berghahn Books.
Williams, K., Haslam, C. and Johal, S. (1994a), *Who's Responsible? BAe: BMW: Honda: Rover*, A Public Interest Report from the Centre for Empirical Research in Accounting and Finance, University of Manchester.

9. Massachusetts Medical Devices: Leveraging the Region's Capabilities

Michael H. Best[1]

9.1 INTRODUCTION

Throughout its economic history, Massachusetts has experienced the emergence and growth to national leadership, and then the decline of regional concentrations of related firms and organisations known as 'clusters', such as in textiles, shipbuilding, footwear and minicomputers. While significant research has been done on the characteristics of fully formed clusters, relatively little work has gone into analysing the rapid growth process between the emergence and establishment of a successful cluster.

The emergence of a cluster right before our eyes – the medical devices sector – creates that opportunity today in Massachusetts. But as we looked into this dynamic sector, we soon realised that NAICS and SIC codes and other ways by which government classifies and measures a sector were unable to historically track the firms or account for the changing product boundaries of the medical devices cluster. Companies are anonymous and company product mix data is not recorded in government economic data. The anonymity requirement obscures growth dynamics and the neglect of product information obscures cluster boundaries and dynamics. For example, the product of Firm X may primarily be computer components, but it also manufactures imaging systems for computed tomography (CT) medical devices. But because that latter activity is not its major activity, Firm X is classified under a non-medical technology code and is thus not categorised as part of the medical device sector.

To understand how the medical devices cluster (which from here onward will be abbreviated as MED) emerged and formed, we need to better quantify and analyse both the companies and the products that constitute it. To this end, we developed a powerful new data base to better measure the scope and substantial growth of the MED sector in Massachusetts. The vTHREAD (Techno-Historical Regional Economic Analysis) database is a longitudinal, historical database of approximately 55 000 public and private high-tech producers classified by a finely granulated taxonomy (see Box 9.1). Because the unit upon which vTHREAD is based is the company itself, it recognises that product boundaries can run through the middle of a company and that a company may have multiple products in multiple industrial categories. vTHREAD allows us to include firms that make medical products even though they are classified in non-medical technology codes.

Box 9.1 Empirical methodology: vTHREAD

To get inside the faceless and ahistorical companies that feature in official data, we have constructed vTHREAD (Techno-Historical Regional Economic Analysis Database), a database of approximately 55 000 public and private, high-tech producers and a set of research tools designed to analyse regional industrial specialisation, growth, decline and reinvention.

The vTHREAD database is populated with a new longitudinal file covering 1989 to the present, based on CorpTech data. The primary purpose of the CorpTech data set is to provide company information on private and public high tech companies in the United States. It is supplied quarterly to subscribers and currently includes approximately 55000 high-tech companies in the United States, including 5000 in Massachusetts. Although the dataset is not constructed for scholarly purposes, CorpTech established sophisticated data collection and research methodology, including quality control systems and consistency checks.[2] The data base is longitudinal; that is firms in the data base are observed and measured over a number of years and their year-to-year records are then linked. The file is organised with a unique, finely granulated taxonomy of companies and products. It can therefore be used quite powerfully for industry studies, being able to identify and measure inter-industry linkages.

The development of the vTHREAD database has been a joint effort. Albert Paquin, the first research assistant, has stayed on from the beginning in various roles. Andrew Frisch did the original programming. Research assistants Hao Xie, MinYu, and John Sharko and colleagues Georges Grinstein and Edward March have all made major contributions to either the research methodology or the development of the database. The Chancellor's Office of the University of Massachusetts Lowell and a CVIP Development Fund award from the President's Office of the University of Massachusetts provided financial support.

In the case of Massachusetts for 2004, we identified 177 companies classified as medical device firms in Massachusetts and another 105 companies that are classified as non-medical device but which make such products. A total of 63 of these firms had more than 200 employees (32 MED, 31 non-MED).

9.1.1 Sector Overview

The annual output of American medical devices firms grew from $10 billion in 1979 to more than $90 billion in 2004.[3] By U.S. Department of Commerce data, Massachusetts is ranked within the top five medical device states in terms of the value of shipments, employment, payroll and valued added by both per capita and absolute size (Clayton-Mathews, 2001, p. 3).[4] An index of Massachusetts merchandise exports shows a growth of medical device exports of 78 percent, compared to growth in total exports of 18 percent between 1998 and 2003 (MISER, 2004).

Why has Massachusetts been so successful in medical devices? The easy answer lies in the region's plethora of research hospitals, which have attracted a disproportionate share of federal research and development (R&D) funding, which in turn has fostered technology transfer, business spin-offs and otherwise created opportunities for medical device companies. But what exactly are the links between medical device clusters and research-intensive medical centres, such as Massachusetts General Hospital in Massachusetts and the Mayo Clinic in Minnesota? What makes these regions different from many others that have successful research-intensive hospitals but lack a substantial medical device industrial counterpart?

Part of the answer lies in Massachusetts' long history of precision engineering and instrument making, which collectively has gone through many design iterations with the transitions from defence to minicomputer to information-communication technologies and now to life science-based industries.

With the vTHREAD methodology in place to identify components of the MED sector, we are now better able to understand how these companies and this cluster have grown in Massachusetts. Much of this chapter examines Boston Scientific, the largest of, and a paradigm for, five fast-growing MED firms with more than 1000 employees, that have grown rapidly in Massachusetts over the past 15 years. Boston Scientific's sales grew from $2 million in 1965 to $5.6 billion in 2004. We then briefly cite fast-growing, mid-sized MED companies and operating units located in Massachusetts. While most are Massachusetts-headquartered companies, we also find fast-growing operating units of companies headquartered in other states.

As noted, the new database and classification system allow us to identify non-medical device companies that make medical device products or provide medical device services. Included here are specialist outsourcing and manufacturing services companies. Such companies straddle industry boundaries and are most important in understanding industrial change and renewal. We have also been able to examine a group of large instruments companies that have transitioned into medical devices. The size of these companies suggests they can play a major role in cluster development. Another category of firms is the large group of foreign-headquartered companies that have operating units in Massachusetts. This group offers clues to the distinctive regional capabilities that cannot be found elsewhere.

9.2 BOSTON SCIENTIFIC: PARADIGM OF THE LARGE, FAST-GROWING COMPANY

To see the rapid growth of the medical devices industry in the United States and in the Boston area, one needs to look at only one firm: the growth behemoth called Boston Scientific.

In 1979, the national industry was worth only $10 billion and only one company, Medtronic Inc., was of significant size. By 2004, the industry, narrowly defined, was worth nearly $94 billion. Over the same 25 years, Boston Scientific's revenues grew from $2 million to $5.6 billion. Employing more than 17 000 people worldwide,[5] it has become the largest medical devices company in the world in the category of minimally invasive therapy.[6]

The Boston Scientific story began in the late 1960s, when co-founder John Abele acquired an equity interest in Medi-tech, Inc, an R&D company focused on developing alternatives to traditional surgery. Medi-tech's first products, introduced in 1969, were steerable balloon catheters that were used in some of the first less invasive procedures. Medi-tech co-founder Peter Nicholas had run the medical products division of Millipore Corp, a large purification equipment company in the biotech and pharmaceutical sectors.[7] Abele and Nicholas are still leaders at Boston Scientific.

9.2.1 Process Integration

Before 1979, the small size of the medical devices industry reflected the passive character of device production. Medical device companies were primarily small instrument companies that built custom devices to the specifications of physicians. The market development challenge was not only technical, but organisational and even political and academic. Power in the medical community was concentrated in the physicians and in cardiology, where surgeons were both powerful and highly skilled in a well-developed, major surgery methodology. The political and intellectual challenge for Abele and Nicholas was to gain physicians' acceptance of angioplasty, an invasive but non-surgical approach to treat diseased arteries. For many physicians at the time, invasive surgery was immoral and unethical because, as Abele put it, it was 'safer to have a big opening so if something went wrong it was easier to fix it' (Swain, 2004, p. 9).[8]

Abele first had to communicate and win the academic argument within the medical community. Forming a partnership with a leading physician was as crucial to the early development of Boston Scientific as it had been to Medtronic. Medtronic's pacemaker technology was an outcome of a long collaboration between Earl Bakken and Dr. C. Walton Lellehie, a pioneer in open-heart surgery at the University of Minnesota's School of Medicine.[9] A similar partnership between Abele and Dr. Andreas Gruentzig was pivotal to the creation of Boston Scientific's revolutionary angioplasty technology. Bakken and Abele cite the communication role of a clinical doctor/researcher to win over the medical community. In the case of the first battery-powered wearable pacemaker, Bakken noted: 'Our friend and collaborator, C. Walton Lillehei of the University of Minnesota, spread the word throughout the worldwide medical community' (Bakken, 1999, p. 63).

But important as such personal partnerships were to the early development of Medtronics and Boston Scientific, they were not enough to build a company. The new product development process involves the lateral integration of a whole range of specialist activities and skills/occupations. Physicians had to be convinced about the benefits of minimally invasive therapy and educated in the procedures. At the same time, device engineers had to be responsive to the physicians' knowledge and use their feedback in the product design process. The challenge was one of integration, of communication across disciplines and specialties. Successful new product development demanded more than inputs from each of these occupational specialities on its own; organisational routines for ongoing dialogue and translation across disciplinary domains had to be institutionalised.

It was also necessary to integrate customers and suppliers into the new product development and production processes. Nicholas led the effort to reorganise business units to interface directly with major customers rather than through specialist marketing/sales offices. This opened up direct lines of communication between customers and product developers/makers. Suppliers to Boston Scientific were encouraged to re-engineer their own organisations along the same lines, which brought the sales function to the shop floor teams. The organisational design was an application of world-class manufacturing practices to the medical device supply chain. The new organisational design would have met the approval of W. Edwards Deming and his principle of system integration (1982).

9.2.2 Technology Integration

Besides organisational process integration, and perhaps fostered by it, Boston Scientific has been a leader in technology integration. Boston Scientific's drug-eluting coronary stent represents the future: drug-device combination products that help the body heal itself. It is a breakthrough technology that is radically changing the cardiovascular field. It reflects Boston Scientific's organisational capabilities in integration, this time with interdisciplinary teams anchored in physics, including fiber-optics, polymeric chemistry, and biologics. Boston Scientific has tapped into the region's leadership in biotech research and is redefining the disciplinary boundaries of medical devices.

Boston Scientific's technology management strategy involves leveraging its technology platform into other speciality markets within the medical field still encumbered by organisational barriers to product development. In the words of Abele:

> ...[W]e were not focused on one marketplace. For example, radiology was big on guide wires, but urology, gastroenterology, and cardiology weren't. But communication between those specialties was, and still is, almost nonexistent. Each of these fields presented an opportunity for us to evolve our technology. In essence, every R&D dollar we spent had benefits in multiple fields, giving us a three-to-four-times value for our spending. (Swain, 2004, pp. 13–14)

While Boston Scientific invested heavily in organisational capabilities in new product development and technology management, it remained a privately held company until 1992, when an IPO was followed by an aggressive acquisition strategy.

Boston Scientific's acquisition strategy had two prongs. First, the firm acquired strategically to deepen its core technology. Two of its biggest acquisitions, Minnesota-based SCIMED and Schneider USA, had polymer technologies important to drug-eluting stent development. Second, it acquired to leverage superior capabilities in new product development in growth markets. In the words of Tom Gunderson, a research analyst for Piper Jaffray: '...[T]here came the imperative to get newer and better products out as fast as you can. In those days, Boston Scientific was way beyond everyone else from a speed standpoint' (quoted in Swain, 2004, p. 18).

Organisationally, Boston Scientific was the business model from which management in other companies in the region must have learned. It is one of a handful of Massachusetts companies that have grown to Fortune 500 size since the demise of the minicomputer giants in the late 1980s.[10] Life sciences replaced computers as the source of growth over the same period. But the business model of the new leaders has also been transformed. The vertically integrated organisations of the past have been replaced by an open-systems business model in which companies focus on core capabilities and partner for complementary capabilities. It represents, as well, the emergence of a new model of innovation and product development. Design has been diffused across networked groups of companies and decentralised within large companies.

9.3 OTHER FAST-GROWING, BIG MED FIRMS

While Boston Scientific is in a class of its own, at least four other companies located in Massachusetts and classified in MED by CorpTech boast more than 1000 employees each (see Table 9.1). These large, fast-growing Massachusetts-headquartered MED companies fit squarely within one of the three related production capability areas in which Massachusetts has historically had a competitive advantage: instrument making, precision equipment, and complex product systems (Best, 2001). In this sense, their success can be considered a process of industrial renewal as fast-growing companies reallocate resources in pursuit of new market opportunities by tapping the region's production and technological capability heritage.

Table 9.1 Fast growing, big medical device companies: employment

	Founded	1990	1995	2000	2001	2002	2003
Analogic / Medical Imaging Division	1969	1275	1400	1700	1850	1800	1800
Haemonetics Corporation	1971	666	1109	1366	1366	1400	1500
Smith & Nephew Inc / Endoscopy Division	1986	250	575	750	1000	1300	1500
DePuy Codman (J&J)	1838		1170	1200	1200	1200	1200
Boston Scientific	1979		5000	13500	13500	13500	14400

Source: vTHREAD.

Indeed, Depuy Codman was founded in 1838 by Thomas Codman, a mechanic in Roxbury.[11] Codman successfully designed a cupping instrument for the application of ether. The company boomed during the Civil War with a range of surgery and amputation instruments. While it had many ups and downs, the company's web site reports that the company maintained a core of skilled instrument makers and an apprenticeship programme.

In 1964, Codman & Shurtleff, Inc. became a member of the Johnson & Johnson 'Family of Companies' while retaining its Codman identity. Codman has a long history of cooperation with surgeons in the development of instruments. In recent years, these have included a hip prosthesis for total hip replacement and a set of instruments that allowed an anterior approach to the cervical spine for treatment of diseased and herniated inter-vertebral discs. In fact, some of the first joint reconstruction implants, marketed under the name Cintor, led to the creation of the Johnson & Johnson Orthopedics division, today named DePuy.

Other big, fast-growing MED companies in Massachusetts include:

- Smith and Nephew Inc./Endoscopy Division is an operating unit of its UK parent. This operating unit, originally Dyonics, was founded in 1986 and acquired by Smith and Nephew in 1996. The Endoscopy Division designs, develops and manufactures endoscopic surgical instrumentation used in minimally invasive surgery. This involves miniature cameras, xenon light sources and arthroscopic surgical instruments.
- Haemonetics Corporation designs and manufactures automated blood processing systems. The company estimates that 60 million blood collection procedures are performed worldwide every year to obtain blood's three major components: red cells, platelets and plasma. Haemonetics designs and manufactures automated blood processing systems to make this possible.
- Analogic Corporation's Medical Imaging Division represents Massachusetts' distinctive regional capability in complex product systems. Analogic is a developmental engineering and manufacturing company that builds medical and security imaging systems for original equipment manufacturers (OEMs). The company claims to supply approximately 75 percent of the data acquisition systems installed in computed tomography (CT) systems worldwide and half of the advanced power systems used in magnetic resonance imaging (MRI). Its product range includes a wide range of imaging systems including digital, laser, phased array, magnetic resonance, and ultrasound. The parent, Analogic Corp. has sales of nearly $.5 billion.[12]

9.4 MID-SIZED, FAST-GROWING MED COMPANIES

There is a large group of fast-growing, mid-size, medical device companies, or operating units, located in Massachusetts (see Table 9.2). Mid-size is defined as between 200 and 1000 employees. These companies can be divided into two groups: independent companies headquartered in Massachusetts and once independent companies that have been acquired by medical device companies headquartered elsewhere in the United States, but which continue to operate in Massachusetts.

When we analysed 12 fast-growing, mid-sized MED companies that remain headquartered in Massachusetts, we found the following:

- Employment in these firms increased from under 2000 in 1989 to between 6000 and 7000 in 2003.
- These are long-established companies.[13] Four were founded in the 1970s (Candela, NMC Diagnostics, Lifeline Systems, and Nova Biomedical) and eight in the 1980s. Inverness Medical Innovations is the newest company, founded in 1992.
- These firms reinforce the region's production capabilities in complex product systems, instruments and equipment found in the fast-growing big companies. Companies adopted a technology-focused strategy based on product development efforts and not mass production.
- They reflect and reinforce the region's technological heritage. At least five are in imaging/scanning, four are in optics, three are in blood processing/diagnostics and two are in cardio equipment. These mirror fairly closely the big medical device companies in Massachusetts. Optics, in particular, is a technology that also goes back to the earliest days of Massachusetts industry.

Five fast-growing companies, founded in Massachusetts, have been acquired by leading MED companies headquartered elsewhere in the nation. Medtronic can be added to this group as this Minnesota company recently established a major presence in Massachusetts partly by acquisition. The acquirers represent two of the biggest and most successful specialist medical device-making companies in the United States, Medtronic and Stryker, and two global giant healthcare companies, Abbott Laboratories and Johnson & Johnson.

The acquisitions by these leading medical device companies of Massachusetts' operating units are a good indicator of the state's distinctive capabilities. These companies, with the exception of Abbotts' acquisition of

Table 9.2 *Fast-growing, mid-size medical device companies: employment*

Companies	Founded	1990	1995	2000	2001	2002	2003
Lifeline Systems, Inc.	1974	250	325	620	790	850	850
Inverness Medical Innovations	1992		78	419	704	800	800
Hologic, Inc.	1985	130	170	600	839	780	750
Nova Biomedical	2001	475	500	500	664	665	665
Cytyc Corp.	1987	25	55	200	495	495	626
Zoll Medical Corporation	1980	150	275	390	430	585	844
Gentex Optics, Inc.	1932	120	250	500	500	500	500
Candela Corp.	1970	174	180	285	285	300	325
Clinical Data	1972	9	80	175	285	151	302
Biopure Corporation	1984	40	110	180	173	240	240
ABIOMED	1981	55	70	182	265	264	238
Aspect Medical	1987			100	200	230	205
American Medical Instruments	1975	60	145	145	149	149	195
Hologic/Lorad	1989	150	280	350	350	275	
Summit Technology	1985	60	211	425	Acquired by Nestle SA		
MediSense	1981	60	850	Acquired by Abbott Labs. In 1996			
NMC Diagnostics	1971	140	250	Acquired by Fresenius AG in 1998			

Source: vTHREAD.

MediSense, have maintained and expanded the operating units they have acquired. Massachusetts' historic strengths in instruments, including optics and imaging, combined with leadership in biotech, were the key elements in many of the acquisitions. These companies must come to Massachusetts to acquire such capabilities and they are not easily removed to other locations.

In some important cases, medical device companies have moved their headquarters to Massachusetts. The UTI Corporation, a 3500 employee, integrated outsourcer of manufacturing services for medical device companies, recently moved its headquarters from Pennsylvania to Massachusetts and renamed itself Accellent. Interestingly, the move of Accellent to Massachusetts represents a return of the remains of at least three once independent private companies to Massachusetts: Brimfield Precision, APEX Engineering, and ACT Medical. All three had been acquired earlier by Minneapolis-based MedSources Technologies, before itself being acquired by UTI.

9.5 NON-MED COMPANIES WITH MEDICAL PRODUCTS

Table 9.3 shows 12 companies that are not classified as medical device companies but which design and make medical device products. Six are classified in biotech, two in instruments, and one each in pharmaceuticals, advanced materials, photonics, computer hardware, and computer software.

The growth in Massachusetts companies as measured by employment is remarkably similar for this group as it is for the big medical device companies. (These 12 companies grew from around 2000 employees in 1990 to nearly 20 000 employees in 2003, compared to the job growth of the six big medical device companies (excluding MediSense), which grew from around 2200 in 1990 to 22 000 in 2003.[14] The closest to a Boston Scientific in employment is a combination of Genzyme, Parexcel and Charles River Laboratories. None of these three, however, are located primarily in medical devices in that all are life science companies and supply products/services to BIO, PHA and MED. But they all offer insights into the dynamics of a rapidly reconfiguring cluster.[15]

Perhaps not surprisingly, virtually all of the non-MED companies with MED products are in the life sciences or healthcare technologies. They represent industrial renewal via technology convergence or new technology combinations. In some cases, companies are in life sciences that are extending into medical devices as a means of drug delivery. In virtually all cases, they imply a regional capability in systems integration: the capacity to redesign or reconfigure the whole in order to take full advantage of design changes in a sub-system.

Table 9.3 Fast growing, non-medical device companies with medical products: employment

Company	Year Established	1990	2003	Class.	Product Codes
Genzyme	1981	450	5500	BIO	MED, PHA
Charles River Labs	1946		5000	BIO	MED
PAREXEL International	1982		4860	PHA	MED, BIO, COM
PolyMedica	1988	4	1679	MAT	BIO, CHE, MED, PHA
Mercury Communications	1982	180	600	COM	MED, SOF, TEL
Harvard Apparatus	1904	65	450	BIO	ENR, MED, SUB, TAM
Alkermes	1987	30	425	BIO	MED, PHA
Perceptive Informatics	2000		331	SOF	MED, TEL
Clinical Data	1972	9	302	TAM	
MJ Research	1985	15	220	BIO	CHE, MED, SUB, TAM
Bruker BioSpin	1972	100	180	TAM	MED, SUB
Organogenesis Inc.	1985	100	180	BIO	MED, TAM
Total Employees		953	19 727		

Source: vTHREAD.

Genzyme, a leader in biotechnology is a major provider of biodiagnostic products ('in vitro diagnostics') and also supplies bioengineered tissue that helps repair damaged cartilage in the implant market. Genzyme's product range straddles biotech and medical devices and represents the convergence of physical and life sciences as the boundaries of the medical device cluster have shifted.

Parexel is one of the world's largest contract research organisations. It specialises in the design and management of clinical research (Phases I-IV), including regulatory, data management and biostatistical services to the pharmaceutical, biotech, and medical device and diagnostic products worldwide. Founded in 1982, Parexel had 5000 employees worldwide and over $600 million in sales in 2003. Parexel achieves economies of scale in clinical trial management, which enables technology development companies to specialise in their core capabilities and outsource for this critically important function. Its rapid growth and location in Massachusetts is explained by the combined number of companies in all three life science based industries in the state. Partnering can drive down new product development time. Parexel's specialist capability developed in Massachusetts has been leveraged globally; it claims to have participated in the development of 23 of the top 25 drugs introduced in the world in 2000.[16]

Charles River Laboratories, formed in 1946, has repositioned itself in recent years from a company that specialised in animal research and diagnostics to human research models required in R&D for new drugs, devices, and therapies. It specialises in clinical trial support and a portfolio of products and services that enable customers to reduce cost and time and increase productivity and effectiveness of product development in the life sciences.

Polymedica, established in 1988, was classified in MED from 1992 to 1997, followed by advanced materials, and now in pharmaceuticals. 'Poly' in the title is short for polyurethane, a material that can be used in long-term implantable medical devices. The rapid growth of the company, however, has been a consequence of becoming a leading provider of healthcare products and services to patients with chronic diseases. Polymedica, with nearly 1700 employees, is the nation's largest provider of blood glucose testing supplies to people with diabetes, most of which are supplied directly to the consumer.

Mercury Computer, a designer and developer of digital signal processing computer systems, grew from 109 employees in 1989 to 600 in 2003. Mercury leveraged its technology platform, which was originally in defence applications, to medical imaging systems.

Clinical Data was in drug delivery systems and more recently has developed transdermal patch drug delivery products. It primarily develops and manufactures scientific and clinical laboratory instrumentation, including blood coagulation analysers, chemistry analysers, spectrophotometric industrial process monitors, and diode ray process spectrometers.

Organogenesis manufactures living organ equivalents, including skin, arterial, and knee ligament replacements. It was classified as testing and measurement in 1992, MED from 1993 to 2001 and since as BIO. It designs develops and manufactures medical therapeutics containing living cells and bioengineered surgical products.

9.6 FOREIGN-HEADQUARTERED COMPANIES

Nearly 8 percent of medical device companies in Massachusetts are foreign headquartered. The behemoth here is Philips Medical Systems with 6650 employees, an estimated 2000 of whom are in Massachusetts (see Table 9.4).

Table 9. 4 Foreign-headquartered medical device companies located in Massachusetts: employment

Company	Founded	1990	1995	2000	2003
HP/Philips (Holland)	1981	200	100	5300	6650
Smith&Nephew (UK)	1986	250	575	750	1500
Instrumentation Lab. Co. (Spain)	1959	400	400	500	500
Gentex Optics, Inc. (France)	1932	120	250	500	500
Bunzl Extrusion MA, LLC (UK)	1949			100	149
The Straumann Co. (Switzerland)	1989		35	68	160
Smiths Medical ASD, Inc. (UK)	1986		160	240	200
EBTEC/TI and Smiths (UK)	1963	80	60	75	80
Schott AG (Germany)	1954	350	265	270	211
Seimens/Draeger Medical Systems, Inc.	1988	360	540	320	370
GSI Lumonics (Canada)	1968	350	399	1500	950
Summit Technology Inc. (Switzerland)	1947	60	211	425	
NMC Diagnostics Services, Inc. (Germany)	1981	140	250		
TUV Product Services Inc. (Germany)	1989			250	149
Spacelabs Medical (Finland)	1987		7	120	120
Bionostics, Inc./Ferraris Group (UK)	1982	40	55	55	55
Pyrosequencing, Inc. (Sweden)	2000				30
LightLab Imaging, LLC (Japan)	1998				12
Tecan Boston, (Switzerland)	1994			26	12
Symfo (USA) Inc. (Belgium)	2000				35
Ophir Optronics, Inc. (Israel)	1985			130	25

Source: vTHREAD.

Philips has a long medical equipment history going back to 1896 when Philips manufactured the first x-ray tubes for medical applications. With the acquisition of Agilent Technologies Healthcare Solutions Group (previously Hewlett-Packard Medical Products Group) of Andover, MA, in 2001, Philips became the number two medical equipment provider in the world.[17] Philips joins a strong group of local and foreign headquartered companies in imaging, optic, and laser technologies. In fact, of the 21 foreign-headquartered medical device companies with operating units in Massachusetts, more than half are in imaging and/or optics.

Another example is Bruker BioSpin Corp., a Massachusetts-based member of the German-headquartered Bruker family of companies, a worldwide leader in nuclear magnetic resonance (NMR) since its development in the mid-1900s.[18] Bruker has specialised on R&D in high-resolution NMR and its extension into mass spectrometry and into in-vivo NMR or medical research MRI. Today, Bruker BioSpin makes nano-scale precision equipment for drug development and gene and protein research.

GSI Lumonics, previously General Scanning, Inc., was founded in 1968 as a laser components engineering company. It grew from 300 employees in 1990 to 1500 in 1999, when it merged with Lumonics Inc. of Canada. Today, it supplies a broad line of turn-key systems, subsystems, and components – most of which leverage laser technology – to OEMs that compete in industrial, medical, imaging, and laboratory marketplaces.

9.7 CONCLUSION

Our dataset of medical device companies serves as a metaphorical laboratory to study the emergence of a high-tech industrial sector in Massachusetts. It includes rapidly growing companies, transitioning companies, relocating companies and foreign-headquartered subsidiaries. The basis for our analysis is that growing, transitioning, and relocating companies drive industrial growth and, as such, they are the carriers, developers, and consolidators of underlying and regionally distinctive technological capabilities.

By applying this research methodology to the medical devices industry of Massachusetts, we find important insights into the processes of regional specialisation. New product development (NPD) capability, for example, is at the heart of technology-driven companies. It is also enormously costly, as is technology leadership. From a regional growth perspective, NPD is the immediate source of innovation. The fear was that with the decline of the computer sector and other large industrial operations in the Massachusetts, NPD would suffer, and the region would go into industrial decline. However, this has not been the case, and the MED sector shows one important reason why: the region's techno-diversity. This diversity has been a key element in the transition to a new, regional system of innovation. If the early emergence and growth of the medical devices industry in Massachusetts was fostered by the plethora of instrument-making companies, the next stages have been marked by the incorporation of photonics and software, and, more recently, the integration of devices and drugs, of physics and biologics. Each shows how the techno-diversity of the region creates enhanced opportunities for NPD. And NPD is the handmaiden of industrial differentiation and growth.

The examination of the medical device industry of Massachusetts suggests that both technological transformation and inter-firm or cluster dynamics are integral to industrial renewal and regional innovation. The sector or cluster is a moving target in which each new product iteration as the underlying technology is altered in the process. Unchanging technologies and static cluster boundaries are a recipe for industrial decline. The emergence and consolidation of the medical devices cluster was, at the same time, a transformation of computer and imaging technologies. Consequently, focusing on pre-given sector or cluster definitions and measures can obscure powerful dynamic forces behind the scenes. This has often been missed by the academic and business press that drew conclusions on the demise of Massachusetts industry based on either the region's traditional industries or the minicomputer industry. Studies of industrial decline and renewal, as well as innovation, must go both inside the black-box of the business enterprise and behind it to examine the historic technology capability and skill trajectories of the region.

NOTES

1. An earlier version of this chapter appeared in *Mass*Benchmarks, a publication of the University of Massachusetts in cooperation with the Federal Reserve Bank of Boston.
2. CorpTech's data collection methodology has eight phases: company identification using a variety of sources, telephone interview after which the companies products are coded, editing by senior researcher, data entry, internal proofing, machine check which applies numerous tests to each record, external proofing, and written verification of the record from the listed company.
3. For these and related statistics and their sources on the size and growth of the U.S. medical devices industry go to www.devicelink.com, the website for the trade publication Medical Device and Diagnostics Industry, and www.AdvaMed.org, the website for the medical technology association.
4. Based on the 1997 Economic Census, U.S. Bureau of the Census. Minnesota is first in all four of the per capita rankings, but between 3rd and 4th in absolute size rankings (Alan Clayton-Mathews, 2001).
5. Company employment numbers in the tables are used to measure company size and growth and therefore can include global operating units of Massachusetts' companies. The extreme example is Boston Scientific which has three major divisions in both Massachusetts and California and one in Galway, Ireland and smaller units elsewhere.

6. Using CorpTech's taxonomy, Boston Scientific's product categories and codes are Minimally invasive cardiovascular diagnostic systems: MED-DG-C (SIC 3845); Ultrasound imaging systems: MED-DG-IU (SIC 3845); Catheters: MED-SU-Q (SIC 3841); and Cardiovascular inflation devices: MED-TH-C (SIC 3841). The value of the CorpTech taxonomy is illustrated by the conversion of over 10 000 catalogue product items in over 50 categories into these four codes.
7. Founded in 1954, Millipore (4000 employees and $800 million in sales in 2003) pioneered the use of membrane technology and purification systems widely used today in research laboratories and in pharmaceutical and biopharmaceuticals manufacturing processes. The infiltration equipment is used to purify DNA and RNA proteins. Waters Corporation (4000 employees and $1 billion in sales) and Mykrolis Corporation (900 employees and $200 million in sales) are two large Massachusetts instruments companies that were once divisions of Millipore. In a sense, Boston Scientific represents 'renewal' of the equipment and instruments industries of Massachusetts in the form of extension into medical devices.
8. Erik Swain is an astute observer and chronicler of the medical devices industry.
9. In the words of Bakken: 'During Medtronic's formative years, I became a more or less regular feature over there...I spent so much time in the offices, surgery suites, and animal labs at the U of M that I was given my own locker' (Earl E. Bakken, 1999, p. 46).
10. The others in high tech are EMC, Genzyme Corp. and Biogen Idec Inc. Massachusetts had 11 companies in Fortune's 500 biggest for 2005, the same as Missouri and far below Ohio's 30 and Michigan's 22. Massachusetts still ranks second highest in per capita income behind only Connecticut (see Charles Stein, 2005, F-1,8).
11. American Optical Lens, established in 1833, may be the oldest medical device company still operating in Massachusetts.
12. The employment numbers are for a family of seven divisions, five of which are in medical devices: Life Care Division, B-K Medical Systems, Medical Imaging, and Sky Computers all in Massachusetts and Camtronics Medical Systems, Ltd., in Wisconsin. Its Test and Measurement Division is co-located with Medical Imaging and Life Care Systems at its company headquarters in Peabody, Massachusetts.
13. Gentex Optics, an even older company founded in 1932, but was acquired by Essilor, a French firm and is not included in this section. Cybex, founded in 1953 grew to 700 employees in 1995 before declining to 450 in the early 2000s.
14. Charles River Laboratories employed 1200 in 1993, the first year for which employment figures are available from CorpTech; likewise PAREXEL employed 750 in 1994.
15. Two other large company candidates for the category of rapidly growing non-MED companies with medical device products are PerkinElmer and Thermo Electron. Instead, they are treated in a separate category below. *The Boston Globe*, for one, classifies both in medical devices but this may be a stretch. Either way, PerkinElmer was not included in the table, although it is a major employer, because it was not a fast growth firm over the period.
16. See www.parexel.com/about_us/history.asp April 2005.
17. The acquisition of Agilent Technologies' Healthcare Solutions Group added cardiovascular ultrasound imaging, patient monitoring, electrocardiography, resuscitation products, and e-care business to Philips portfolio. This reflects the region's technological capability in both cardiovascular and imaging technologies. In the same year (2001), Philips acquired Marconi Medical Systems from GEC, a medical imaging innovator in the U.K. since 1915.
18. The Bruker family of companies specialises in spectrometers for the pharmaceutical, oil, petrochemical, and polymer industries as well as MRI systems. For a discussion of Bruker Instrument's role in the early development of MRI, see Annetine Gelijns and Nathan Rosenberg (1999).

REFERENCES

Bakken, Earl, E. (1999), *One Man's Full Life*, Minneapolis, Minn.

Best, M. (2001), *The New Competitive Advantage*, New York: Oxford University Press.

Clayton-Mathews, A. (2001), 'The Medical Device Industry in Massachusetts', University of Massachusetts Donahue Institute, www.donahue.umassp.edu.

Deming, W. E. (1982), *Quality, Productivity and Competitive Position*, Center for Advanced Engineering Study, MIT, Cambridge, Mass.

Gelijns, A. and Rosenberg, N. (1999), 'Diagnostic Devices: An Analysis of Comparative Advantages', in D. Mowery and R. R. Nelson (eds), *Sources of Industrial Leadership*, UK: Cambridge University Press, 328–30.

Massachusetts Institute of Social and Economic Research (MISER) (2004), foreign trade database, http://www.misertrade.org/.

Stein, C. (2005), 'State seeks a few good giants: After Fleet and Gillette, local corporate titans are scarce', *The Boston Globe*, May 17.

Swain, E. (2004), 'Boston Scientific: Making the Most of its First 25 Years', *Medical Device and Diagnostic Industry*, August.

10. Economic Restructuring, Regional 'Visioning' and the Role of Universities: The Outcomes of an Automotive Plant Closure in Southern Adelaide, Australia

Andrew Beer and Holli Thomas

10.1 INTRODUCTION

Across the globe manufacturing regions in developed economies have been confronted by ongoing restructuring as mature industries have entered a period of decline or sluggish growth, and as manufacturing investment has been concentrated in low-wage nations, such as China. This trend is evident within the European Union, North America and Australia. Nations and regions have attempted to respond to this challenge in a number of ways: some have sought to encourage investment in alternative manufacturing industries, supporting supplier diversification away from auto-sector dependence and towards growing sectors such as medical equipment and nano-technology (Bailey, 2003; MG Rover Taskforce 2005b); others have promoted the growth of the service sector and knowledge-based industries; while a third group has looked to the resources sector and/or primary production as a way of establishing a new niche within the global market place. Commonly these strategies involve the development of a new 'vision' for the region (McKinsey and Co, 1994), the identification of industries whose growth is to be encouraged and the implementation of policies and programmes aimed at fostering economic growth.

Universities are increasingly viewed as an economic asset for regions as a consequence of their capacity to generate new technologies and intellectual property that supports new enterprises; through the skills passed on to graduates, and through the expenditures of the university, its staff and students (Saxenian,1994). The growth of the 'North Carolina Triangle', for example, is frequently linked to the presence of a number of top tier universities which has encouraged firms to locate within the region in order to gain access to highly-qualified graduates (Stephen *et al.*, 2004). Similarly, the growth of the bio-technology industry in San Diego is seen to be an outcome of the presence of research institutions and the University of Southern California. This chapter considers the conflation of these two trends by examining how a manufacturing region in Adelaide, South Australia has responded to the challenge of economic restructuring, and the role assigned to universities within that response. Particular attention is focused on how the sole university located within the region is considered within the statements and policy documents produced in the aftermath of the announcement of a major plant closure. The chapter argues that in southern Adelaide, while universities and the education sector as a whole are recognised in plans for the region's economic revitalisation, the discussion of education's potential contribution is limited when compared with more conventional industries, such as manufacturing. It is argued that this reflects limitations within industry policies overall; a degree of path dependency, which is reflected in the ongoing emphasis on manufacturing and related industries; the absence of a long history of engagement between the region – including the custodians of its future – and universities; and, the failure to fully apprehend the potential of the higher education sector to transform a region's economy.

10.2 ECONOMIC RESTRUCTURING IN SOUTHERN ADELAIDE

In April 2004 Australia's then Prime Minister, John Howard – flanked by South Australian Premier Mike Rann and Mr Tom Phillips, the serving CEO of Mitsubishi Motors Australia Limited (MMAL) – announced that the Lonsdale plant of MMAL would be closed with a loss of 700 jobs, with a further 400 voluntary redundancies from MMAL's Tonsley Park assembly plant. The factory had been in operation since the mid 1960s and performed a number of roles, including a foundry where engine blocks were cast, as well as hosting the assembly of some components, such as brake knuckles. The Tonsley Park plant has remained a site for vehicle assembly, but functions such as inventory and upholstery manufacture have been further outsourced.

The loss of just over 1100 jobs in the southern part of metropolitan Adelaide was recognised as a major setback to the regional economy. The Federal Government responded by announcing a $45 million assistance package for the region – called the Structural Adjustment Fund (SAF) – as well as enhanced employment assistance for retrenched workers. This assistance was to be delivered via the Jobs Network, Australia's network of federally-funded labour market providers. In addition, the South Australian Government committed $10 million of assistance to displaced workers, mainly in the form of enhanced access to services. The loss of employment from MMAL in 2004 can be seen to be part of the longer-term restructuring of the automobile industry, and manufacturing more generally, in Australia (Government of South Australia, 2006). In the mid 1970s manufacturing employment accounted for 25 per cent of the workforce but by 2001 it had declined to 12 per cent of the labour force, even though the value of production had increased (Forster, 2003). Where once car-making plants could be found in all state capitals except Perth, by the year 2000 motor vehicle production had consolidated into a limited number of sites with Toyota and Ford assembling vehicles in Melbourne, Mitsubishi and General Motors Holden building cars in Adelaide and General Motors Holden building engines in Melbourne.

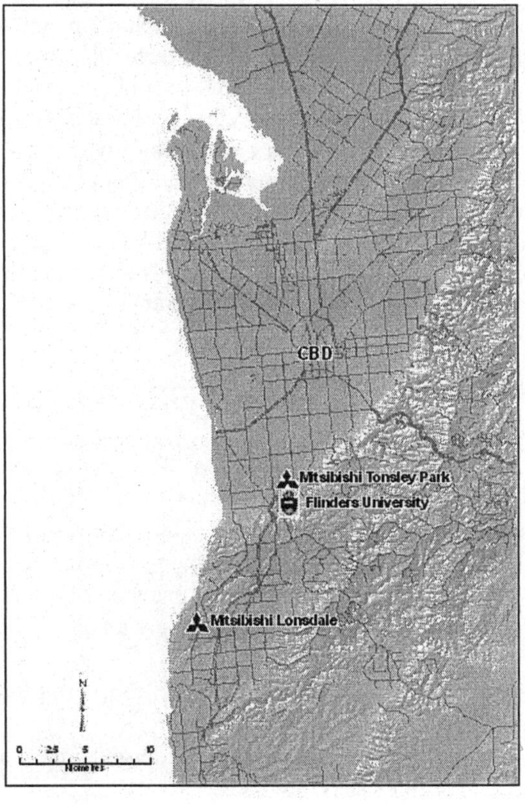

Figure 10.1 *The South Adelaide Region*

The southern region of Adelaide – defined as the jurisdictions of the City of Onkaparinga and the City of Marion (Figure 10.1) – was perceived to be at risk economically as a consequence of the MMAL job losses. Its potential vulnerability reflected a number of structural factors, including the fact that the region has a relatively unskilled and under-qualified workforce (City of Marion and City of Onkaparinga, 2006); regional incomes are lower than the national average (ABS, 2001); and, a significant proportion of the workforce is employed outside the region. The further loss of local employment had the potential to undermine the viability of the region's small businesses. More specifically, the workforce being made redundant was mature and tended to be concentrated in neighbourhoods close to the MMAL factories. There was, therefore, a real prospect that those who left Mitsubishi would not find paid employment and that the consequences of employment loss would be concentrated in a relatively small area. In addition, the region as a whole has lagged behind the expansion of manufacturing – and especially advanced manufacturing – in other parts of the metropolitan area, as the majority of new manufacturing enterprises have established in northern Adelaide. Businesses within the southern region of Adelaide tend to be small scale and relatively mature (Kearins, 2002). The Mobil (Exxon) oil refinery at Lonsdale had closed two years previously with a significant loss of employment; and, the wine industry in the southern part of the City of Onkaparinga (Southern Vales/McLaren Vale) faltered in 2004 and 2005 as the national supply of grapes for wine production exceeded demand. Finally, it is worth noting that the southern region of Adelaide is relatively poorly served in terms of access to infrastructure, with transport, power and telecommunications of a poorer standard than in its competitor regions.

10.3 THE RESPONSES OF GOVERNMENTS

The announcement of job losses from MMAL was accompanied by the establishment of a new institutional structure to deal with the consequences of the plant closure and employment loss. A new body was established – the South Australian Government Advisory Group – to provide the government with industry-relevant advice under the chairmanship of a former President of General Motors Japan. Four sub-committees were also established:

- Lonsdale Facility Assets to advise on the best possible use of the vacated Lonsdale Plant;
- Outplacement Opportunities to provide guidance on labour market programmes and issues;

- The Southern Suburbs Industry Development Working Group (SSIDWG) to assist with the further development of the southern region economy; and,
- Tonsley Park Utilisation which was charged with identifying strategies to ensure the ongoing financial viability of MMAL's remaining factory.

From its inception considerable importance was attached to the work of SSIDWG as it was the only sub-committee to involve representatives of the region and it had a mandate to shape a new future for the region. SSIDWG commenced meeting fortnightly in May 2004 and began to address an ambitious programme of work including: planning for a Southern Summit to raise the profile of the challenges confronting the Southern Region; preparation of a regional economic development strategy that embraced the two council areas – The Blueprint for the Future; research into the availability of land for further industrial development; contact with businesses and other organisations interested in investing in southern Adelaide or in applying for money from the Structural Adjustment Fund (SAF); and, planning for an Innovation Centre in the south.

SSIDWG's role needs to be viewed within the context of the broader processes of government, as well as the overall response to the closure of the Lonsdale plant. It is important to emphasise that the $45 million of funds made available through the SAF was by far the most significant response to economic restructuring. Limited, or not, resources were committed by the State and Australian Governments to other initiatives such as SSIDWG. This stands in stark contrast to policy responses in other regions – such as Birmingham, UK – where governments have invested substantial resources in the institutions managing the processes of change at the regional level. The MG Rover Task Force set up when MG Rover announced that it had entered administration had £175 million allocated to deal with the impact of the MG Rover closure and to assist further modernisation and diversification of suppliers (MG Rover Task Force, 2005a).

The SAF was the most significant response by central governments to the loss of employment at MMAL and we must recognise that it does not fit easily within contemporary paradigms of regional development (see, for example, Beer *et al.,* 2003a; OECD, 2001; Rainnie, 2003) because the programme consisted of grants – effectively capital subsidies – to firms willing to invest in South Australia. The SAF supported firms that were able to make a 'business case' that the injection of additional capital would allow for the expansion of business and would result in a significant number of new jobs. Firms had to match the grant awarded to them and complete a substantial application which was assessed by the advisory group. Two issues are critical here: first, SAF monies were not targeted exclusively to the southern region of Adelaide and, as Table 10.1 shows, as many grants were awarded to firms outside the region as within southern Adelaide. Second, grants of this nature are a relatively blunt policy instrument and one which has fallen out of favour in most developed economies (Haughton *et al.*, 2003). Commonly, subsidy programmes of this nature do not achieve the employment outcomes forecast – and committed to – by businesses (Beer *et al.*, 2003b) and the diverse firms able to take up these opportunities effectively precludes the targeting of those industries considered to have the best long-term prospects. In this instance, the single largest grant announced at the time of writing – to Ingham Processing – was awarded to a chicken processing plant in northern Adelaide.

Table 10.1 Grants awarded under the structural adjustment fund

Company	Cash Grant	Location and Region	Number of Jobs Created
Redarc Electronics Expansion of an electronics Business	1.6 million	Lonsdale, Southern Adelaide	62
Alloy Technologies International Project to develop state of the art light metals foundry	1.8 million	Wingfield, Northern Adelaide	100
Resource Co Expansion of an established waste processing company	3 million	Lonsdale, Southern Adelaide	120 direct 40 indirect
Cubic Pacifc New plant for the construction of motor vehicle components	0.95 million	Edinburgh Park, Northern Adelaide	75
Fibrelogic Pty Ltd Establishment of a fibreglass reinforced pipe manufacturing plant	5.9 million	Lonsdale, Southern Adelaide	140
Ingham's Enterprises Chicken processing	$7 million	Edinburgh Park, Northern Adelaide	245
Intercast and Forge Expansion of cast metals moulding at the Wingfield Cast Metals Precinct	2.5 million	Wingfield, Northern Adelaide	68
PBR Australia Brake components for the automotive industry	1.5 million	Lonsdale	Not Stated

The SAF and the failure of the South Australian Government to commit significant resources to other regional adjustment initiatives is indicative of neoliberal policies in Australia more generally. Australia has been profoundly influenced by the philosophies and practices of neo-liberalism (McGuirk, 2005; O'Neill and Moore, 2005; Pusey, 1991) with government services either reduced or restructured to meet the needs of a market-oriented economy. Peck and Tickell (2002, p. 380) argue that it is possible to identify a general model of neoliberalism based on free market economic theory and enacted through the processes of globalisation and contemporary state reform. Peck (2001, p. 445) has argued that this model is international and characterised by the purging of obstacles to the functioning of free markets, restraint in public expenditure, the weakening of social transfer payments and 'the "inclusion" of the poor and marginalised into the labour market, on the market's terms'. Researchers writing in this area, however, note that neoliberalism is not uniform and that its expression at the national, regional or local level is contingent upon historical, political and geographical circumstances (Peck, 2004), or as O'Neil and Argent (2005, p. 5) have argued '...neoliberalism presents an agenda of possibilities for those with the power to enact change. But this agenda must be devised and played out in historical and geographical circumstances'. Critically, neoliberalism also generates a strong philosophical and practical argument for denying direct action, as the ethos of 'fiscal responsibility' and allowing markets to operate unencumbered can be used to negate calls for action.

Neoliberal policies have been a prominent feature of the Australian policy landscape since the mid 1980s. In South Australia the philosophy of market-based solutions to questions of social and economic well-being has been prominent since the mid 1990s. Critically, the Rann Labor Government, which was elected in 2002, restructured and reduced its industry department – the Department of Trade and Economic Development (DTED) – in reaction to the substantial subsidies offered to private businesses by the previous Liberal government. Critically then, government responses to the loss of employment at MMAL included a high-profile grant programme from the Australian Government that was not targeted to the affected region and which the evidence base suggests is likely to be ineffective in the long term. The South Australian Government response was restricted and did not include ongoing measures to assist the adjustment of regional businesses. Local governments – the junior tier of government within Australian federalism (Stilwell and Troy, 2000) – were left with the responsibility for developing and implementing a more strategic approach to advancing the well-being of the southern region. SSIDWG emerged as an important avenue for local government to articulate its vision for the future of the region as it adopted the roll of a 'clearing house', whereby ideas and issues were raised and tested against the opinions and attitudes of central government agencies. SSIDWG was the catalyst for a number of regional development strategies that attempted to present a new 'vision' for southern Adelaide and these are discussed in more detail in the section below.

10.4 REGIONAL VISIONS FOR A NEW SOUTHERN ECONOMY

From the first announcement of the loss of jobs at MMAL there was a flurry of activity as individual agencies and coalitions of agencies sought to articulate their strategy for the future development of southern Adelaide. At least 15 strategic planning documents were initiated between April 2004 and April 2006. Two background papers were also commissioned (Table 10.2). Critically, many of these reports were instigated but not completed, or completed but not released to the public. There is a high degree of overlap and repetition between the plans – even when the authorship differs – and this reflects a process of institutional searching, as entities have sought to identify and define their role in responding to large-scale employment loss, and as they have sought to map out their relationship with other actors in the field.

A number of common themes are evident in the documents that have been developed in response to the closure of MMAL Lonsdale. First, and at a broad level, all the documents have adopted a sectoral approach to encouraging the development of the region. That is, they have attempted to identify industries with the potential to grow and have sought to encourage their more rapid development. These key sectors are seen to be: the environment and environmental industries; advanced manufacturing; knowledge/innovation industries; and, the food and wine industry. The identification of these sectors reflects a number of factors including recognition of existing capacity – or the potential for capacity development – in the region; existing resources; trends in the State and national economy; and, the desire to generate employment in high wage sectors. It is worth noting, for example, that policy documents eschew further development of the retail sector because of the low wages, insecure employment and limited multiplier effects of that industry.

The multiple economic development strategies are indicative of the broader dynamics surrounding the formulation of policy responses produced as an outcome of the closure of the MMAL plant at Lonsdale and voluntary redundancies at Tonsley Park. Effectively the production of (draft) development strategies has provided both a conduit and a focus for the establishment of new visions for the future of the region.

Table 10.2 Economic development strategies and background documents, April 2004–2006

Report Title Strategy Documents Developed in Response to Economic Restructuring	Released to the Public	Author/Common Agent
Southern Region Economic Diversification Plan	No	City of Onkaparinga City of Marion on behalf of SSIDWG
Blueprint for a new Economy for the Southern Region of Adelaide – Draft Outline	No	City of Onkaparinga City of Marion on behalf of SSIDWG
Southern Region Economic Diversification Plan – Marketing Plan	Yes	City of Onkaparinga City of Marion on behalf of SSIDWG
Centre for International Innovation and Creativity	Launched 2006	Flinders University
Outcomes of Strategic Planning Sessions with Industry Sector Representatives	Yes	City of Onkaparinga City of Marion on behalf of SSIDWG
Southern Region Economic Diversification Blueprint (SREDB)	No	City of Onkaparinga City of Marion on behalf of SSIDWG
Creating Our Future	Yes	City of Onkaparinga
Southern Suburbs Economic Development Plan	Yes	City of Onkaparinga City of Marion on behalf of SSIDWG
Southern Region Economic Diversification Plan – draft 2	Yes	City of Onkaparinga and City of Marion on behalf of SSIDWIG?
Economic and Employment Development Strategy – draft	Yes	City of Onkaparinga
Major Project Brief for Strategic Investment Plan	Yes	City of Onkaparinga
Southern Suburbs Industry Attraction Analysis – Industry and Market Overview	Yes	Department of Trade and Economic Development

General Strategic Plans		
Draft Strategic Plan	Yes	City of Marion
Corporate Plan	Yes	City of Marion
Background Papers		
Southern Region Economic Diversification Plan – Background Paper	No	Andrew Scott, Commissioned by the City of Onkaparinga and City of Marion
Southern Region Economic Diversification Plan – Background Paper	No	Richard Blandy, City of Onkaparinga and City of Marion

In process terms, the production of an economic development strategy permits individual agencies, and coalitions of agencies, to present their vision of the future and test it against the attitudes and positions of stakeholders both internal, and external, to SSIDWG. It is therefore important to examine some of the key documents and what they envisage. These statements of future intent are important for understanding the region's prospects and for understanding the position of universities within regional development efforts.

10.4.1 Taking Southern Adelaide Forward

The most significant of the documents that articulate a vision for southern Adelaide is the draft version of the Southern Region Economic Diversification Blueprint (SREDB) released in April 2006. In many respects the SREDB represents an end point to regional economic development strategies in southern Adelaide. The SREDB was produced by the City of Onkaparinga and the City of Marion for SIDWIG, having been in development since December 2004. All previous documents make reference to the future SREDB document, and in many vision documents detailed responses have been delayed until the SREDB was produced. The other significance of the SREDB is that it builds upon all previous documents and is based upon consultations with targeted industries in the southern region, business leaders external to the region, and a review of the economic development programmes conducted in the region over the last five years. The SREDB also clearly states that it has been developed within the context of relevant state and local government plans.

SREDB (City of Onkaparinga and City of Marion, 2006) focuses on the priority areas of food, wine and tourism; advanced manufacturing (including automotive); environmental industries; knowledge and innovation industries; and small to medium-sized enterprises (SMEs). The report articulates a number of issues requiring urgent action. The first order issue is seen to be workforce development. SREDB acknowledges that there is a shortage of both unskilled and skilled labour in all industries and that multiple and overlapping responses are required to ensure the achievement of the region's vision. The responses suggested include: process automation, increased skilled immigration, up-skilling existing workforce and succession planning (p. 7). Interestingly, the region's only tertiary education institution – Flinders University – is not identified as part of the workforce development strategy (pp. 14–15). Developing and implementing a broadband strategy is also seen as vital to the region establishing and maintaining competitive advantage (p. 9). Emphasis is also placed on marketing the region. It is suggested that with the closure of Mobil and Mitsubishi the region has the opportunity to reposition itself as the home of emerging industries (p. 9). The need to develop an 'icon brand strategy' is suggested, together with a market emphasis on the lifestyle advantages of the region to potential investors. The fourth issue seen to require urgent intervention was the shortage of industrial and commercial land. The SREDB recommends preparing a Mobil Site Master Plan; identifying potential sites for industrial expansion in the southern region; establishing a Southern Eco Precinct at the Mitsubishi Lonsdale site and positioning Edwardstown/Melrose Park as a 'start-up/niche' area for commercial and industrial development. The emergence of a specialist precinct in the north-west of the metropolitan area is held up as an example of the type of development that could be pursued (pp. 16–17). Transport is seen as the final policy priority, with actions to include the improvement of the north–south corridor as well as enhancements to transport into and out of the southern region.

The SREDB draws from, and builds upon, the second draft of the Southern Region Economic Diversification Plan (SREDP) which was jointly produced by the City of Onkaparinga and the City of Marion. The SREDP highlights six strategic directions that the local governments need to pursue to ensure a robust local response to the Mobil and Mitsubishi Lonsdale closures. They are:

- Nurturing the local growth momentum;
- Augmenting the business skills base to achieve the momentum;
- Mobilising the community resource;
- Attracting new industry and investment;
- Meeting infrastructure needs;

- Marketing/branding the region.

Strategic planning sessions were held as part of the preparations for the SREDB and these consultations have resulted in a summary document entitled *The Southern Region Economic Diversification Blueprint – Preliminary Outcomes of Strategic Planning Sessions with Industry Sector Representatives.* Industry consultation focused on the four targeted industry sectors. It was argued that the SREDB needed to be industry led and the Cities of Onkaparinga and Marion engaged industry leaders in each sector to outline a long-term vision for their sector and the region as a whole. The consultation with industry highlighted that early steps were needed to make this vision of the region achievable in these four targeted sectors. These steps included:

- Addressing the poor transport linkages within, and to and from, the region;
- The improvement of telecommunications;
- The preparation of a Master Plan for the future economic development of Port Stanvac;
- A strategy to address the shortage of industrial land in the medium to long term;
- Leveraging housing growth in the south to foster the development of innovative and sustainable technology; and,
- Waterproofing the south through effective management of water resources.

Two important reports were commissioned by the cities of Onkaparinga and Marion in the development of the SREDB. The report produced by Dr Andrew Scott (Scott, 2004) identifies those industry sectors with the greatest potential for growth in the region and how they should be targeted to maximise development. The report suggests that several steps need to be taken. First, Scott suggests that a strong and effective leadership needs to be appointed. The report noted that while there are many stakeholders and agencies involved in economic development in the south, these groups have differing priorities and jurisdictions. The report suggests that a single senior person with time and authority be appointed to drive this process. Second, Scott (2004) notes that there is a significant skills base, particularly in the area of manufacturing, which is at risk of being lost to other manufacturing regions, such as northern Adelaide. Scott suggests that programmes to retrain workers and find them suitable jobs in the south have merit (p. 15). The Scott Report notes that component suppliers to the automotive industry depend upon car manufacturers for orders and that these same skills and production technologies could be adapted to produce components for defence or electronic equipment (p. 9). Scott suggests that because most components for the growing defence industry are brought in from interstate or overseas there is an opportunity for competent manufacturers in the south to capture these markets. This point was not developed further in the SREDB. The Scott Report suggests that because the service sector and the knowledge economy (including the Green Industry) should be high priorities in the region, there needs to be training and education programmes in order to ensure that there are workers with appropriate skills in the region. The Scott Report encourages improvements in access to training through the Technical and Further Education (TAFE) network, which would involve bringing together several smaller colleges into a single campus. Scott also notes that there is a shortage of well-located and serviced sites where investors could look to establish a business. Despite this, he suggests that the success of the development of Edinburgh Park and the cast-metal precincts at Wingfield could be emulated in the south. Opportunities in locations such as the Mitsubishi Lonsdale site; the former Mobil refinery – which Scott argues 'could be a potent symbol if used as a location for Green industry' (p. 16); the former Glenthorne Research Station; as well as a site further to the south called Pedler Creek, which could be used as a recycling plant, are identified. Access to broadband telecommunications is also seen as a high priority in attracting investment in the knowledge economy sector. Finally the report stresses the need to deal with transport problems in and around the south.

Professor Richard Blandy was also commissioned to write a report as part of the Southern Region Economic Diversification Plan. The Blandy Report notes that manufacturing is the largest economic sector in terms of employment (Blandy, 2004), with 18 per cent of employment in the southern region. Blandy calculates the loss to the southern region economy from the closure of the Mobil refinery to be $36 million. The closure of Mitsubishi at Lonsdale is estimated to result in a loss of $303 million to the southern region. Perhaps most significantly, Blandy estimates that if Mitsubishi at Tonsley Park were to close it would result in a loss to the southern region of a further $606 million (p. 8). This sees an overall annual projected loss to the southern region of over $900 million by the end of 2008/09. The closure projections imply a loss of just under $1 billion annually to the southern region and the departure of just over 15 000 jobs by 2013 (p. 9). Blandy (2004) makes a number of suggestions as to how an injection of expenditure into the region can off-set the impact of the closures. These initiatives include investment in an expanded freeway, increasing the population of the southern suburbs, improving health facilities, accelerating the development of wine tourism in the Southern Vales region, improving broadband, providing a 'green' focus to industrial development in the southern region economy and finding replacement activities for the Mobil and Mitsubishi Lonsdale sites. Most of these suggestions were developed within the SREDB as 'urgent initiatives'.

10.5 THE 'IMAGINED' AND 'REAL' ROLE OF FLINDERS UNIVERSITY IN RESPONDING TO REGIONAL STRUCTURAL CHANGE

The economic development strategies developed in response to the closure of the MMAL plant at Lonsdale in large measure have articulated a consistent vision for the future of the region: an ongoing emphasis on manufacturing; the development of new industries based around environmental technologies; the continuing promotion of the food processing and wine industries; and, innovation industries – broadly defined. There is an ongoing emphasis on the established industries within the region and this focus is moderated by the recognition that new industries and new ways of doing business must be identified and encouraged. The strategies and approaches outlined in these policy documents can usefully be examined with reference to the role assigned to universities and the education sector. Universities and the education sector generally are potentially an important source of innovation within the region – as well as labour force productivity – and it is critical to understand the imagined role assigned to the tertiary education sector in regional planning documents and compare this with the outcomes.

10.5.1 The Education Sector and Universities within Plans for Economic Development

Universities can contribute to regional economic growth and prosperity in two fundamental ways: first, they can foster the emergence of a knowledge-based economy, which Forbes and Cutler (2006, p. 177) argue is characterised by an emphasis on innovation and technological change; an efficient infrastructure, especially in information and telecommunications; a business environment which stresses innovation and enterprise; and, a strong commitment to education and human resources. McKeon and Weir (2001) argue that nations such as Canada, Singapore and Australia 'are developing knowledge-based economies with around half their GDP attributable to knowledge-based industries, while just under a third of their workforce could be labelled "knowledge workers" ' (Forbes and Cutler, 2006, p. 178). Universities and other education providers have a direct impact on the knowledge economy through their research, consulting, and perhaps most importantly, through the development of skills amongst graduates

(W.K.Kellog Foundation, 2000). Second, the education sector can have a direct impact on the level of economic activity within a region. The British Council (2003) reported that the global demand for higher education is forecast to rise from 988 000 students in 2003 to 3.4 million students in 2025 (Forbes and Cutler, 2006). Australian higher education exports in 2003 were valued at $4.1 billion, compared with $7 billion in higher education exports from the UK (Forbes and Cutler, 2006, p. 179). Significantly, however, South Australia has a smaller share of international students than its per capita population indicates and within South Australia, the education sector in the southern region of Adelaide is underdeveloped with the majority of international students living and studying in the city centre.

Clearly universities can make an important contribution to the well-being of regions and it would be expected that the education sector would feature prominently in the planning for the economic revitalisation of southern Adelaide. Data on the prominence of education and education providers within the key policy documents are presented in Table 10.3. The topic of 'education' generally was discussed 33 times across the 15 documents, and was raised as a major focus of activity on 16 occasions. However, vocational education and training was awarded greater priority, being mentioned 39 times in total and on 20 occasions it was examined as an important theme in plans for future economic development.

The role of Flinders University, as the sole institution based within the region, was considered within the policy documents on 24 occasions, and was addressed in a substantial way on 14 separate occasions. However, it should be recognised that Flinders University has been an active participant in SSIDWG and related activities and that one of the key policy documents was written by people from Flinders University. The Southern Innovation Centre – a pivotal focus for a knowledge-based economy – was discussed on four occasions, while new technology/information technology was discussed 12 times and research and development was discussed on 11 occasions. Schools and the university sector as a whole were mentioned 11 times and discussed in a substantial way only once.

Education and universities have been addressed in the development of a new vision for the economy of southern Adelaide but they have been awarded relatively little priority when compared with the development of specific industry sectors, or the need to provide additional infrastructure. For example, the vision articulated in the 'Blueprint for a New Economy for the Southern Region of Adelaide' suggests that by 2020 the region will have 'enviable strategic alliances between industry, research institutions, educational bodies and governments which result in an integrated research capability, colliding ideas and new emerging skills' (City of Onkaparinga and City of Marion, 2004, p. 2).

Table 10.3 Universities in vision statements

KEY WORDS	Number of Times Referred to in Key Policy Documents	Number of Times Awarded Priority in Key Policy Documents
Education	33	16
Flinders University	24	14
Universities	7	1
Schools	7	1
Vocational Education and Training	39	20
New technology (technology)/ Information Technology	12	8
Southern Innovation Centre	4	2
Research and Development	11	5

However, this was only one of 14 key characteristics the region was expected to exhibit. The content of the 'Blueprint for a New Economy for the Southern Region of Adelaide' is instructive because it was an early response to employment loss and it illustrates the ways in which the education sector, and universities in particular, have been incorporated within economic planning for the region. As noted above, the emergence of *alliances between industry and research institutions* was a central concern, while elsewhere in the same document education and research are presented as a critical success factor in developing a new skills base for the region. The 'Blueprint for a New Economy for the Southern Region of Adelaide' argues that four strategies are needed to achieve its vision – nurturing the local growth momentum; augmenting the business skills base to achieve the momentum; mobilising the community resource; and, attracting new industry and investment. Education is seen to be part of the third strategy, with encouragement offered to apprenticeships and traineeship opportunities, and the fourth strategy, where the region should seek to 'build an international reputation for innovation and research and development in the southern region through the work of a Centre for Innovation and Creativity' and 'alliances with industry and education' (p. 8). Critically then, the education sector was not ignored, but it was seen as an adjunct to industry, that is a pathway to a more productive manufacturing sector. Promotion of education, or the education sector, was not seen to be a goal in itself.

Later strategy documents have expressed similar sentiments, and awarded comparable priority, to those expressed in the 'Blueprint for a New Economy for the Southern Region of Adelaide'. The Southern Region Economic Diversification Blueprint, which was produced in April 2006, has an emphasis on both workforce development and knowledge innovation industries. In the latter case, the development of industries around medical devices, where the Blueprint recognises that Flinders University is an acknowledged research leader, is foreshadowed, although it is noted that 'there has been some difficulty in engaging private industry participants and this is reported to be due to the limited scale of the industry and uncertainty regarding the value of engagement' (City of Marion and City of Onkaparinga, 2006, p. 24). This response suggests that industry in the region is not equipped to participate in the development of new technologies and leading-edge products. The Workforce Development Plan within this document, by contrast, simply notes the looming shortage of workers and the need to act quickly.

10.5.2 The 'Real' Role of Universities in Responding to Economic Change

The documents developed in response to the loss of jobs at MMAL have set out a particular role for education and the education sector and in large measure it is one which is supportive of ongoing manufacturing production through workforce training and the introduction of new technologies into production processes. Intuitively, we would also expect that university-led research would be central to the development of new 'green industries' in the region but this relationship is not clearly expressed in the documents. The role allocated to universities and the education sector in policy statements can usefully be compared with the part played by representatives of the sector since the announcement of the Lonsdale plant closure.

Participation in SSIDWG has been one of the most significant roles played by the education sector in responding to the closure of the MMAL Lonsdale plant. From the first meeting of SSIDWG one representative of Flinders University was a board member and by September 2004 another Flinders staff member had joined.[1] In addition, a third member of SSIDWG held a dual appointment with an automotive components manufacturer located in southern Adelaide as well as the University of South Australia. University sector participation in SSIDWG is important because that involvement has affected the development of plans for the region. While it could be argued that education and research have a relatively minor role in the regional vision statements developed in response to the Lonsdale closure, that part is greater than it would have been if there had not been University participation in SSIDWG.

The Innovation Centre has been the second significant form of engagement between regional development planning and the tertiary education sector. The need for an Innovation Centre – or a Southern Innovation Centre – was identified in the early regional planning exercises. At one stage an independent Southern Innovation and Creativity Centre was mooted, but in the end a state government-funded Innovation Centre was established in September 2005. Crucially, while staff funding has been provided by the state government, the Centre is located in a Flinders University property and the accommodation costs are met by the university. The university is therefore an active partner and is one of the pivotal responses to employment loss within the region. This outcome needs to be examined in light of earlier research that showed that key decision makers sitting on SSIDWG believed the most substantial contribution universities can make to a region's economy is through the exploitations of Intellectual Property (Beer and Cooper, 2008). The Innovation Centre awards the university a central role in the provision of services to enterprises in the region and potentially offers a further conduit for the exploitation of knowledge.

The Innovation Centre and participation in SSIDWG have been the only concrete responses to the job losses at MMAL that universities have been involved in. However, there have been very few measures enacted by any of the stakeholders working to secure the future of the southern region, and in relative terms the university sector has performed well. The loss of employment at MMAL did not result in a large-scale shift of workers into either technical or higher education, although there are critical shortages of both skilled trades labour and graduates in areas such as nursing. A survey of 374 workers who left MMAL found that almost 60 per cent intend to undertake further training or education but that potential demand has not yet been expressed in enrolments in either a university or TAFE. In part this reflects the buoyant labour market in Adelaide overall. To a certain degree it is a function of the older age profile of the workers and it is also an outcome of a working culture that does not value education highly. This has meant that one of the most direct and straightforward contributions universities and other education institutions could make to the region has not been realised.

10.6 CONCLUSIONS

The relationships between universities and the regions within which they sit are often complex (Garlick, 1998). As nationally or globally focused institutions universities do not necessarily have a strong relationship with their surrounding communities and they may not be well placed to respond to economic shocks, such as the loss of major manufacturing plants. It is worth noting that Birmingham University and the other universities in that region were not actively involved in addressing the closure of Rover's Longbridge plant. Conversely Flinders University has been an active participant in formulating and implementing a response to the job losses at MMAL and in many ways this built upon relationships that had already begun to emerge. The fact that Flinders University is the only university within the region both helped and hindered this process. On the positive side, the institution was recognised as an important stakeholder. However, debates about what the education sector could deliver for the region often became embroiled in discussions around the capacity of the university to meet those needs.

We would conclude that the plans that have been articulated for the future of the southern region of Adelaide do not pay sufficient attention to the role of universities and the education sector overall. As this chapter has shown, the vision statements for the southern region suggest that education is critical for providing – often basic – training for skilled labour; for working with and for manufacturers in developing new products; and, in establishing 'green industries'. In many ways these are limited roles and we should ask, what other strategies could have been adopted, what could have been done better?

The development of education as an industry was an option for the future development of the region that was not acted on. The southern region of Adelaide has relatively poor transport infrastructure, has a limited supply of land for manufacturing and has no competitive advantages in terms of access to infrastructure. On the other hand, it is an attractive and scenic environment, with low housing costs and reasonably good access to secondary, technical and tertiary education facilities. The global education market is growing and the region could have positioned itself to participate in its further expansion. The failure to do so, and to instead continue to focus on manufacturing, reflects entrenched attitudes to the nature of economic development – and what constitutes economic development activities (Loveridge, 1996) – rather than an objective assessment of the long-term prospects of service-based industries *vis-à-vis* traded goods. This path dependency in economic development planning effectively precludes universities from a major role because they were not important historically. The University of Wollongong has been the largest employer in the Illawarra region of New South Wales since restructuring at the city's steel mill. The southern region of Adelaide may well present a similar profile in the near future. Secondly, the region could have embraced a wider strategy of skills development and education. Of the 7000 workers who lost their jobs at Longbridge, Birmingham some 3200 were offered a training opportunity relevant to their needs. Over 2000 individuals took up this opportunity within six months of Rover closing. Virtually all 7000 workers who lost their jobs at Longbridge in Birmingham received some skills training – albeit of a basic nature (MG Rover Task Force, 2005b). There was a very limited take-up of skills training or further education amongst the MMAL workers, and while the closure of the Lonsdale plant released some skilled workers into the labour market, few additional skills were acquired and a significant number of workers appear to have left the labour market entirely. The regional vision statements developed in response to the job losses at MMAL could have, and should have, made education and training a centrepiece of economic planning.

NOTE

1. The first named author of this chapter was the Flinders University staff member to sit on SSIDWG from its inception. The second Flinders staff member to join the Working Group was the Deputy-Vice-Chancellor (Academic).

REFERENCES

Australian Bureau of Statistics (2001), Census of Population and Housing, 2001, ABS, Canberra.

Bailey, D. (2003), 'Globalisation, Regions and Cluster Policies: The Case of the Rover Task Force', *Policy Studies*, **2** (3), 67–85.

Beer, A. and Cooper, J. (2008), 'University-Regional Partnership in a Period of Structural Adjustment: Lessons from Southern Adelaide's Response to an Automobile Plant Closure', *European Planning Studies*, **15** (8), 1063-84.

Beer, A., Haughton, G. and Maude, A. (2003a), *Developing Locally: Lessons in Economic Development from Four Nations*, Bristol: Policy Press.

Beer, A., Maude, A. and Pritchard, B. (2003b), *Developing Australia's Regions: Theory and Practice*, Sydney: University of NSW Press.

Blandy, R. (2004), 'A Quantitative Assessment of the Southern Economy and the Effects of the Closures of the Mobil and Mitsubishi Plants at Lonsdale and Tonsley Park together with an Outline of a Strategic Response', Commissioned by the City of Onkaparinga and the City of Marion.

British Council (2003), *Educating UK: Positioning For Success*, London: British Council.

City of Marion and City of Onkaparinga (2006), 'Southern Region Economic Diversification Plan Draft', Unpublished.

City of Onkaparinga and City of Marion (2004), 'Blueprint for a New Economy for the Southern Region of Adelaide- Draft Outline', Unpublished.
City of Onkaparinga and City of Marion (2006), 'Southern Region Economic Diversification Blueprint (SREDB)', Unpublished.
Forbes, D. and Cutler, C. (2006), 'The Global Knowledge Economy, the University and the South East Asian City', in T. Wong, B. Shaw, and K. Goh (eds), *Challenging Sustainability: Urban Development and Change in South East Asia*, Singapore: Marshall Cavendish, pp. 175–97.
Forster, C. (2003), *Australian Cities, Continuity and Change*, 3rd edition, Melbourne: Oxford University Press.
Garlick, S. (1998), 'Creative Associations in Special Places: Enhancing the Partnership Role in Building Competitive Regional Economies', Canberra: Department of Employment, Education and Youth Affairs.
Government of South Australia (2006), 'Submission to the House of Representatives Standing Committee Inquiry into Employment in Automotive Component Manufacturing'.
Haughton, G., Beer, A. and Maude, A. (2003), 'Understanding International Divergence and Convergence in Local and Regional Economic Development', in A. Beer *et al.* (eds), *Developing Locally: An International Comparison of Local and Regional Economic Development*, Bristol: Policy Press, pp. 15–36.
Kearins, B. (2002), 'Exporting Locally: A Strategy for Regional Small Business Growth', *Sustaining Regions*, **2** (1), 17–28.
Loveridge, S. (1996), 'On the Continuing Popularity of Industrial Recruitment', *Economic Development Quarterly*, **10** (2), 151–58.
McGuirk, P. (2005), 'Neoliberalist Planning? Rethinking and Recasting Sydney's Metropolitan Planning', *Geographical Research*, **43** (1), 59–71.
McKeon, R. and Weir, T. (2001), 'Preconditions for a Knowledge Based Economy', *B-HERT News*, **11**, 4–5.
McKinsey and Co (1994), *Lead Local, Compete Global*, McKinsey and Co, Sydney.
MG Rover Task Force (2005a), 'Closure of MG Rover: Economic Impact Assessment', July 2005.
MG Rover Task Force: Six Months On (2005b), Paper Prepared for Submission to the Department for Trade and Industry, November 2005.
OECD (2001), *Best Practices in Local Development*, OECD, Paris.
O'Neill, P. and Argent, N. (2005), 'Neoliberalism in Antipodean Spaces and Times: An Introduction to the Special Theme Issue', *Geographical Research*, **43** (1), 2–9.
O'Neill, P. and Moore, N. (2005), 'Real Institutional Responses to Neoliberalism in Australia', *Geographical Research*, **43** (1), 19–28.
Peck, J. (2001), 'Neoliberalizing States: Thin Policies, Hard Outcomes', *Progress in Human Geography*, **25**, 445–55.
Peck, J. (2004), 'Geography and Public Policy: Constructions of Neoliberalism', *Progress in Human Geography*, **28**, 392–405.
Peck, J. and Tickell, A. (2002), 'Neoliberalising Space', *Antipode*, 34-380-404.
Pusey, M. (1991), *Economic Rationalism in Canberra: A Nation Building State Changes its Mind*, Cambridge: Cambrige University Press.
Rainnie, A. (2003) 'New Regionalism in Australia', *Sustaining Regions*, **3** (2), 13–18.
Saxenian, A. (1994), *Regional Advantage: Culture and Competion in Silicon Valley and Route 128*, Cambridge: Harvard University Press.

Scott, A. (2004), 'A Qualitative Assessment of Opportunities for Infrastructure and Industry Development to Counter the Effects of the closure of Mobil and the Mitsubishi Plant at Lonsdale On the Southern Region of Adelaide', Report Commissioned by the City of Onkaparinga and the City of Marion.

Stephen, P., Sumell, A., Black, G. and Adams, J. (2004), 'Doctoral Education and Industry: The Flow of New PhDs into Industry', Economic *Development Quarterly*, **18** (2), 151–67.

Stilwell, F. and Troy, P. (2000), 'Multi-level Governance and Urban Development in Australia', *Urban Studies*, 37 (5–6), 909–30.

W.K.Kellogg Foundation (2002), 'Home Grown: Community – University Partnerships Can Create the Workforce of the Future', http:www.wkkf.org.

11. Maturity or Decline of Italian Industrial Districts

Ivana Paniccia

11.1 INTRODUCTION

Since the second half of the 1990s, Italy has lost large market shares of international trade, GDP has increased by less than in the other European partners, while productivity has stagnated.

Economic observers have engaged in debate as to the causes of what some of them call 'Italian industrial decline', where industrial districts (IDs) are inevitably questioned, given that they account for much of the Italian model of international specialisation (Ciocca, 2003; Faini and Sapir, 2005).

The industries in which IDs are specialised have suffered from intense competition raised by emerging countries, and from low rates of productivity compared to other industries or countries, and they have not benefited from the present buoyant demand for technology-based products. Similarly, the small-sized firms that characterise IDs have struggled not only in providing new products but also in reaching new international clients and markets. The most significant effect of these difficulties is the decreasing market shares of Italian IDs in international trade.

Opinions differ on the implications of all this for the viability of the ID, either as an empirical construct or as a theoretical model. On the one hand stand those who regard IDs as the main culprits for Italian decline and maintain that they (or, at least, a large proportion of them) risk being rapidly swept away by current technological and market changes. Measures in support of IDs, therefore, are likely to be a waste of public resources, whereas industrial policy should be targeted on innovative industries that ensure high productivity and competitiveness.

On the opposite side stand those who argue that even within industries characterised by low productivity rates or negative export variations, firms belonging to IDs achieve performances better than the average. Becattini and Dei Ottati (2006) provide evidence of such positive performances, compared to non-IDs areas, in past years, while some recent signs of recovery in the variation of exports (Ice, 2006a) suggest that the interpretation of current difficulties is merely conjunctural. IDs therefore warrant a targeted industrial policy aimed at guiding them through this difficult phase while awaiting their resurgence, perhaps on new bases, but without altering their essential nature.

Other commentators maintain that certain IDs have already altered their structure, which is no longer based on a large number of similar small-sized firms, but presently rotates around a number of leading firms, generally of larger-than-average size and much better equipped to face international competition and technical change. These IDs and their larger firms are the appropriate targets of an industrial policy aimed at prompting Italian industry recovery.

This chapter keeps the theoretical validity of the model distinct from its historical or empirical manifestations, preserving the former but referring to a conceptualisation which envisages a larger number of theoretical variants of the model which I have already proposed elsewhere. The chapter then assesses the capacity of each of their empirical instances to resist current competitive challenges. In what follows, therefore, it will emerge that Italian IDs have already undergone a process of internal change which has generated a variety of forms subject to further evolution.

The analytical scheme adopted here draws on previous empirical investigations by the author on the various types of IDs in Italy (Paniccia, 2002), and enables identification of their likely evolutionary patterns – though not their points of arrival – and the drawing of some policy implications.

We shall see that the difficulties of certain forms of IDs, such as the canonical and craft-based ones, correspond partially to those of mature regions in Europe. IDs, however, differ as regards both their specialisation in industries with limited economies of scale and also a governance and institutional system which is very specific to Italian history.

Here, the term ID is used generically to denote any agglomeration of small to medium-sized enterprises (SMEs) specialised in one or a few industries in a geographically bounded area. This definition coincides with that used by the Italian national institute of statistics for the identification of Italian IDs (Istat, 1997 and 2005).[1] If the requisites of the prevalence of SMEs and small spatial scale are relaxed, the terms 'local productive agglomerations' (LPAs) or 'economies of agglomeration' may be more appropriate; the term 'cluster' of firms is similarly appropriate for these cases, but it will be used sparingly here because of its association with Porter's work.

Section 11.2 presents a classification scheme of LPAs which may be applied to the Italian cases. The scheme is built as a combination of different structural dimensions of analysis, which also explain the evolution of the different types. Section 11.3 discusses the capacity to react to external change and the associated performance of each type, while the final section discusses some likely scenarios of change and policy implications consistent with the structural analysis of the Italian types previously conducted.

11.2 A CLASSIFICATION SCHEME FOR ITALIAN INDUSTRIAL DISTRICTS

Agglomeration may be analysed along different dimensions. By 'de-constructing' the various definitions of LPAs put forward in the literature into their constitutive elements or factors, four sets of classification criteria can be obtained (Table 11.1).

By definition, agglomeration is a distinctive feature of IDs or LPAs, but the values assumed by its measuring variables, such as the size of the industry and the degree of specialisation (or, symmetrically, diversification), describe a continuum of forms with different implications for the model of labour division between firms and performance. The nature of the LPA's industry of specialisation, in terms of degree of complexity, fungibility or diffusiveness and its technological paradigm, set the technical constraints for the way in which specific activities may be organised and interact with each other and with other external institutions, and they define the within- and between-industry externalities that may be generated. Real forms of agglomeration differ in their ranges of horizontally- or vertically-related industries, depending, among many factors, on their history and age. The localisation of the most value-added activities similarly depends on the relevant competence, whether technical or commercial, exhibited by local actors. Competencies are related to the (prevailing) ownership and governance model(s) of localised firms and are also fostered by external institutions (apart from the firm), such as knowledge ones. The access to qualified services and public or private support structures as well as to a high-ranked labour market for managerial or creative competencies also depends either on whether the localisation is in a high-ranked urban centre or on its accessibility.

Using the criteria summarised in Table 11.1, the following typology may be drawn up:

- (Semi)canonical IDs;
- Diversified or urban IDs;
- Satellite platforms or hub and spoke agglomerations;
- Co-location areas;
- Concentrated or Integrated Agglomerations or IDs;
- Science-based or Technology Agglomerations.

Table 11.1 Descriptive factors of a typology

Factors of agglomeration

- Size of the industry (no. of firms and employees)
- Degree of specialisation/diversification

Industry factors

- Industrial specialisation (complexity, fungibility, etc. nature of the technological base: analytical, synthetic, intuitive)
- Range of horizontally-related industries. These include industries related by common technologies, end/users, distribution channels and other non vertical relationships
- Range of vertically-related industries. These include all the stages of activity of a *filière*
- Type of interdependence: 'pooled', 'sequential' 'reciprocal' (or two-way) and 'intensive' (or 'organic') interdependence
- Activity set

Companies (and institutional) characteristics

- Size distribution of firms and presence of business groups;
- Competencies: technical and organisational (allocative, transactional, administrative and financial)
- Ownership:
  - Local
  - Non local (national or multinational; multinational: a) *Truncated Miniature Replicas*; b) *Specialised Product Subsidiaries*; c) *Rationalised Product Subsidiaries*; d) *World Product Mandates or Regional Product Mandates*)
- Governance:
  - Family firms
  - Entrepreneurial firms
  - Managerial firms

Urban and socio-institutional context

- Urban setting or environment
- Economic institutions and knowledge institutions
- Social institutions

Source: Adapted from Paniccia, 2006.

After describing the above types, I shall discuss the rationale of each type and its reaction to different competitive scenarios and exogenous conditions of change, drawing on my first attempt to operationalise the above-mentioned explanatory factors on a sample of 37 Italian IDs using the 1991 Istat classification (1997).[2] More precisely, following a multivariate analysis, I classified the IDs in the sample in canonical, craft-based, concentrated and satellite types. While the canonical IDs' group included only a minority of areas (LLSs), the other types groups were found to be more numerous.

11.2.1 The Canonical Type

The *canonical* type of ID is characterised by an extended division of labour mostly based on family-owned and family-run firms with few employees and which rely on family members or relatives. Entrepreneurship is based on on-the-job training and has low formal educational levels, with an artisan or factory background.

The structure of the industry (generally fashion or design-intensive industries) is highly fragmented with a highly-decomposable nature, in that it consists of independently owned firms scattered among small units.

The network of relationships among the firms is dense and is characterised by a two-way sequential interdependence whereby firms exchange information (bi-directional flows) and goods (one-way or bi-directional flows). This does not exclude, however, the existence of more integrated firms, networks of firms or a few subsidiaries of transnational corporations (TNCs) with a 'world mandate' to produce specialised intermediate or final products; but their presence does not alter – at least in their mature stage – the flat model of labour division of canonical IDs. While the internal ties are numerous, the opposite is the case regarding external ties, since these IDs have a limited degree of international openness in terms of global chains of production and knowledge. Relationships with external markets and technologies are ensured by just a few specialised actors.

The large number of firms and the long-standing manufacturing tradition jointly ensure a high degree of self-containment, since all the specific activities in the leading industry's *filière* are carried out locally. These include the manufacturing of investment goods (machinery) and the provision of auxiliary services (transport, banks, etc.) specifically dedicated to the leading industry.

Local institutions, like local banks, trade unions, trade associations and technical or vocational schools, tend to offer services tailored to the needs of local firms and workers. However, their performance, activism and extent of membership (or in other words, the local 'institutional thickness') is not homogeneous and depends on a variety of factors including the past history and the national and local political context.

Workers are on average highly skilled. They are trained in technical schools or within families where, in the first phase of development, the proximity between workplace and home favours a process of industry socialisation and then processes of learning and imitation.

The complete range of cooperative agreements along vertical and horizontal lines, can be observed when this canonical type of ID is in the growth stage. In its maturity, when demand shrinks and competitive pressures increase, horizontal cooperation in terms of the sharing of orders becomes rarer, and opportunistic behaviour such as labour poaching, unfair fees for subcontractors and free-riding on quality becomes evident.

Cultural and income disparities are compressed because people live and work locally. This also explains why canonical IDs cannot develop in metropolitan cities with their large peripheries and often conflicting sub-cultures. The Marshallian triad of external economies is contained within a single urban area of a small-to-medium size, which also explains the limited extent of these areas. The most prominent examples of this canonical type of ID are to be found in Italy, and at least in the 1990s, they include the cases of Prato, Porto Sant'Elpidio, and Santa Croce.

On the positive side, canonical IDs are very able to absorb fluctuations in the economic cycle and they are well suited to markets characterised by variable and differentiated demand (Langlois and Robertson, 1995). Their modular organisation also gives them a high degree of operational and market flexibility that permits rapid reactions to changes in tastes, allowing for savings on the costs that would otherwise be incurred in the rearrangement of production lines. Productivity is expected to be high because of the relatively low costs of organisation, but also because of the high expertise of workers and technicians. Evidence of this may be found in the both case studies of IDs and also quantitative studies, although these refer to a generic list of IDs defined according to the Istat methodology (Fabiani and Pellegrini, 1998; Signorini, 2001).

On the negative side, as will be shown here, canonical IDs are vulnerable to structural changes in technology, intermediate or final markets, and they may similarly struggle when new standards or organisational procedures are introduced.

11.2.2 Diversified IDs or Urban IDs

The *diversified* or *urban* type of ID is characterised by the location and the embeddedness of manufacturing activities in an urban setting, and renders the economic structure more diversified than in the case of canonical IDs.

It may include either small-to-medium sized towns in which the manufacturing activity takes place within the town or large metropolitan area, and often where productive activities are confined to distinguishable neighbourhoods. In its many variants, the urban settlement offers business services and expertise, larger and more diversified labour markets and cultural amenities, and it is also well equipped to foster design and craft-based activities, which are very typical in the Italian case. As a consequence, if their urban environment is rich with cultural and knowledge institutions, at least three typical types of firms and industries prosper in this type of ID: i) firms specialised in creative industries (i.e. the film industry) or activities (i.e. designers), ii) artisan art firms (i.e. reproduction furniture, jewellery, art ceramics) and iii) technological firms specialised in high-tech industries or activities, as with the case of science-based clusters.

The variety within this category is linked to the differing nature of the urban settlement, which may perform different roles according to the functions exerted, its size, its international connections, and also its cultural, historical and institutional traditions.

As regards the structure of the localised industry in these diversified or urban IDs, the number of firms specialised in the dominant industry is smaller than in canonical IDs, *ceteris paribus*. The division of labour is less extensive, and so too is the range of complementary activities and industries (such as those providing investment goods). Firms are mainly final output producers,[3] and are surrounded by a high proportion of subcontractors with commercial autonomy and a high degree of diversification in other manufacturing and service industries. Related to the less extensive division of labour is a slightly larger average size of firms, *ceteris paribus*. In this respect, as compared to canonical IDs, these types of IDs are less able to adapt to short-run market and (non-radical) technological changes. Their lower level of 'internal variety' (Maskell, 2001) impedes experimentation with new organisational or technical solutions within the dominant specialisation and industry.

The diversification of economic activities is an asset for the area. This type of ID also has its own independent logic that implies that it is not merely a preliminary version of the canonical type with a lesser degree of specialisation. In Italy most IDs have arisen in small-to-medium urban centres. Some of them have built on the artisanal and artistic content of their production, exploiting the links between local products and the historical tradition or the artistic reputation of the city (i.e. Bassano del Grappa, Marostica, Murano, all in the Veneto region). In other cases the local production has lost some of its more artisanal content in order to embrace a pattern of greater standardisation, as in the cases of Civitanova Marche or Pesaro, which specialised in footwear and furniture, respectively.

11.2.3 The Satellite Type(s)

Satellite local production agglomerations (LPAs) generally comprise a limited number of small firms as compared to the canonical type of ID, but not in absolute terms and not compared to the diversified type. These usually operate as subcontractors to larger commissioning firms, which may be localised in the same area or outside. The degree of labour division is therefore not particularly marked, and a horizontal pattern tends to prevail. Firms' size, however, is not necessarily small. The relationships between suppliers, subcontractors and final firms are characterised by a sequential interdependence of one-way flows of goods. There are not the types of structural conditions appropriate for the emergence of horizontal cooperation.

Knowledge institutions are very rare in these areas, which also tend to host only a small proportion of relatively limited business services. The ownership of commissioning firms tends to be local; but for a very small proportion of them, it may be national-domestic or foreign-owned. According to the presence of larger firms, their ownership, and the amount of work subcontracted locally, there may be two variants of this type resulting in different degrees of self-containment:

a) Where the local commissioning firms are foreign-owned, a sub-type of *'TNCs-led' (satellite) LPAs* may be distinguished: a few TNCs supported by clustered supply chains of rather dependent SMEs specialised in modular and complex activities, such as in the automotive industry.
b) When the presence of large firms is negligible and the industry is not modular or complex, these areas may be better defined as *external subcontracting areas*. The subcontractors are dependent on external commissioning firms and are specialised in just a small number of phases of the production cycle, and in particular, those with more labour content. Examples of this type can be seen in many southern Italian IDs, but also some northern ones, which appear more to be the peripheral rings of canonical IDs.

Satellite LPAs can be considered to be a form of agglomeration which is largely independent of urban scale and with a weak potential for development, although they may experience high growth rates of employment and exports during the formation stage. They are in competition with less favoured regions, given that their typical production features are close to mass or standardised production facilitated by low-cost location advantages. A problem with this type of agglomeration, as underlined in the literature on international value chains, is that even when the commissioning firms are involved in forms of vertical cooperation, upgrading does not coincide with innovation. Commissioning firms push their suppliers to introduce cost-saving technologies, to reduce delivery times and to enhance quality reliability, without necessarily contributing to the upgrading of technology and skills (Gereffi et al., 2005). Therefore, if the relocation of high added value activities occurs, it militates against the local formation of the competencies which would be required for future development.

11.2.4 The Concentrated/Integrated Type

The *concentrated* or *integrated* type of ID is characterised by the integration of a small number of parallel technological *filières* which leads to the discovery of new markets where local firms occupy leadership positions. This integration derives from the development of a small number of the Marshallian auxiliary industries, which acquire their independence and autonomy over time either by placing their products in different markets, or from technological spillovers associated with applying a variant of the principal industry's technology to a different product or technology.

The first case is manifested by the Italian districts in which a predominant specialisation, whether in fashion or design-led and highly-specialised industries, is integrated with a strongly-developed mechanical industry which makes the tools and machines used in the production process. Included in this group would be the areas with a strong mechanical tradition such as Cento or Guastalla in Emilia Romagna or Suzzara in Lombardy, which have developed expertise in the production of agricultural machines. Another example is the integration of an already established mechanical engineering tradition with electronics and ICT in order to produce automated packaging machines and electrical engines for marble working. Unlike the canonical type, where a wide range of complementary activities similarly exist, in this type of ID complementary industries have developed autonomous final markets. This allows local firms to escape the trap of having rely only on local demand. This combination is not a natural outcome or evolution of canonical IDs, since it requires the localisation of an already strong complementary industry. Moreover, on closer inspection, a mechanical tradition in many IDs, which is now classified as being either of a canonical or integrated type in regions such as Emilia Romagna and some areas of Northern Italy, actually pre-existed their development.

The second case is demonstrated by three examples; the application of the local expertise in ceramic materials (also thanks to collaboration with the nearby universities of Modena and Bologna) to aerospace and electrical transmission industries in Sassuolo; innovations in the manufacturing or design of traditional ID products that have created new markets for 'technical textiles' in the IDs of Cossato and Biella, and customised laser-designed shoes in Vigevano; Montebelluna, in the Veneto region, which has developed a range of products, from ski boots to leisure-time shoes, that integrate plastic and leather technologies.

In this sense, concentrated types of IDs are an evolution of the canonical type, but the transition to this form configures a different kind of interdependence between firms. In particular, the development of higher and diversified competencies within firms and more open and urbanised settlements, are all features that, if not locally present, may make this transition very difficult.

Reciprocal interdependence is common in these areas. It may come about between firms, but it is much more frequently observed among the internal R&D departments of the more innovative firms. The IDs specialised in mechanical engineering industries include cases which involve the joint application of complementary resources to a common activity in integrated manner. Overall, the firms located in this type of ID are less interconnected than in canonical IDs, with interdependence focused within selected user-producer interactions for innovative activities. Owing also to the complex nature of the industry of specialisation, these areas may foster the development of high-level technological competencies and related knowledge institutions.

The Italian IDs falling within this category contain two groups of firms which interact with each other. On the one hand, there are the relatively larger and vertically integrated firms, often at the head of formal or informal groups of firms and with a market or technological leadership. On the other hand, there are small firms specialised in more specific tasks and supplying components or services to the former group of firms, or operating as final demand firms. The leading firms are either national or multinational. In the latter case, subsidiaries usually have a 'world' or a 'regional product mandate' and they oversee the entire range of activities (production, R&D, finance, marketing, etc.) related to one line of production or one product, which is then marketed on the global market.

My typology also includes *Co-location areas* and *Technology or Science-based agglomerations or clusters* (Paniccia, 2006). Evidence of these types is still scanty in present-day Italy because co-location areas may be seen as forms of organisation of production typical of the early stages of a process of industrial agglomeration from which some of the existing canonical or craft-based IDs have evolved. On the other hand, evidence of technology clusters may be found, such as the case of Catania microelectronics and nanotechnologies or the high-tech mechanical engineering in Emilia Romagna. These, however, are agglomerations of a few firms and institutions which are either confined to a small area or spread across an entire region. As such, they do not yet exhibit all the features of a typical technology cluster.

As already shown, the various Italian IDs types also exhibit different performance profiles. The recent literature aimed at measuring the 'district effect' generally analyses the performance of IDs as an undifferentiated whole, coinciding with the Istat list (see note 1). On the basis of the Istat listings of IDs, the growth of employment, export and value added in IDs was found to be above the national average over a long period (Fabiani and Pellegrini, 1998; Signorini, 2001). Paniccia (2002) makes an attempt to differentiate among the various types and performances of a sample of IDs. This study shows that the concentrated type exhibited the best performance in the 1990s, and this is confirmed when updating data to more recent years. Moreover, this is the case, even allowing for the fact that this group included a higher proportion of the industries which were more severely affected by the dramatic increase in the cost of energy, such as textiles and ceramics.

A structural feature of the typology presented here is also their institutional setting. Each type has its own typical institutions, but these are also an expression of the national system of economic and knowledge organisations, public policies and infrastructures that make up the 'national system of innovation' (Freeman, 1987; Lundvall, 2002). Among these IDs, the commonalities are the governance model of the firms and the associated entrepreneurial pattern. The governance model of firms generally in all Italian SMEs is built on family ties, a low degree of ownership contestability and concentration, scant investment in personnel training, and a weakness of control mechanisms. Entrepreneurship is based on on-the-job training, self-employment, low educational levels, artisan or factory backgrounds, personal savings, work ethics and family support alone.

More recently, it has emerged that the typical knowledge institutions of the past do not work with the same efficiency: technical schools attract students in lesser proportions, apprenticeships in crafts firms have vanished in many areas, the mechanisation of industrial processes and the detachment of young generations from parental jobs have altered the patterns of socialisation, which now tend to take place outside of the mill or the factory. Moreover, the shortage of skilled labour induces employers to ask for extra hours of work and cases of labour poaching occur. Overtime now makes it less necessary to hire new personnel and therefore impairs the socialisation of young workers in the production process. In addition to these factors, the national system of innovation and its local variants exerts an obvious influence on the quality of human capital and generally on the nature and diffusion of technological capabilities. A general issue of human capital deficiency arises in the Italian scenario, with obvious consequences for productivity and growth. Data on education levels for a sample of IDs examined by Paniccia (2002) displays levels of higher schooling which are below the national average and an education profile of the local population which is concentrated in the lower levels of schooling.

11.3 RISKS AND OPPORTUNITIES FOR ITALIAN INDUSTRIAL DISTRICTS

The previous sections have described the structural features of Italian IDs using a comprehensive analytical scheme that identifies the major types. In this section the resilience and the likely evolutionary patterns of IDs are analysed.

The present competitive scenario is characterised by growing international market integration, with the emergence of major competitors, such as China and India. These emerging countries increasingly occupy the low and medium-quality bands of the personal and household goods markets. Goods produced in emerging countries are increasingly reaching international markets thanks to improvements ICTs and decreased transport costs, thereby reducing the market shares of countries with similar patterns of specialisation, such as Italy. This loss of competitiveness can be related to either supply or demand factors.[4] On the supply side, the competition raised by the newly-industrialised countries produces low labour cost challenges to the efficiency and organisation of IDs. On the demand side, this current trend, which is likely to continue for the foreseeable future, could be counterbalanced by the rise of a large middle class in these emerging economies anxious to buy traditional goods with a high fashion or design content. If the near future matches this scenario, the prospects are good for the firms specialised in luxury goods, although not for IDs as a whole.

The nature of these changes in consumption patterns should be monitored by organisations which are also able to control the distribution chains of 'made in Italy' products. Some of the larger ID firms have already invested in Eastern European or Mediterranean countries or Far Eastern markets for cost-seeking or market-seeking reasons, but the extent of this trend has been found so far to be limited (Federico, 2004; Basile and Giunta, 2005).

Most of the constitutive features of Italian ID types, and specifically the small, often micro, size of firms prevailing in canonical and craft-based types and their governance patterns, are unsuitable for active internationalisation strategies. The small size of firms impedes direct connections with the types of new markets which would be able to convey information about emerging consumption tastes and styles. In addition, many of the traditional local intermediaries (exporters, traders or generally wholesalers) that used to link producers in Italian IDs to markets are no longer able to perform this function. The closed governance model of most Italian firms, combined with other financial constraints, hampers their growth to sizes sufficient for foreign investment. On the other side, if Italian IDs' firms do invest abroad, there will inevitably be effects on the structural features of IDs themselves. These changes, together with the other ongoing social transformations regarding mechanisms of training, education and work socialisation of young generations, are unlikely to preserve IDs in their canonical form. These changes therefore raise the question as to whether there would conversely be greater opportunities for the more concentrated firms of integrated clusters or IDs. According to the typology presented above, larger firms and concentrated IDs are better equipped than small firms and canonical IDs in general, to deal with most of the external challenges. Many economic observers in Italy are therefore very optimistic about the abilities of concentrated groups of firms, and larger firms, to rejuvenate the competitive evolution of IDs, due to their greater propensity for internationalisation and achieve better-than-average indicators of productivity or profitability and higher levels of performance (Mediobanca-Unioncamere, 2005). However, if groups of firms embrace a pattern of standardisation rather than 'mass customisation', the distinguishable content of the product will weaken along with the reputation of the firm and area in which they are located. This will have negative effects on the relationships between firms and will lead to the destruction of those tacit capabilities that are at the root of customised production in canonical or craft-based districts.

Besides these risks, which are mainly specific to Italy, *integrated agglomerations* have great opportunities of development since their rationale resides in that 'contamination' of co-located *filières*. This is typical of processes of technological change in which traditional products have acquired new functions incorporating innovations in their design or in the production methods and materials used in a way that also redefines markets' boundaries. Examples of this are new textile products for improved human performance, paramedical textiles, smart textiles and clothing, or ceramics for innovative technical applications. The hybridisation of *filières* may also derive from the adapting towards environmentally-friendly technologies and markets. These tendencies, however, may not necessarily imply the physical proximity of the technology and capabilities employed. But the outcomes depend on the likelihood that the entrepreneurial class of existing Italian IDs respond to such technological upgrading.

The arguments in this analysis imply that there is likely to be a reduction in the number of IDs as a whole and a gradual transition from canonical to urban or integrated types. Within either canonical or diversified districts there will be a slimming down of the number of firms due to the disappearance of fashion-led production linked to low brands, unless it is countered by a set of new product brand innovations. At the same time, the near future may also see the advent of firms or networks organised in extensive worldwide value chains with weak ties with their original regions. This will lead to the breaking up of the complex system of traditional interdependences between local firms and local institutions and the affirmation of new open-governance models of firms. The few remaining canonical IDs will enlarge their territorial boundaries, and this process will continue even further in integrated agglomerations due to the replacement of local ties with international market relationships. On the other hand, where the reciprocal user–producer patterns of interdependence between local firms is likely to continue is in terms of the activities of the *filières* in integrated agglomerations which still require proximity in order to foster innovation.

Satellite IDs risk losing their potential as a model of local development, becoming thinner networks of subcontractors linked to a few commissioning firms, or changing into peripheral concentrations of activities, primarily located in the southern regions of Italy.

A recent account of the evolution of Italian IDs based on Istat 1997 listings provides evidence of the evolution of IDs along the lines of (a) concentration by integration among firms or group formation; (b) diversification into other manufacturing industries, including machinery; and (c) tertiarisation of the ID into intermediate and advanced services (Bracalente and Cossignani, 2005). In addition, the publication of the 2001 list of IDs by Istat also confirms a process of change is taking place. The number of IDs, identified on the basis of 2001 census data, has fallen relative to the number identified on the previous census year of 1991 (from 199 in 1991 to 156 in 2001) (see note 1).

11.4 CONCLUSIONS AND POLICY IMPLICATIONS

The competitiveness of IDs in many ways is now based on factors and processes that lie outside the boundaries of the canonical ID model. These include the hybridisation by industries into completely different technological fields or the movement towards advanced business services. Similarly, largely outside the control of IDs are issues such as changes in communications and transportation technologies, organisational changes, and developments in finance, as well as issues such as the level of local policy governance, and the formation and delivery of education policies,

If the analysis in this chapter is correct, there are therefore only weak reasons for advocating a specific or targeted industrial policy for IDs in the Italian context, given that most of their weaknesses are common to a large part of the Italian industrial structure. On the other hand, all IDs may benefit more from general policies at the national level not relating to territorial constraints, such as those designed to improve educational standards, policies encouraging company mergers, immigration policies focused on attracting high-ranked competencies, and liberalisation policies in network industries such as energy, transport and logistic networks which are aimed at promoting competition. Most of these policies will stimulate structural change, but more targeted devices are also possible, such as measures favouring the hybridisation of firms' technology bases, increasing the turnover and mobility of creative personnel such as technicians and scientists, designers, architects, artists, and the promotion of public-private technological partnerships. However, the heterogeneity of IDs in terms of their structural features, evolutionary patterns and classification difficulties,[5] together with the practical difficulties of designing a targeted industrial policy for IDs, militates against the search for a single policy rationale specific to IDs. Those who maintain that the causes of IDs' competitive weakening are mainly exogenous prefer to propose policies specifically targeted on IDs, while those recognising that IDs are changing, although not necessarily declining, assign much more value to broader structural and competition policies. The arguments here suggest that specific support for all IDs may actually delay the necessary processes of adjustment change which have been taking place in many other mature regions of Europe.

NOTES

1. Istat has identified IDs by means of a multistep procedure which starts from of the concept of the 'Local Labour System' (LLS) in order to define their territorial boundaries. LLSs are aggregations of two or more contiguous municipalities identified on the basis of the self-containment of daily commuting flows between the place of residence and the place of work. The concept is close to the British ones of 'Travel-to-Work-Market-Areas' or 'Local Labour Market Areas'. Istat provides the list of Italian LLSs for the last three population censuses (1981, 1991 and 2001). The LLSs are used as the basis for the identification of IDs, which are defined as those LLS with a higher-than-average proportion of manufacturing activities and a higher-than-average proportion of SMEs in the dominant industry. See Paniccia 2002 (Chapter 2) for details. At present there are two lists of 'Istat IDs', one derived from 1991 LLSs and listing 199 IDs (Istat, 1997) and one from 2001, listing 156 IDs (Istat, 2005). These two lists of IDs are not perfectly comparable since the number of IDs has diminished and their LLSs are not the same, because either their municipalities or number have changed in many cases.
2. In my previous work (Paniccia, 2002) the taxonomy of IDs was obtained inductively by considering a number of variables not perfectly coinciding with those implied in the analytical framework here resumed.
3. Final firms are those at the head of a value chain or *filière* specialised more in the design of products than in their manufacturing.
4. I deliberately do not discuss macroeconomic factors such as the adoption of a fixed exchange rate which acted as powerful shocks for Italian firms.
5. One of the main problems is that different forms of agglomeration operate on different geographical scales, and these boundaries can be only endogenously determined when the structural features of the area are known. This requires an operationalisation procedure (different from identification) incompatible with a suitable method for identifying firms entitled to public incentives, which, by its nature, must be parsimonious and objective.

REFERENCES

Basile, R. and Giunta, A. (2005), 'Things change. Foreign market penetration and firms' behaviour in industrial districts: an empirical analysis', *Documento di lavoro ISAE*, (48), Rome: Isae.

Becattini, G. and Dei Ottati, G. (2006), 'L'economia italiana negli anni novanta: un confronto tra aree di grande impresa e distretti industriali', *Economia e Politica Industriale*, (1), 5–24.

Bracalente, B. and Cossignani, M. (2005), 'L'evoluzione della specializzazione nei sistemi locali manifatturieri italiani', paper presented at the *XXVI Conferenza italiana di scienze regionali*, Naples, 17–19 October 2005.

Ciocca, P. (2003), 'L'economia italiana: un problema di crescita', *Banca d'Italia-Bollettino Economico*, (41).

Fabiani, S. and Pellegrini, G. (1998), 'Un'analisi quantitativa delle imprese nei distretti industriali italiani: redditività, produttività e costo del lavoro', *L'Industria*, **XIX** (4), 811–31.

Faini R. and Sapir, A. (2005), 'Un modello obsoleto? Crescita e specializzazione dell'economia italiana', in T. Boeri, R. Faini, A. Ichino, G. Pisauro and C. Scarpa (eds), *Oltre il Declino*, Atti della Conferenza organizzata dalla Fondazione Rodolfo Debenedetti, Bologna: Il Mulino.

Federico, S. (2004), 'L'internazionalizzazione produttiva italiana e i distretti industriali: un'analisi degli investimenti diretti all'estero', in *Economie locali, modelli di agglomerazione e apertura internazionale.-Bologna, 20 novembre 2003*, Banca d'Italia, marzo 2004.

Freeman, C. (1987), *Technology Policy and Economic Performance: Lessons from Japan*, London: F. Pinter.

Gereffi, G., Humphrey J. and Sturgeon, T. (2005), 'The Governance of Global Value Chains', *Review of International Political Economy*, **12** (1), 78–104.

ICE (2006a), 'Osservatorio sull'internazionalizzazione dei distretti industriali', no. 1, March, Rome.

ICE (2006b), 'Osservatorio sull'internazionalizzazione dei distretti industriali', no. 0, March, Rome.

Istat (1997), *I Sistemi Locali del Lavoro*, F. Sforzi (ed.), Rome: Istat.

Istat (2005), *Distretti industriali e sistemi locali del lavoro 2001*, Rome: Istat.

Langlois, R. N. and Robertson P. L. (1995), *Firms, Markets and Economic Change*, London & New York: Routledge.

Lundvall, B. A. (1992) (ed.), *National Systems of Innovation. Towards a Theory of Innovation and Interactive Learning*, London: F. Pinter.

Maskell, P. (2001), 'Towards a knowledge-based theory of the geographical cluster', *Industrial and Corporate Change*, **10** (4), 921–43.

Mediobanca-Unioncamere (2005), *Le Medie Imprese Industriali Italiane (1996-2002)*, Milano-Roma.

Paniccia, I. (2002), *Industrial Districts: Evolution and Competitiveness in Italian Firms*, Cheltenham, UK and Northampton, MA, USA: Edward Elgar Publishing.

Paniccia, I. (2006), 'Cutting through the chaos: toward a new typology of industrial districts and clusters', in B. Asheim, P. Cooke and R. Martin (eds), *Clusters in Regional Development*, Routledge, 90–114.

Signorini, L. F. (2001), *Lo Sviluppo Locale: Un'indagine Della Banca d'Italia sui Distretti Industriali*, Roma: Meridiana Libri.

12. Knowledge Spillovers and Industrial Transformation: The West Midlands and Saxony Automotive Clusters

Odile E. M. Janne and Mahtab Akhavan Farshchi

12.1 INTRODUCTION

Although significant spatial clustering of industrial activity is observed, the sources and dynamics of agglomeration economies vary across locations and sectors and are difficult to isolate empirically (Martin and Sunley, 2003; McCann, 1995; Nelson, 1996). Empirical studies on the success of regional economies such as the Silicon Valley, Route 128, Baden Württenberg and Emilia Romagna have focused much of their recent work on the geographical dimension of innovation processes and their dynamics for explaining localised growth. While the geographical clustering of innovative activities has been established, the most important mechanisms and types of externalities underlying localised innovation processes are fiercely debated. The main controversy lies in the role of localised knowledge spillovers (LKS) versus Marshallian pecuniary externalities as the most important sources of agglomeration economies in innovative clusters (Breschi and Lissoni, 2001a&b; Beaudry and Breschi, 2003; Caniels and Romijn, 2005). This controversy can be characterised as between the roles played by dynamic versus static externalities, in the sense that knowledge technology spillovers can be considered to be 'dynamic' externalities, whereas the pecuniary externalities explicitly modelled by early new economic geography frameworks (Krugman, 1991) can be considered as being 'static' externalities. Economic geographers (Martin and Sunley, 1996) criticise new economic geography models for ignoring the role of knowledge technology spillovers. As a result of these debates, however, the prominent role given to knowledge spillovers in the economic and new industrial geography literature, along with a focus on high-technology clusters, is also criticised for being conceptually overly limited (Breschi and Lissoni, 2001a; Capello and Faggian, 2005), in that these approaches underestimate the significance of pecuniary externalities (Breschi and Lissoni, 2001a&b; Caniels and Romijn, 2005), and also ignore many of the sectoral aspects of industrial clusters (Iammarino and McCann, 2006; von Tunzelmann and Acha, 2005).

The first aim of the chapter is to address these theoretical contentions by examining the distinction between the different sources and mechanisms of externalities which drive innovation in industrial regions. We develop an analytical framework distinguishing static and dynamic pecuniary externalities and real knowledge spillovers in order to explain the co-location and accumulation of innovative activities. Our framework allows for the specific innovation patterns in the automotive industry and suggests the different possible channels for the realisation of externalities. We argue that sector-specific dynamic externalities, including real and rent knowledge spillovers, are more localised (or less prone to geographical dispersion) than the static ones; and that the relative salience of these externalities in regions is key to understanding their innovative performance in a changing environment.

The second aim of the chapter is to investigate the relevance of that debate for understanding the growth and restructuring of mature industrial regions. While much research has been targeted at successful or relatively new regions in high-tech or other such dynamic industries, very much less research has investigated the evolution and transformation of mature industrial regions. Innovation in mature industrial regions is often underestimated, as the assumption is made that they have reached the end of their life cycle, with fewer prospects for growth and development in a context of fierce global competition, fewer opportunities for innovation (Utterback and Abernathy, 1975), and fewer agglomeration tendencies for innovative activity (Audretsch and Feldman, 1996). Knowledge spillovers are argued to be less relevant in mature industries compared to young industries, as firms hold more established routines and independent large-scale research and development (R&D) departments (Doring and Schnellenbach, 2006). However, the process of industrial maturity can be reversed (Abernathy and Clark, 1985) and the specific nature of the innovation-inducing externalities can evolve over the industry life cycle (Caniels and Romijn, 2005). We propose that the ability of a mature industrial region to remain innovative and grow or survive depends on the predominant type of externalities supporting innovative activities in the region, and particularly the presence of dynamic externalities, which cannot be explained solely by the mature stage of evolution of its dominant industry. This type of analysis of mature industrial regions is therefore primarily concerned with their adaptation, learning and innovative capabilities in response to negative shocks.

Finally, we empirically apply our framework and discuss the role of externalities for the cases of the West Midlands (WM) and Saxony clusters so as to better understand the contrasting innovative performance of these two mature regions, both of which are dominated by the automotive industry. We find that the differing nature and sources of local externalities that are present in the two regions offer strong insights regarding their apparent ability to innovate, grow and restructure. While the WM seems to benefit mainly from conventional static Marshallian externalities, Saxony exhibits more dynamic knowledge spillovers. In turn, innovation capabilities in Saxony appear more promising for the development of the region compared to those of the WM, despite its recent privatisation and dependency on investment from West Germany and Europe.

The next section critically reviews the main arguments regarding the nature and sources of local external economies for innovation, in relation to the debate on the importance of LKS for explaining the co-location and agglomeration of innovative activity. The third section discusses the sectoral dimension of innovation and the main sources of innovation in the automotive industry. Our analytical framework for identifying innovation externalities and dynamics in clusters is developed in section four. Section five outlines the methodology and data, and section six presents an empirical analysis and comparison of the innovation externalities and evolution of the WM and Saxony automotive clusters respectively. Conclusions and policy implications are drawn in the last section.

12.2 THE NATURE AND ROLE OF EXTERNALITIES FOR INNOVATION IN CLUSTERS

The concept of industrial clusters rests on the implied benefits of externalities.[1] These external economies can be related to Marshall, who identified three possible reasons why firms gain from locating in proximity to each other: the development of local labour markets for specialised skills, local markets for specialised intermediate goods, and information and knowledge spillovers. While Marshall focused on location externalities with reference to production activities, we here discuss his insights referring more particularly to innovation activities. We then address the debate about the nature and relative importance of different externalities, including LKS, as the main source of the clustering of innovative activity.

The economic benefits of static externalities are traditionally described in terms of economies of scale and scope in the transaction costs achieved by firms through labour pooling (i.e. reduced search costs) and the sharing of tangible and intangible assets (e.g. more accessible and efficient local services and more rapid supplier–firm communication). Lower transaction costs for local interaction also lead to pecuniary gains from collaborations between actors, through the sharing of the costs and risks of R&D. These static supply-side pecuniary externalities may also be supplemented by demand-side externalities (i.e. market effects), which allow for access to more sophisticated local markets that lower the search costs for customers for better products and processes (Swann, 1998).

Static Marshallian pecuniary externalities can be distinguished from pure knowledge spillovers. Pure or real knowledge spillovers refer to new knowledge or technology developed by one institution (e.g. a firm or university) becoming potentially available to other institutions. This is the notion of LKS employed by Breschi and Lissoni (2001a&b). The arguments developed by Breschi and Lissoni (ibid.) suggest that influential new industrial geographers (e.g. Becattini and Saxenian) and many economic geographers (e.g. Feldman, Audretsch and Jaffe) have overemphasised the importance of pure knowledge spillovers over more traditional Marshallian pecuniary economies as the main driving force in the agglomeration of innovation in clusters. Their critical review of the literature contends that the notion of LKS is conceptually and methodologically confusing, and that the dominant focus on LKS has obscured other important kinds of pecuniary externalities which contribute to innovation and learning in clusters. The arguments of Breschi and Lissoni (2001a&b) therefore suggest that the analyses of these geographers provide misleading policy prescriptions.

Drawing on insights from the evolutionary theory of the firm, Caniels and Romijn (2005) suggest a theoretical framework that allows for a broader notion of LKS than the conventional description of LKS as employed by Breschi and Lissoni (2001a&b). In the analysis of Caniels and Romijn (2005) we distinguish between the concepts of rent knowledge spillovers and pure knowledge spillovers. This approach posits that rent knowledge spillovers constitute the (dynamic) pecuniary cost advantages benefiting purchasers which derive from quality improvements embedded in products, when the quality increase is not fully reflected in prices (Griliches, 1979). As such, these arguments imply that the benefits from knowledge spillovers may take a pecuniary rather than a pure form. This distinction between static pecuniary externalities, rent (dynamic) pecuniary externalities and finally pure knowledge spillovers, provides a novel conceptual background for identifying the different mechanisms underlying the clustering of innovative activity. The LKS debate may then be redefined accordingly in terms of the relative importance of static cost advantages versus (broader) dynamic externalities (Caniels and Romijn, 2005).

On the basis of evolutionary economics arguments, our proposition is that dynamic externalities, defined to include both rent knowledge spillovers and pure knowledge spillovers, are a key source of the innovation and learning in clusters. In a globalising context, dynamic externalities are argued to be geographically more specific, i.e. less ubiquitous, than static cost-based externalities. In the evolutionary framework, dynamic externalities remain more localised than static externalities because they concern tacit knowledge transfers reliant on close interactions and relationships between firms and institutions. As such, they are not easily transferred or reproducible from one place to another. While traditional static economies of scale, scope and transactions are still recognised, our argument therefore primarily relates to the recent innovation literature which emphasises the role played by dynamic knowledge spillovers.

Spatial clusters, however, may also experience negative externalities as they mature, and may be locked in outdated technological trajectories (Baptista and Swann, 1998; Beaudry and Breschi, 2003; Swann, 1998). Negative static pecuniary externalities relate to local congestion and competition constraints. Negative dynamic externalities entail the exclusion or destruction of previous relational knowledge ties with other organisations and adverse technological 'lock-in' trajectories (Beaudry and Breschi, 2003). On balance, we would expect that mature industrial clusters in which dynamic agglomeration advantages for innovation dominate any static externalities are less likely to be adversely affected by changes brought about by globalisation, new technologies and/or new regulatory regimes. The resilience and growth of an industrial cluster will therefore depend on its ability to promote and benefit from positive dynamic externalities, while avoiding negative static externalities.

12.3 SOURCES OF INNOVATION IN THE AUTOMOTIVE INDUSTRY

The analysis of innovation and clusters goes beyond the regional level to the firm level, as firms are key actors of regional dynamism (Caniels and Romijn, 2003, 2005). Firms can enhance their innovation and learning by appropriating the innovation-stimulating externalities of the region in which they are located. In turn, the future prosperity of regions depends on how firms perceive and exploit the location-bound advantages that they offer. For firms in different industries, the relative importance of interaction between the different sources of technological knowledge, technological threats and opportunities, differs considerably (Archibugi, 2001; Pavitt, 1984; von Tunzelmann and Acha, 2005). More recently, it has been argued that the type of knowledge base that underlies each industry shapes the innovation processes of firms in clusters (Asheim and Coenen, 2005; Moodysson et al., 2006).

Countries have long regarded the automotive industry as a driver of industrialisation and economic development. Although the industry has matured, substantial innovations in the technological and organisational development of vehicle manufacturing have challenged it. The possibility of the de-maturity of the industry was first discussed by Abernathy and Clark (1985). The phenomenon of de-maturity occurs where technical, demand and regulatory changes mean that the renewed opportunities for innovation pose significant managerial challenges to the established firms. Such changes are currently occurring in many areas of the automotive industry, and the survival and performance of firms in this industry depends on transformation in the sector's technologies, skills and ways of working (Fine and Raff, 2002).

Following the relationship between industry structure types and the sources of innovation which was first defined by Pavitt (1984), firms in the automotive industry belong to the *production intensive* type of industry organisation, which includes the scale-intensive and specialised supplier's categories. In this schema, the key sources of innovation for large-scale firms are the firms' own innovations through its R&D and production engineering departments, as well as innovations by their supplier firms. R&D in the automotive industry is important for both production, and sales and marketing, since a car design needs to take into account the production process and the appeal to customers. Yet, innovation in the industry reflects the incremental nature of production engineering and the significance of process innovation (Fine and Raff, 2002). Significant technology linkages in the British automotive industry come from electrical machinery and communications equipment industries (Drejer, 1999), while Bergman and Feser (2000) found that the US automotive industry obtains inputs from a range of sectors like the refrigeration and heating equipment, or rubber and plastic hose and belting. In particular, from the 1980s onwards a major feature of technical change in the automotive industry has been the wide diffusion of computer-related process equipment. These developments include the adoption of computer-controlled machine tools, robotics and flexible manufacturing systems, and other measuring and testing equipment (Bessant and Haywood, 1991).

Meanwhile, the key sources of innovation for the smaller specialised supplier firms are own design and development and the larger user firms. The contribution to industry innovation of the supplier firms goes beyond the purchase of goods embodying new technology. These firms embody specialised knowledge and experience and are also directly involved in the development of new labour skills. As such, in this type of technological environment, education and training, and engineering talents in particular, are crucial for the supply of a competent workforce.

The two-way relationships between the large assembly firms and the smaller supplier firms are therefore both possible sources of industry innovation. However, if the industry is to take advantage of these various sources of innovation, innovation in the organisation, quality and flexibility of the supply chains and networks is also of fundamental importance (Fine and Raff, 2002; Schamp et al., 2004). The capacity of the WM and Saxony automotive clusters to continue generating and exploiting innovation will therefore depend on their ability to foster these two-way upstream and downstream sources of innovation.

12.4 ANALYTICAL FRAMEWORK

Our framework helps in identifying the different types and sources of externalities for innovative activities, that firms located in an automotive cluster can potentially gain access to and appropriate (Table 12.1). The interplay between the industrial and regional patterns of innovation will define the specificity of agglomeration processes within clusters. Examples of the channels through which the externalities may be realised for firms are given for each case. We classify externalities for the clustering of innovation into traditional Marshallian static pecuniary economies, based on the reduced transaction costs for co-located firms (column 1), and LKS, based on the interactions between co-located organisations. Those include rent spillovers founded on inter-organisational relationships (column 2), as well as pure knowledge spillovers derived from face-to-face contacts, the mobility on the local labour market, and/or intended exchanges (column 3). The rows distinguish the nature and sources of knowledge for innovation in the automotive industry that emanate from 1) corporate R&D departments, 2) corporate product engineering, design and development departments, and 3) external suppliers and/or demand/users.

Although actual clusters may combine different types of forces behind spatial clustering, the analytical distinction of the various elements at work is important in understanding the effects and implications of spatial clustering and designing regional policies (Gordon and McCann, 2000). This framework provides a useful analytical tool for better understanding the factors underlying the advantages (and disadvantages) of a cluster to generate innovation, and their relative importance.

The need for a dynamic capability of mature industrial clusters to restructure, grow or regenerate commands a shift of emphasis from static to dynamic externalities. Complex regional interdependencies for innovation associated with (rent and pure) knowledge spillovers are at the core of long-term comparative and competitive advantages, because they are unique and exclusive to the environment in which they are created and developed. LKS are the dynamic externalities supporting the learning and evolution of industrial clusters that are sector specific and necessary in a context of ongoing technological, regulatory and globalisation changes. While different types of externalities are likely to overlap, our proposition is that the role of LKS is crucial to explain the uneven innovative performance between clusters and regions.

Table 12.1 Region- and industry-specific sources of positive externalities for innovation in an automotive cluster

Automotive-specific sources of knowledge for innovation	Region-specific sources of location economies for innovation		
	Static pecuniary	Knowledge spillovers	
		Rent spillovers	Pure spillovers
In-house R&D	e.g.: access to R&D services, pool of scientists/engineers, access to partners for sharing cost and risk of R&D	e.g.: improved price-quality availability of scientists and engineers (from interaction with universities and research centres)	e.g.: face-to-face exchanges, mobility of local labour, demonstration effect, collaborative exchange of knowledge
In-house product engineering / design & development	e.g.: access to training/technical services, pool of engineers/designers, access to partners for sharing cost and risk of engineering projects	e.g.: improved price-quality availability of engineers/designers (from interaction with universities and training centres)	e.g.: face-to-face exchanges, mobility of local labour, demonstration effect, collaborative exchange of knowledge
Local suppliers / users	e.g.: access to input-output transactions (local supply and demand) and partners for cost and risk sharing	e.g.: improved price-quality transactions of capital goods (from interaction within supply chains)	e.g.: face-to-face exchanges, mobility of local labour, demonstration effect, collaborative exchange of knowledge

Some types and sources of external economies in Table 12.1 may be absent or have a negative influence on innovation in the cluster. At the same time, the relative importance of different types of externalities depends on the industry characteristics and structure, and may change over time. For example, a consequence of oligopolistic competition in mature industries is that (both static and dynamic) pecuniary economies may be more positively perceived by firms than pure knowledge spillovers, which may be regarded as knowledge 'leaks' to competitive rivals (Iammarino and McCann, 2006). Hence, oligopolistic firms in the automotive industry will prefer to formally contract and cooperate with suppliers, customers and employees over knowledge transfers and exchanges in order to prevent their own knowledge from unintentionally spilling over to competitors.

Over time, static advantages may be declining (or become negative) in mature regions that suffer from congestion costs or skilled labour shortages, particularly if there are many non-innovative firms in the region (Beaudry and Breschi, 2003). Static advantages may also be eroding due to increased global pressures that make geographical proximity less relevant for the search for suppliers or markets. Indeed, improvements in transport and communication processes, particularly in a mature industry where large-scale routine flows of transactions may prevail mean that vertical relationships may increasingly become global, as opposed to local. This indicates that positive knowledge spillovers from interactions with suppliers and customers are likely to be weaker where market relations alone prevail. Likewise, collaborative relationships that facilitate pure knowledge and/or rent knowledge spillovers will better support the clustering of innovative activities, than those that mainly provide the sharing of costs and risks of investments and transactions. Therefore, the role of cooperation on cluster innovation is expected to vary according to the nature of its roots in static or dynamic advantages.

12.5 METHODOLOGY AND DATA

We use our framework to investigate the nature and relative importance of LKS compared to static externalities for firms established in the automotive industry of the WM and Saxony. These two regions are both dominated by the automotive industry, supported by cluster policies, and yet have different patterns of local economies, innovation drivers and prospects.

The roots of the WM automotive industry in the UK go back to the end of the nineteenth century. Today, the regeneration of the WM as a mature automotive industry cluster depends on how innovation can be promoted and used as a key source of sustainable comparative advantage. The automotive cluster of the WM, its restructuring and crisis in the context of the collapse of MG Rover and its reactive cluster policies are also discussed in Chapter 8 by Bailey and Kobayashi, and in Chapter 7 by Clark. The car manufacturing industry has existed in Saxony since the beginning of the twentieth century, but after the Second World War it was under the East German government control and was only officially organised following the German reunification (Lichte, 2003). After Volkswagen started producing cars in Mosel in the beginning of the 1990s, Saxony has continued to develop and grow through the establishment of subsidiaries of large, mainly West German Original Equipment Manufacturers (OEMs), and has focused its own efforts to develop networks and sub-networks for a large number of activities. Furthermore, there have been some recent investments by suppliers in the region. In contrast to the WM, subsidiaries of foreign-owned multinational corporations (MNCs) are the suppliers to locally-established, West German and other European OEMs, and are observed to be practising cooperation in production and development with indigenous firms, using and developing their know-how (Lichte, 2003).

Our empirical analysis draws from an in-depth study of the West Midlands in the UK (Janne and Farshchi, 2004), and Saxony in Germany (Lichte, 2003) over the period 2001–2004 as part of the EU-founded WEST-EAST-ID project.[2] Face-to-face interviews were conducted with the top management of 26 and 24 component suppliers, and 10 and 11 local actors in the WM and Saxony, respectively, between April 2003 and March 2004. The use of a uniform protocol facilitated the analysis of the data and offered improved replicability and reliability of data collection across the two regions. Local actors included, for instance, representatives of local chambers of commerce, regional agencies, local banks, sectoral associations. Our data in the tables and discussion refer to the period of the interviews unless otherwise specified. Other secondary sources of evidence were used to increase the construct validity of our study (Yin, 1991).

The WM sample includes 4.8 per cent of all component suppliers in the West Midlands employing 12.5 per cent of the total workforce in the region's component suppliers. The range of firms interviewed relatively over-represents the West Midlands' components suppliers in the ranges of (10–19), (100–499) and (500+) employment categories and under-represents firms in the (1–9) and (20–99) employment categories. The Saxony sample covers 4 per cent of the region's component suppliers. Comparing the average size of firms in terms of employment shows that the sample of firms in Saxony offers a considerably smaller average size than that in the WM (71 against 147 in 2001). The same feature becomes evident considering that the

Saxony sample covers 4 per cent of the district's firms but only 2.7 per cent of its employment.

Most firms in Saxony were set up after the 1990 reunification and there is only one firm which dates back to the beginning of the twentieth century. Most firms in both regions do not have any other branches in the same locality but there are more firms in the WM than in Saxony with at least one other branch in the same locality or nationally. Some of the WM firms interviewed have branches in other countries (eight for WM, none for Saxony). Most firms in both countries are private limited companies (21 in WM, 22 in Saxony) and there are no public limited companies in the Saxony case while there are three in the WM. On the other hand there is one cooperative in the Saxon sample. More firms in Saxony are family-owned but fewer firms choose their top management from the owner family compared with those in the WM. Independent family firms are only observed in the Saxony case (seven firms in Saxony, none in the WM). There are also more affiliates of European firms in the WM than there are in the Saxony sample.

12.6 A COMPARISON OF INNOVATION IN THE WM AND SAXONY

12.6.1 Innovation Activities

The market and industry environments in the two countries are rather different. In the UK, the automotive sector is only second to IT Hardware in the list of the top ten sectors by total R&D internationally (DTI, 2003). However, only 12 per cent of UK automotive R&D has been carried out by UK-owned companies (DTI, 2003). On the other hand, relatively more firms in the Saxony automotive cluster have dedicated budgets to R&D and register patents internationally than firms in the WM automotive cluster (39 per cent in Saxony, 23 per cent in the WM in 2001). Moreover, this is despite the smaller average size of the firms in the case of Saxony (31 for WM, 21 for Saxony). In terms of R&D there are more firms with dedicated R&D budgets in the Saxony case than in the WM (17 vs. 9 in 2001). However, the total budget for R&D investment and average number of R&D employees are larger for the WM firms than those in Saxony (e.g. 7.8 R&D employees on average in the WM against 0.6 in Saxony). This difference may be explained by the fact that the average size of firms in the WM is larger. However, the difference between the overall number of international patents in 2001 is less between the two samples, despite the smaller size of firms interviewed in Saxony.

Table 12.2 compares innovation in the two regions for different types of innovation during a period of up to three year prior to the interviews. Product and process innovations are more prevalent in Saxony compared to the WM, in which 50 per cent or more firms have not declared any major innovation. Most innovations have been carried out within the firm, although innovations sourced at the national level play a significant role in Saxony, which may be explained by the strong presence of German firms in that region. The majority of firms have no major innovations in the organisation of production or the organisation of sales and distribution.

Table 12.3 shows that there is a significant difference in the composition of skilled labour between the two cases. Whereas the WM heavily relies on employees with no qualification or intermediate qualification (>13.2 per cent qualified employees), the Saxony firms enjoy a much more educated workforce (>38.9 per cent qualified employees). Similarly, more entrepreneurs and CEOs in Saxony have gained a university degree compared with those in the WM (15 vs. 8 and 14 vs. 9 respectively).

The WM seems to suffer from a lack of skills and R&D (Donnelly et al., 2005). In 2000, 37 per cent of employers in the WM had provided off-the-job training in the previous year, which is one of the lowest proportions among the English regions (Matheson and Babb, 2002, Table 3.26). According to the same survey, employers in the WM were also the least likely in the UK to have reported a training plan and a training budget. Figures from the WM Regional Innovation Survey of Business (1994–98) showed that up to 30 per cent of firms had no R&D expenditure, and only 40 per cent claimed to have introduced a new product, which is low relative to the UK as well as international standards (Advantage West Midlands, 2003). Many companies in the WM automotive sector depend on traditional low value-added manufacturing and the sourcing decisions of foreign-owned vehicles manufacturers (BEIC, 2002; Donnelly et al., 2005). The region underperforms in terms of productivity and competitiveness when compared to possible competitor regions elsewhere in Europe, such as Stuttgart, Franche-Comté and Piemonte (Rover Task Force, 2000).

12.6.2 Characteristics and Sources of Innovation

Table 12.4 reports the significant sample differences regarding the sources of technical knowledge, using a Mann Whitney U test for non-parametric data. Data is shown to indicate the number of responding firms in each region, as well as the total score for each question. The total score includes all responses from ‘of little importance’ to ‘very important’.

Table 12.2 Characteristics of innovation

Type of innovation	Region	No. major innovations %	Innovation					Total firms %
			In the firm %	In the district %	In the country %	Abroad %	Not applicable/Missing %	
Product	WM	50	34.6	3.8	0	11.5	0	100
	Saxony	12.5	50	0	20.8	16.7	4.2	100
Process	WM	61.5	30.8	3.8	0	0	3.8	100
	Saxony	16.7	45.8	0	25	8.3	4.2	100
Change in organisation of production	WM	53.8	34.6	7.7	0	0	3.8	100
	Saxony	54.2	29.2	0	12.5	0	3.8	100
Change in organisation of sales and distribution	WM	53.8	42.3	0	0	3.8	0	100
	Saxony	58.3	16.7	0	0	8.3	20.8	100

Table 12.3 Distribution of employees by qualification and employment size classes

	West Midlands (%)	Saxony (%)
Employees with university qualifications	9.9	15.1
Employees holding diplomas	3.7	4.5
Qualified employees	13.2	38.9
Employees with intermediate qualification	12.5	3.1
Employees without qualifications	24.1	8.6

Firms in the WM rely relatively more on continuous improvements to the production processes, as well as knowledge from their (foreign-owned) parent/subsidiaries than the Saxony firms, which are more reliant on innovations emerging from R&D expenditures. The most significant difference between the two cases is in the level of knowledge sourcing from interactions with clients and/or suppliers, where the WM firms score higher. However, our interviews demonstrate that these interactions in the WM are dominated by concerns regarding control over quality, prices and delivery, rather than joint innovation efforts. While not shown in the table, it is worth noticing that Saxony firms scored particularly high with regard to the knowledge they derive from interactions with public institutions, such as universities, public research centres and local government agencies (19 firms associated with a total score of 60 compared to 13 firms and 46 for the WM).

Table 12.4 Sources of technical knowledge[3]

Internal Sources of Knowledge		West Midlands		Saxony	
Significance of difference between samples		No. of firms	Score[4]	No. of firms	Score
*	Knowledge gained from continuous improvement of production processes	25	104	23	80
*	Knowledge derived from parent or subsidiary companies	16	59	7	21
Local Sources of Knowledge					
***	Knowledge derived from interactions with clients and/or suppliers	22	96	9	21
**	Knowledge derived from cooperation with other companies	21	80	11	35
*	Knowledge gained from interactions with semi-public institutions (e.g., chambers of commerce; industry associations, trade unions, etc.)	17	58	14	34

Note: Mann Whitney U Test results:
* marginally significant ($P < 0.05$, two-tailed test);
** significantly different ($P < 0.01$, two-tailed test);
*** highly significant ($P < 0.001$, two-tailed test).

Table 12.5 indicates that Saxony firms are generally more involved with other local or national firms, especially in terms of the joint development of new products or processes. These firms have also more joint distribution agreements at the national level than firms interviewed in the WM.

Table 12.5 Nature of inter-firm relationships

With local firms	West Midlands (No. of firms)	Saxony (No. of firms)
Joint distribution in national/international markets	1	2
Joint export of a set of products (e.g. through a consortium)	0	1
Joint development of a new product/process	2	7
Common training programmes (through the auspices of an institution)	1	3
Other	1	2
With national or international firms		
Joint distribution in national/international markets	1	7
Joint export of a set of products (e.g. through a consortium)	0	3
Joint purchasing of raw materials or services (e.g. through a consortium)	1	0
Joint development of a new product/process	2	6
Common training programmes (through the auspices of an institution)	1	1
Other	1	1

Table 12.6 shows the perceived advantages and disadvantages of the region by firms in each sample and the significance of samples' difference. Results confirm that firms in the WM perceive the local advantages as deriving from the more traditional external economies, described in terms of benefits from the local cost and mobility of labour, cost of housing, availability of competent suppliers, local competition and demand (significant differences between the responses in the two samples). Although the local cost of housing and employees' mobility are regarded as advantages

Table 12.6 Locational advantages and disadvantages

Characteristics of the area	Saxony (No. of firms)			West Midlands (No. of firms)				Difference between the WM and Saxony responses (Mann-Whitney U test results)
	Advantage	Disadvantage	Not relevant	Advantage	Disadvantage	Not relevant	Not known	
Availability of competent suppliers	4	1	19	21	2	2	1	***
Competition	3	1	20	19	1	5	1	***
Cost of labour	3	0	21	14	2	9	1	***
Housing cost	1	1	22	8	6	10	1	***
Imitation (between contiguous firms)	18	0	6	7	0	18	2	**
Local demand for the company's products/services	5	0	19	16	2	7	1	**
Leaving employees	2	0	22	8	6	11	1	**
Public support for training	1	0	22	4	5	16	1	*
Availability of land or industrial premises	6	0	18	10	7	7	1	*
Access to technological competence	17	0	7	11	0	14	1	*
Access to capital loans or grants, risk capital and venturing	11	1	12	17	1	7	1	
Availability of effective business services	7	1	16	12	2	11	1	
Availability of buyers or distributors	11	0	13	15	2	8	1	
Availability of qualified employees	2	1	21	1	5	18	2	
Physical infrastructure	9	4	11	9	1	15	1	
Regulatory environment	12	0	12	10	2	13	1	
Propensity to innovate	14	0	10	16	0	9	1	
Opportunity to enact common strategies /actions with similar firm	10	0	14	4	9	12	1	
Other	16	0	8	5	1	19	1	*

Note: Mann Whitney U Test results:
* marginally significant ($P < 0.05$, two-tailed test);
** significantly different ($P < 0.01$, two-tailed test);
*** highly significant ($P < 0.001$, two-tailed test).

compared to Saxony, they are also perceived as disadvantages by many firms in our sample. A similarly balanced view also exists regarding the provision of land and public support for training. The claimed advantage of the cost of labour in the WM can therefore be weighted against the disadvantage of staff turnover (i.e. employees leaving). Indeed, our company and interviews with local actors also hinted at issues of skilled labour shortages in the region. By contrast to the WM, firms in Saxony significantly perceive locational advantages of their region in terms of increased opportunities for imitation and the access to technological competence. Other types of innovation-related externalities, such as those associated with formal and informal networks, are claimed to become increasingly important for regional prosperity in Saxony, which explains the difference in 'other' as a locational factor in the Table (Lichte, 2003).

Table 12.7 Factors limiting innovative activities

	West Midlands	Saxony
	No. of firms	No. of firms
Shortage of specialised staff within the company	15	11
Lack of appropriate incentives and commitment towards innovation within the company	11	2
Nature of product/service (new products/services are not in demand)	12	4
Rapid imitation by neighbour firms	13	3
Shortage of knowledge infrastructures (research centres, science parks, universities, etc…)	8	3
Lack of market information	8	5
Other (e.g. high turnover of skilled workers, inappropriate legal framework, etc.)	2	0

Table 12.7 finally shows the specific factors that are claimed to reduce the innovative capacity of firms in the WM and Saxony. Firms in the WM and Saxony are both constrained as a result of shortages of skilled labour, which is particularly important for innovation in the automotive industry. However, this shortage seems to reflect, on the one hand, rapid growth and expansion in Saxony, whereas it reflects a structural crisis in the WM. Furthermore, a lack of incentives within the firm, rapid imitation by neighbours, and shortages of

infrastructure all contribute to limiting innovative activities in the WM, while they are not perceived to be major issues for firms in Saxony.

12.6.3 Discussion

The traditional supply- and demand-side characteristics of externalities identified for the automotive cluster of the WM indicate that the WM is dominated by externalities of the static Marshallian type. Very little or no evidence is found of advantages in the region that are derived from (rent or pure) knowledge spillovers through collaboration or other exchanges for innovation. While relationships between suppliers and clients are important, they are mostly dominated by control issues over costs, quality and delivery, rather than joint R&D associated with LKS. Technology centres, consulting firms, universities and higher education centres were also perceived by our sample firms to be the least relevant institutions for generating advantages to the region.

This pattern may explain the apparent lack of innovative activities and difficulties for the WM in transforming and regenerating. The benefits from static pecuniary externalities that prevail in the WM are declining not only because some of these have become negative, but also because the region suffers competition from other regions in a global context. As the region has matured, location advantages due to labour and housing costs, the availability of land and industrial premises, are deteriorating as the region endures skilled labour shortages and increasing congestion costs, such as the rising costs of labour, housing, industrial land and traffic congestion. In addition, advantages due to the local proximity of suppliers and clients are gradually diminishing as the foreign-owned final assembly OEMs increasingly develop pan-European or global suppliers networks. The dominance of foreign-owned MNCs in the WM automotive industry may lead to a redirection of high-value activities away from the region and the UK, and towards other locations in Europe, the US and Asia (Donnelly et al., 2005), thereby endangering the future technological capability and competitive position of the region (Farshchi and Janne, 2003).

Endogenous institutional factors of the WM and the UK also help to understand the issue at hand. In particular, the relative decline of the UK in the automotive and other engineering industries has been discussed in relation to deficiencies in the training of engineers, financial and social systems (Freeman and Soete, 1997; Lazonick, 1994; Mowery and Rosenberg, 1989; Patel and Pavitt, 1994). Furthermore, the inability of the British domestic car manufacturers to rationalise and organise the proliferation of models and assembly plants inherited from various mergers since the 1950s has been associated with their decline, especially since the 1970s (see also Clark, Chapter 7, and Bailey and Kobayashi, Chapter 8).

The Saxony case exhibits different characteristics from those of the WM and which reflect more predominant localised knowledge spillovers. Dynamic externalities can indeed be identified for the different key sources of innovation in the automotive industry. In contrast to the WM, close relationships with universities and research institutes and good cooperation with them are also perceived by many firms to be major sources of knowledge. The role of learning for knowledge generation and innovation in Saxony is not only revealed in the presence of highly skilled managers and employees, the significance of continuous improvement and internal training programmes, but also in the perceived strong relevance of local institutions such as universities, service centres (including training associations) and the local chambers of commerce. The German educational and training system has been described as the foundation for the superior skills and productivity of the German workforce in many manufacturing and engineering industries and a strong basis for cumulative learning (Patel and Pavitt, 1994). In addition, Saxony institutions have proactively supported the local 'automotive cluster' (e.g. the network initiative 'AMZ Sachsen 2005') which is claimed to have successfully developed special trust relationships between local firms (Lichte, 2003). For many firms then, innovation plays an important role in the relationship with subcontractors and clients, and technological learning is derived from interaction with suppliers and clients, as well as from other sources such as trade fairs, publications and bought technology and components.

12.7 CONCLUSIONS

This chapter has discussed the different types of externalities behind the clustering of innovative activities. Our analysis highlights the significant role of dynamic externalities, including rent pecuniary and real knowledge spillovers, in the clustering of innovative activities. While different sources and institutional foundations of innovation will depend on the region's sectoral specialisation, structure and spatial concentration, localised knowledge spillovers can provide the innovation capabilities for regions to adapt and renew themselves in the face of technological and market changes. Our survey-based empirical evidence highlights the role of localised knowledge spillovers in explaining the different innovation activities and prospects of the WM and Saxony automotive clusters. Our findings imply that the ability of a mature industrial region to remain innovative and grow depends on the predominant type of externalities supporting innovative activities in the region, and is not predetermined by the mature stage of evolution of its prevailing industry. Sectoral and regional sources of innovation are intertwined, and any determinant of sectoral innovation interacts and has to be supported by the relevant regional environment in order to sustain regional innovation and growth. As we have seen, the propensity to innovate in the automotive industry is associated with the capability for firms to develop internal engineering, design and R&D activities, and communication and interaction capabilities in value chains. We show that of the two mature automotive clusters, innovation externalities in Saxony seem to be more dynamic and promising for the development of the region compared to those of the WM. The reason is that whereas the Saxony cluster appears to benefit more from knowledge spillovers, the WM cluster appears to have been based rather more on the traditional external economies derived from local labour pooling, local services and the proximity of demand. As these (primarily) pecuniary externalities are eroding and congestion costs rising, little evidence is found of renewed agglomeration benefits associated with localised knowledge spillovers in the WM. The difficulty of the WM automotive sector to restructure itself therefore appears to be explained rather more by its reliance on an established path of routines and institutions than because of industry maturity. As such, the lack of knowledge spillovers nowadays appears to define the weaknesses of the WM automotive cluster in a time of change. On the other hand, local firms in the Saxony automotive cluster claim that the presence of local knowledge exchanges and regional networking opportunities are a key source of competitive advantage for that region. Moreover, many of the Saxony institutions are actively involved in promoting such local networking.

NOTES

1. While a cluster broadly refers to a set of firms and/or activities that exist in the same local environment, a region is associated with more specific geographical (often administrative) boundaries. Here, we focus on the factors driving the clustering of innovative activity in an industry, and not the definition and broader phenomenon of clusters or regions.
2. The data and information used in this chapter were collected as part of an EU-funded project entitled 'WEST-EAST-ID. Industrial Districts' Relocation Processes: Identifying Policies in the Perspective of EU Enlargement' (Contract no. HPSE-CT2001-00098). For Saxony it draws from the work by our colleagues in Germany, which is summarised in the project reports. We are nevertheless responsible for the use and interpretation of data and reports, any errors or omissions.
3. Technical knowledge is defined as the set of notions, skills and abilities related to the industrial process or product.
4. Score refers to the sum of values given to the variable by respondents. These include 5 = Very important; 4 = Important; 3 = Moderately important; 2 = Of little importance.

REFERENCES

Abernathy, W. J. and Clark, K. B. (1985), 'Innovation: Mapping the winds of creative destruction', *Research Policy*, **14**, 3–22.

Advantage West Midlands, (2003), 'West Midlands. Regional Innovation Strategy and Action Plan, shaping our future,' Birmingham: Advantage West Midlands, www.wmda.co.uk/work/strategies.

Archibugi, D. (2001), 'Pavitt's Taxonomy sixteen years on: A review article', *Economics of Innovation and New Technology*, **10**, 415–25.

Asheim, B. and Coenen, L. (2005), 'Knowledge bases and regional innovation systems: Comparing Nordic clusters', *Research Policy*, **34**, 1173–90.

Audretsch, D. B. and Feldman, M. P. (1996), 'Innovative clusters and the industry life cycle', *Review of Industrial Organization*, **11**, 253–73.

Baptista, R. and Swann, G. M. P. (1998), 'Do firms in clusters innovate more?', *Research Policy*, **27**, 525–40.

Beaudry, C. and Breschi, S. (2003), 'Are firms in clusters really more innovative?', *Economics of Innovation and New Technology*, **12** (4), 325–42.

Bergman, E. M. and Feser, E. J. (2000), 'National industry cluster templates: A framework for applied regional cluster analysis', *Regional Studies*, **34**, 1–19.

Bessant, J. and Haywood, B. (1991), 'Mechatronics and the Machinery Industry', in C. Freeman, M. L. Sharp and W. Walker (eds), *Technology and the Future of Europe: Global Competition and the Environment in the 1990s*, London: Frances Pinter.

Birmingham Economic Information Centre (BEIC) (2002), 'Birmingham and the Sub-Regional Economy: Technical Report', Economic Development Department, Birmingham City Council.

Breschi, S. and Lissoni, F. (2001a), 'Localised knowledge spillovers versus innovative milieux: Knowledge 'tacitness' reconsidered', *Papers in Regional Science*, **80**, 255–73.

Breschi, S. and Lissoni, F. (2001b), 'Knowledge spillovers and local innovation systems: A critical survey', *Industrial and Corporate Change*, **10** (4), 975–1005.

Caniels, M. C. J. and Romijn, H. A. (2003), 'Firm-level knowledge accumulation and regional dynamics', *Industrial and Corporate Change*, **12** (6), 1253–78.

Caniels, M. C. J. and Romijn, H. A. (2005), 'What drives innovativeness in industrial clusters? Transcending the debate', *Cambridge Journal of Economics*, **29**, 497–515.

Capello, R. and Faggian, A. (2005), 'Collective learning and relational capital in local innovation processes', *Regional Studies*, **39** (1), 75–87.

Department of Trade and Industry (DTI) (2003), Innovation.co.uk: http://www.innovation.gov.uk/

Donnelly, T., Barnes, S. and Morris, D. (2005), 'Restructuring the automotive industry in the English West Midlands', *Local Economy*, **20** (3), 249–65.

Doring, T. and Schnellenbach, J. (2006), 'What do we know about geographical knowledge spillovers and regional growth?: A survey of the literature', *Regional Studies*, **40**, 375–95.

Drejer, I. (1999), 'Comparing patterns of industrial interdependence in national systems of Innovation – a study of Germany, the UK, Japan and the US', *Economic Systems Research*, **12** (3), 377–99.

Farshchi, M. A. and Janne, O. E. M. (2003), 'Industrial Transformation in the West Midlands Automotive Cluster: International Networks and Cooperative

Behaviour', paper presented at the 29th EIBA Conference, 11–13 December, Copenhagen Business School, Denmark.

Fine, C. H. and Raff, D. M. G. (2002), 'Automobile', in B. Steil, D. G. Victor and R. R. Nelson (eds), *Technological Innovation and Economic Performance*, Princeton University Press.

Freeman, C. and Soete, L. L. G. (1997), *The Economics of Industrial Innovation*, third edition, London: Pinter Publishers.

Gordon, I. R. and McCann, P. (2000), 'Industrial clusters: Complexes, agglomeration and/or social networks?', *Urban Studies*, **37** (3), 513–32.

Griliches, Z. (1979), 'Issues in assessing the contribution of research and development to productivity growth', *Bell Journal of Economics*, **10**, 92–116.

Iammarino, S. and McCann P. (2006), 'The structure and evolution of industrial clusters: Transactions, technology and knowledge spillovers', *Research Policy*, **35**, 1018–36.

Janne, O. E. M. and Farshchi, M. A. (2004), 'Case study report: The West Midlands Automotive ID', as part of the EU-funded project entitled 'WEST-EAST-ID (WEID) "Industrial Districts" Relocation Processes: Identifying Policies in the Perspective of EU Enlargement', 2001–04 (Contract no. HPSE-CT2001-00098).

Krugman, P. (1991), *Geography and Trade*, Cambridge MA: MIT Press.

Lazonick, W. (1994), 'Social Organization and Technological Leadership', in W. J. Baumol, R. R. Nelson and E. N. Wolff (eds), *Convergence of Productivity: Cross-National Studies and Historical Evidence*, Oxford: Oxford University Press.

Lichte, R. (2003), 'Case study report: The automotive suppliers' network Saxony 2005 in the Freistaat Sachsen/Germany', as part of the EU-funded project entitled 'WEST-EAST-ID (WEID) "Industrial Districts" Relocation Processes: Identifying Policies in the Perspective of EU Enlargement', 2001–04 (Contract no. HPSE-CT2001-00098).

Martin, R., and Sunley, P. (1996), 'Paul Krugman's geographical economics and its implications for regional development theory: A critical assessment', *Economic Geography*, **72** (3), 259–92.

Martin, R. and Sunley, P. (2003), 'Deconstructing clusters: Chaotic concept or policy panacea?', *Journal of Economic Geography*, **3**, 5–35.

Matheson, J. and Babb, P. (eds) (2002), 'Social Trends', n.32, National Statistics, UK, Table 3.26.

McCann, P. (1995), 'Rethinking the economics of location and agglomeration', *Urban Studies*, **32**, 563–77.

Moodysson, J., Coenen, L. and Asheim, B. (2006), 'Explaining Socio-Spatial Patterns of Innovation: Analytical and Synthetic Modes of Knowledge Creation in the Life Science Industry', paper presented at the Fifth Proximity Congress, June 28–30, Bordeaux, France.

Mowery, D. C. and Rosenberg, N. (1989), *Technology and the Pursuit of Economic Growth*, Cambridge: Cambridge University Press.

Nelson, R. R. (1996), 'National Innovation Systems: A Retrospective on a Study', in G. Dosi and F. Malerba (eds), *Organization and Strategy in the Evolution of the Enterprise*, Macmillan Press Ltd.

Patel, P. and Pavitt, K. L. R. (1994), 'National innovation systems: Why they are important, and how they might be measured and compared', *Economics of Innovation and New Technology*, **3**, 77–95.

Pavitt, K. L. R. (1984), 'Sectoral patterns of technological change: Towards a taxonomy and theory', *Research Policy*, **13**, 343–74.

Rover Task Force (2000), 'Technical Report To the Secretary of State for Trade and Industry', prepared by Pieda Consulting on behalf of the Rover Task Force Executive, Birmingham, Advantage West Midlands, www.rovertaskforce.co.uk.

Schamp, E. W., Rentmeister, B. and Lo, V. (2004), 'Dimensions of proximity in knowledge-based networks: The cases of investment banking and automobile design', *European Planning Studies*, **12** (5), 607–24.

Swann, G. M. P. (1998), 'Towards a Model of Clustering in High-Technology Industries', in G. M. P. Swann, M. Prevezer and D. Stout (eds), *The Dynamics of Industrial Clustering*, Oxford: Oxford University Press.

Utterback, J. and Abernathy, W. (1975), 'A dynamic model of process and product innovation', Omega, *The International Journal of Management Science*, **3** (6), 639–56.

von Tunzlemann, N. and Acha, V. (2005), 'Innovation in "Low-tech" Industries' in J. Fagerberg, D. C. Mowery and R. R. Nelson (eds), *The Oxford Handbook of Innovation*, Oxford: Oxford University Press, pp. 407–32.

Yin, R. (1991), *Case Study Research: Design and Methods*, revised edition, Newbury Park: Sage Publications.

PART III

REGIONAL POLICY AND MATURE INDUSTRIAL REGIONS

13. A Cost-Benefit Approach to the Assessment of Regional Policy

J. Kim Swales

13.1 INTRODUCTION

This chapter explores the implications of thinking about regional economic problems as being the result of some form of market failure. The main advantage of adopting this approach is that the extensive literature concerning welfare economics can be used as a lens through which to view regional policy issues. Further, the applied arm of welfare economics, cost-benefit analysis (CBA), can be employed as a framework for the assessment of regional policy.

The chapter proceeds in the following way. Section 13.2 discusses the nature of regional economic problems. Section 13.3 investigates how these regional economic problems should be tackled. Section 13.4 outlines the application of cost-benefit analysis to regional problems. Section 13.5 gives the role of shadow prices in such an approach. Section 13.6 is a schematic application of the cost-benefit approach to determining the optimal level of discretionary regional aid. Section 13.7 gives some problems for the use of cost benefit, particularly the problem of 'gaming' the system. Section 13.8 provides some conclusions.

13.2 WHAT IS THE NATURE OF REGIONAL PROBLEMS?

The approach taken in this chapter is to treat regional problems from the perspective of welfare economics. I begin with the two fundamental theorems of welfare economics. These are as follows (Mas-Coll et al., 1995):

- All perfectly competitive economic outcomes are Pareto optimal.

- In a perfectly competitive economy all feasible Pareto optimal outcomes can be achieved through an initial redistribution of resources.[1]

These fundamental theorems immediately identify the two key elements of regional policy concerns. The first is that the existing spatial operation of the economy might be inefficient, in the sense that Pareto improvements are possible. The second is that the present outcome, whilst efficient, might be unsatisfactory in terms of the distribution of benefits, in particular their distribution across space. In terms of the distribution of benefits, it is conventional to represent the trade-off between individuals using a Social Welfare Function notionally employed by the government (Bergson, 1938).

In practice regional policy concerns are likely to be made up of a combination of these two. However, a key point is that either of these problems can be identified as a market failure in that the unaided market is failing to reach the position that maximises the government's Social Welfare Function. To begin, each potential source of market failure will be considered in more detail.

13.2.1 Perfect Competition Generates Pareto Optimality

One possible source of regional problems is that the economy does not operate efficiently across space. The notion that, under certain conditions, a perfectly competitive economy will reach Pareto optimality is fine but the perfectly competitive ideal is one that can never be fully achieved. In so far as the economy will otherwise deviate from the perfectly competitive paradigm, government intervention might be justified.

One major problem is externalities, or what are sometimes called spillovers. This is where the actions of one agent have positive or negative impacts on other agents that are not regulated through markets. Essentially the issue is one of missing markets, either because the appropriate property rights are not in place or because transactions costs are too high. Relevant examples here are agglomeration economies and congestion costs.

We can start with an example of agglomeration economies. An expansion in a particular sector in a specific region might improve the labour market for all firms operating in that sector in that location. A thicker labour market might increase skill levels, reducing recruitment costs and improving job matching. In these circumstances, an expansion in activity by one firm would have labour market benefits to all firms in that sector and location. However, the firm making the initial increase in activity receives no payment for these benefits and therefore the level of activity would be inefficiently low. The marginal social benefit in this case is higher than the marginal private benefits. If the provision is driven only by private gain, then this provision will be too low.

With congestion costs the argument operates in the opposite direction. An expansion in one activity in a particular location will increase the costs of travel and transporting goods borne by other individuals and organisations. In this case the marginal social cost is greater than the marginal private costs and there will be over-expansion. The level of activity will be, *ceteris paribus*, too high.

A second problem is the exercise of market power or some other form of market imperfection. The classic example here concerns monopolistic or oligopolistic situations in the product market, so that price is greater than marginal cost and output is sub-optimally low. However, in regional analysis imperfections in factor markets appear more relevant. First, the regional problem is often perceived as a problem of unemployment or low activity. This suggests some problem in the operation of the local labour market. Regional subsidies are also often given on investment, suggesting failure in the capital market. These issues are dealt with in greater detail in Section 13.5.

A third general problem concerns public goods. These are goods and services that are non-excludable and non-rival. This means that their provision by one agent generates benefits that automatically accrue to a number of other agents. These other agents cannot be excluded from enjoying the benefits of the public good and their enjoyment of the benefits does not directly affect the benefit that any other agent receives from the good or service.[2]

Examples of public goods are defence, street lighting and clean air. The nature of these goods and services are such that they generally have to be supplied by the public sector and many will be local public goods which primarily benefit individuals living in, and firms operating in, a particular geographical location. For Pareto optimality it is important to be able to value the benefits to individuals and deliver the public goods at minimum cost. It could be that regional problems at least partly reflect sub-optimal decisions concerning the provision of these goods.

The fourth major problem for the achievement of Pareto optimality is the future. Many decisions – essentially decisions concerning the investment in physical and human capital – require knowledge of, or conjectures concerning, costs and prices in the future. In the perfectly competitive ideal, the market supplies this information. However, nothing like a complete set of futures markets for goods, services and factors of production exists.

One problem is that exogenous events, such as the weather, or partly endogenous events, such as technological developments, cannot be predicted perfectly. Given that any efficient market system would need to take into account these events once they had occurred, an efficient set of futures markets would have to include contingent elements. However, agents cannot even list all possible future circumstances, and even if they could, drawing up appropriate contracts would be prohibitively costly. Moreover, these difficulties have to be overlaid with the asymmetric information problems.

Missing futures markets, when combined with any form of externality, generate coordination problems. The profitability of an investment in one sector in a given region might depend on the existence of a low-cost (high volume) local source of supply for a key input. However, the investment in the capacity to generate this low-cost input, plus the demand to fully utilise that plant and therefore realise the economies of scale in production, will depend on decisions made by the supplying firm and other firms that use its product. These decisions are not coordinated through the market and initial cost advantages achieved in particular locations might generate polarised growth.

13.2.2 Redistribution plus Perfect Competition can Maximise Social Welfare

This second theorem of welfare economics is perhaps even more troublesome than the first. The operation of a completely unrestricted perfectly competitive system – abstracting from the problems raised in the previous subsection – will generate a Pareto efficient outcome. However, this outcome might have very undesirable distributional characteristics. But the second theorem states that an *ex ante* lump sum redistribution can generate any feasible Pareto optimal position. This suggests that a combination of perfectly competitive markets, plus the appropriate lump sum redistribution, can get the economy to the point where social welfare is maximised.

One major problem is that it is difficult in practice to disentangle the efficiency and redistribution functions. For example, individuals have different capacities to earn income. If we want an egalitarian outcome, workers with a low earning capacity would get favourable initial income or resource transfers. But how do we determine each individual's earning potential? The best measure is likely to be his or her actual earnings. But if this is the basis for the redistribution, moral hazard difficulties arise: an individual's incentive to maximise earned income (through labour or capital accumulation) is blunted if there is egalitarian redistribution based on *ex post* earnings.[3] This is not to say that redistribution should not take place on this basis, but simply that there are likely to be efficiency implications.

A second issue concerns the specific redistributional characteristics of regional policy. One critique is that in the standard welfare economics approach, welfare is linked to the preferences and utility of individuals.

Further, the redistributional function of the government is again typically seen as one in which the inequality amongst individuals is reduced. However, in justifying regional policy, an appeal is often made to territorial equity. This implies that there should be a reduction in inequality across regions. Examples would be that the interregional unemployment rate or per capita GDP differential should be reduced. An immediate issue here is that reducing territorial inequality might not reduce individual inequality. Another way of putting this is to pose the question: should policy be orientated towards place or person prosperity?

Two comments are appropriate here. First, in so far as the redistributional issues relate to the provision of local public goods, the most appropriate way to deal with this will be through spatial policy of some sort. For example, low incomes might be related to inappropriate skills. Clearly, retraining provided locally is likely to be an effective policy response. In this case, reducing the inequality across space also reduces the inequality between individuals, and regional policy is an effective tool to bring about this redistribution. However, this is not always the case.

Second, political power and influence often has a natural spatial character as politicians typically represent geographically defined constituencies. These representatives will press the interests of the particular geographical areas that they represent. Therefore, actual redistribution across space might occur simply as a result of political economy considerations (Persson and Tabellini, 2000, Chapter 6.3).

13.3 HOW SHOULD POLICY ADDRESS THESE PROBLEMS?

Imagine a country with two regions, 1 and 2, where the physical and economic characteristics of the two regions might or might not differ. Within each region the inhabitants are identical, although again they might differ across regions. The utility of an individual inhabitant of region i is given as u_i and combinations of the individuals' utilities in the two regions, u_1, u_2, are plotted in Figure 13.1. The initial position is U_0. The efficiency frontier, which indicates all the feasible Pareto Optimal outcomes, is given by *EF*. SW_0 is the iso-Social Welfare curve that passes through U_0. This shows all the combinations of u_1, u_2 that give the same level of social welfare. The optimal feasible social welfare is given at U_{max} where SW_{max} is tangent to the efficiency frontier. In this approach, the aim of policy is to get the economy from U_0 to U_{max}.

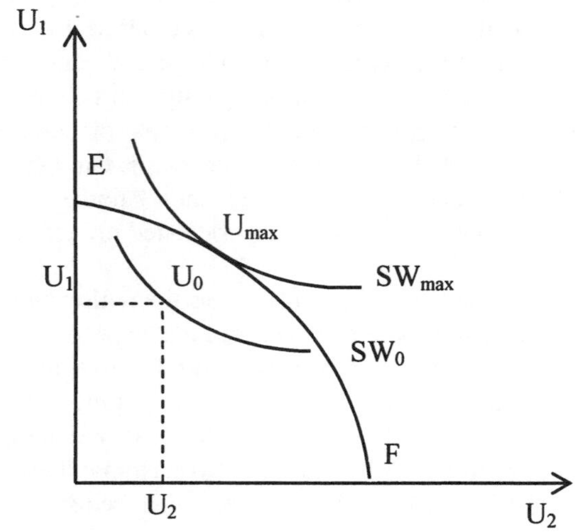

Figure 13.1 Social welfare allocation

The economy is shown initially as operating within the efficiency frontier: that is to say, Pareto improvements are available. It is important to state that some of the inefficiencies might lie outside of the scope of regional policy. Some might be the result of nation-wide market failure, such as the monopoly power in national product markets or the inappropriate provision of national public goods. For our purposes, it is therefore appropriate for regional policy to identify the constrained feasible efficiency frontier as E_1F_1. This falls within the unconstrained feasible efficiency frontier, *EF*: the constraint is defined by the existing national market failures.

When the constrained efficiency frontier is constructed it could be that the initial position lies on this frontier. This is shown as U_2 in Figure 13.2. If this is the case, then the aim of regional policy would be to attempt to move the economy from U_2 to U_1. In this example, regional policy would serve a purely redistributive role: the utility of individuals in region 1 falls and the utility of those in region 2 rises. Alternatively, the economy might initially lie within the constrained efficiency frontier, such as U_3 in Figure 13.2. In this case, regional policy can raise the utility in both regions, which would be the outcome if the optimal feasible outcome is reached, U_1.

The exposition up to now has been purely conceptual. If a benign government knows exactly where the economy is operating relative to the feasible efficiency frontier and its own Social Welfare Function, it could

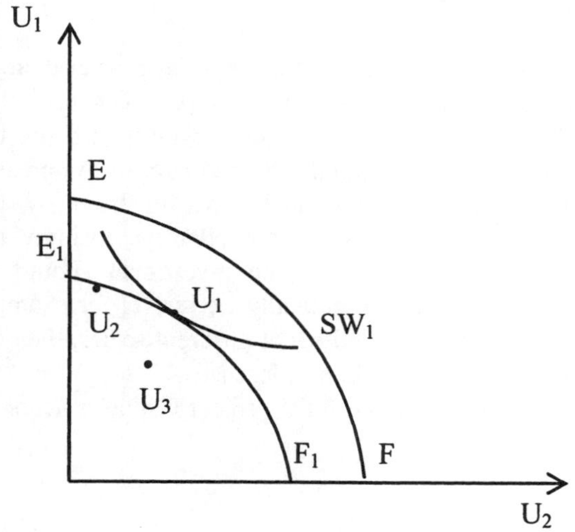

Figure 13.2 Social welfare and regional policy

simply redirect the scarce resources within the economy to alternative uses, in order to reach the social optimum. However, even the most efficient and intrusive government lacks the information to carry out such a policy. There are two basic problems: information and incentives.

These informational and incentive problems are always tackled primarily in a decentralised manner and, for developed economies, with extensive resort to the market. Competitive markets are able to process the disparate information from producers and consumers concerning the relative costs and demands for particular goods and services. Markets therefore act to coordinate these diverse information sources (Hayek, 1937). Perhaps more importantly, market mechanisms provide the incentives for individuals to truthfully reveal private information concerning their production capabilities and consumption preferences (Stiglitz, 1996).

However, we know that the existing market operations and provision of public goods have brought about a sub-optimal outcome. In an attempt to operationalise the policy implications of welfare analysis, cost-benefit analysis maintains the market framework but uses it to reach alternate ends. It does so primarily by replacing the actual prices determined within the market by shadow prices. Decentralised decision making using these shadow prices can, in principal, lead the economy to the social optimal position.

13.4 COST-BENEFIT ANALYSIS

Cost-benefit analysis can be used as an *ex ante* appraisal or *ex post* evaluation tool. We begin with *ex ante* appraisal where a private company is considering an investment decision. The standard approach is to calculate the net present value (NPV) of this investment, which is the discounted sum of the present and future revenues minus similarly discounted costs incurred by the company from undertaking this investment. If the NPV is positive, the investment should go ahead; if negative then the investment should be rejected. Also, if there are projects that are mutually exclusive, for example such as a particular project that can be carried out at different scales, then the investment that gives the highest NPV should be chosen.

The private investment net present value for a project lasting *n* years, is calculated as:

$$NPV_{PI} = \sum_{t=0}^{n} \frac{Q_t p_t - F_t w_t}{(1+i)^t} \tag{13.1}$$

where Q and F represent vectors of output and inputs, p and w are the corresponding vectors of output and input prices, i is the interest rate, and the PI and t subscripts stand for private investment and time respectively. Note that only costs and benefits that go through markets enter into the decision process, and these costs and benefits are valued at their market prices.[4]

If there are no market failures and the initial distribution of resources amongst individuals is optimal, then this procedure is optimal. The price of outputs represents the marginal benefit to the individual consumer and the cost of an input accurately measures the opportunity cost of its marginal unit (that is, the value of other commodities or services given up in the use of the last unit of the input). For labour inputs, the opportunity cost is ultimately the leisure given up to supply additional labour.

However, from the discussion in Section 13.2, we know that market failures occur. Actual markets often don't match up to the perfectly competitive ideal and some key markets simply don't exist. Also the initial distribution of resources is likely to be sub-optimal. Cost-benefit analysis attempts to adjust the disaggregated decision-making process to correct for these deficiencies. The cost-benefit approach to the same decision given by expression (13.1) is represented as:

$$NPV_{CBA} = \sum_{t=0}^{n} \frac{Q_t p_t^S - F_t w_t^S}{(1+r)^t} \tag{13.2}$$

where p^S and w^S are the vectors of the 'shadow' prices for inputs and outputs, r is the discount rate and the CBA subscript indicates a cost-benefit calculation.

The first point to make is that it would be optimal if private companies could use shadow prices, rather than market prices, in making all their output and resource use decisions. The most straightforward way to achieve this is to adjust market prices to shadow prices through appropriate taxes and subsidies and, where possible, to generate markets for presently non-marketed goods and services. For example, in the past the UK government has subsidised all labour and capital inputs in manufacturing in Development Areas (through the Regional Employment Premium and Regional Development Grants). In this framework, when private sector firms use these inputs, the subsidies in effect mean that the market price has been replaced with a shadow price. Similarly, in the environmental field, the creation of tradeable permits for carbon emissions produces a market for an environmental cost, air pollution, which would previously not have been incorporated into the financial decision rules of the firm.

Where output and resource use decisions are made by the public sector, again CBA should be used. If the public sector is supplying a public good, the demand for the good, and therefore its marginal benefit, is derived by summing the individual demand curves vertically rather than horizontally, as with private goods (Samuelson, 1954). In this case, the major problem is likely to be deriving a credible measure of these individual demand schedules. An example would be in decisions concerning infrastructure provision, such as roads. However, the use of shadow prices in the valuation of other outputs and inputs should also be used.

Of course, regional policy might be carried out through other means. Examples would include physical controls, such as zoning, or the provision of training and business services that are supplied free but are quantity rationed. In these cases, CBA could still be used as an *ex ante* appraisal or *ex post* evaluation mechanism. For *ex post* evaluation, a formulation identical to expression (13.2) would be used, except actual, rather than expected, values for the outputs and inputs would be employed. The same decision criteria should be adopted: a positive net present value indicates a policy that has improved efficiency.

The major problem is how to operationalise the CBA approach. The actual physical effects that policy produces and the resources that it uses are clearly key inputs: these are the output and input vectors Q and F in expressions (13.1) and (13.2). The appropriate identification of these vectors is not discussed in this chapter as it is common to any appraisal or evaluation procedure. However, this is not to minimise its significance: without good estimates of the expected or actual impacts of policy, an evaluation or appraisal procedure cannot get to square one.

For CBA, the key issue is how the appropriate shadow prices are to be determined where the market price is thought to be inappropriate or where there are missing markets. It is important to remember we start from the position that the government is, for some reason, unhappy with the present outcome, and wants to alter or distort this. However, the government needs to be aware of the resource, technology and informational constraints embedded in the economy. The problem will be discussed in very general terms initially, and then more specific issues will be dealt with.

We can tackle the problem of the appropriate shadow prices through a top down or a bottom up approach. The top down method is to explicitly model the regional economy, incorporating the relevant market failures. Then by experimenting with the model, using appropriate tax and subsidy price wedges, the feasible outcome for the economy as a whole that maximises the social welfare function can be found. A central problem here is that the modeller needs to know a lot about the operation of the economy, and the decentralised nature of the decision making that is integral to generating the market equilibrium is lost.[5]

The alternative is to take a bottom up approach. Here a partial equilibrium micro-economic perspective is adopted to identify the appropriate opportunity cost of using resources and the marginal social benefit of producing output. These costs and benefits need to be calibrated against a money scale and then used as the shadow prices in a cost-benefit calculation. In practice, a bottom up approach is almost invariably used, though the shadow prices must be estimated to begin with. This is because they will, in general, be affected by system-wide factors, but the shadow prices can be adjusted through a trial and error procedure if the desired outcome is not initially achieved. Ideally a combination of bottom up and top down procedures should be used to calculate the optimal shadow prices.

13.5 SHADOW PRICES FOR REGIONAL POLICY

To use CBA in the evaluation of regional policy, the central issue is to identify which market failures regional policy is attempting to correct. The most obvious candidate would be spatial inefficiency or rigidity in the labour market. Disparities in unemployment or labour market participation rates are often taken to be indicators of regional disequilibrium and the need for policy intervention. A second concern is the capital market. Although the capital market is often characterised as being perfectly competitive, problems of asymmetric information are likely to occur across space. Also, regional aid to companies is frequently given as some form of investment grant, suggesting some inefficiency coming through the capital market. A third important market failure concerns environmental and congestion costs. The effects of some of these will be primarily local and they are appropriately dealt with as local 'public bads'. As such this requires shadow prices that vary according to location. Fourth, there might be spillovers of a technological kind that have some distance decay element, so that they will benefit primarily the productivity of other local firms. Fifth, income going to inhabitants of some geographical parts of the country might be valued more highly from a distributional perspective. The issue of territorial equity arises here. Finally, there is the question of the shadow price for taxation. This does not have a specifically spatial character but might vary with the form of aid typically given for regional development.

13.5.1 Shadow Wages

Traditional input–output or Keynesian demand-driven multiplier analysis is frequently used to calculate the employment effects of spatially targeted policy on the recipient region (Armstrong and Taylor, 2000; Miller and Blair, 1985). This is especially the case where the recipient area is within a problem region. The implication is that there is demand deficient (involuntary) unemployment at the regional level: that there are workers who are presently unemployed or otherwise outside the labour market, for example such as those who are registered as sick and who would like to work at the present wage levels, but who cannot get a job. This means that their shadow wage – the value of the leisure foregone if employed – is less than the wage. It is this shadow wage that should be included as a cost of labour in the appraisal and evaluation of regional policy.

The first question of course is why should it be that the local labour market does not clear. However, there is no shortage of theories. Three will be sketched out here but there are numerous others. First, the effective impact of national policies pursued for other reasons, such as minimum wage or unemployment support programmes, might operate differentially across regions. Second, wages might be set by national agreements or at the national level by multi-regional organisations at rates that fail to clear some local labour markets. Third, imperfect competition might generate wage rigidity in local labour markets.

The fact that demand-deficient unemployment can be explained by a variety of factors does not mean that it is not important for policy evaluation to attempt to identify which are the more significant causes. For example, in the UK, regional variations in unemployment and participation rates for unskilled workers are greater than for skilled workers (Gibbons et al., 2005). It is argued that in the UK interregional migration is more restricted for unskilled workers. If this is the case, there might be grounds for putting a shadow wage on unskilled work but not skilled. Other disaggregations of the labour force, for example by age and sex, might also be important.

It might also be the case that some other market failures, whilst not occurring in the labour market, can be most conveniently linked to changes in employment, and therefore incorporated in the shadow wage. A straightforward one would be congestion and environmental costs. If employment is linked with commuting, and commuting with differential congestion in different parts of the country, then a convenient way to treat this is through adjusting the shadow wage. Similarly, if population is linked to employment (through migration for example), then environmental effects that increase with population could be reflected in an appropriate shadow wage. Note that these sorts of considerations suggest that the shadow wage might be greater than the market wage in congested regions.

It is argued (Gibbons et al., 2005) that there are also public bads that are linked to unemployment, in particular crime, but also other lifestyle characteristics, such as health. If this is so, then the cost of employment is less than the cost to the individual worker, because there are benefits to other members of society when a person becomes employed. Again some of these impacts, such as changes in the level of crime, have a local impact and should be reflected in the allocation of economic activity across space.

Another issue that might effectively be incorporated into adjustments in the labour market concerns income distribution. Imagine that the government values increases in income in some regions more than others. Given that the spatial flows of other forms of income are difficult to track, then this might be practically dealt with by incorporating it into the shadow wage. Again, regions where income is to be favoured would have, *ceteris paribus*, a lower shadow wage, whereas regions whose income is to be less highly valued would have a shadow wage above the market wage. In the UK, the latest guide for government appraisal and evaluation gives explicit recommendations as to how changes in income across different ranges should be valued, with monetary benefits to low income families being valued more highly than to high income families. Such guidelines could be incorporated into regional shadow wage calculations (HM Treasury, 2003).

Employment is a variable that at the regional level is relatively straightforward to measure (as against value added, for example). Moreover, as we have seen above, a number of possible important sources of market failure can be linked to employment and therefore reflected in the shadow wage. This maybe accounts for the popularity of the exchequer cost per job criteria that has been extensively used in the UK as both an *ex ante* appraisal and *ex post* evaluation tool. We will return to cost per job later, but clearly the discussion up to now focuses on a key practical problem: at what level should the shadow wage be set?

13.5.2 Shadow Price of Capital

In the CBA literature one important question is the appropriate rate at which to discount future costs and benefits. This is the parameter r in equation (13.2). If the rate of interest determined in the market does not equal the appropriate marginal rate of time discount, then a shadow discount rate should replace the market one. This is a very important issue when projects have extremely long life spans, particularly if there are sizeable costs and benefits at the end points. An example would be the evaluation of a nuclear power station, where there are high decommissioning costs at the end of the plant's working life and then a long period of storage for the radioactive material.

One issue is that investments in the future involve risk and individuals are typically more risk averse than the state. This is an argument for subsidising investments in both physical and human capital. The same argument makes it appropriate to use a shadow discount rate for evaluating projects using CBA. However, is there an argument for using a shadow price for capital (or the time discount rate) that varies across regions?

There are at least three arguments that could be used here. The first is that although there is a literature that sees financial markets as operating perfectly across space, there is another stressing the asymmetric information problems associated with lending and borrowing money (Stiglitz and Greenwald, 2003). This literature implies that financial markets will never be anything close to perfectly competitive and there are many reasons why these imperfections might vary across space and be particularly acute in problem regions.

The second links capital investment to technological (non-pecuniary) externalities that have a spatial decay. In this argument, the existence of one plant in a region will affect the efficiency with which other plants in the same region will operate. For example, in the past, UK policy towards foreign direct investment (FDI) has been justified at least partly on these grounds. Domestic plants having to interact as customers or suppliers with technologically or organisationally more efficient foreign-owned plants were thought to see an improvement in their own efficiency.

The third concerns the notion that there are missing futures markets which makes optimal investment decisions uncertain. For regional policy, a particular problem relates to the notion of pecuniary externalities. In this case, the local provision of a good or service produced under conditions of economies of scale but with high transport or transactions costs over space, will reduce the costs, and/or increase the real incomes, of individuals in the local area. Whilst these pecuniary externalities are captured by the market process, their optimal exploitation might require corresponding investments by other local firms and individuals. However, the futures markets that are supposed to coordinate these activities in the standard welfare economics analysis typically do not exist. There is therefore a potential for coordination failure which regional policy can address. It is this type of argument, combined with the presence of non-pecuniary externalities, that underlies the analysis of Marshallian industrial districts and Porter's (1990) cluster analysis.

Finally, for a shadow price on capital, as with a shadow price for labour, environmental externalities or income distribution issues might be linked to investment, and these might affect the appropriate cost of capital.

13.5.3 The Shadow Price of Taxation

Where regional development projects require public funding, that public funding needs to be financed ultimately through taxation. From a welfare economics perspective, increased taxation, *ceteris paribus*, reduces welfare. There are two reasons for this. First, resources are required to collect, process and administer tax collection. These not only cover the costs born by the state but also the compliance costs imposed on private companies and individuals. Also, when evaluating subsidy programmes the additional administration and compliance costs of the subsidy programmes should be included. Second, taxation drives a wedge between the marginal cost of producing a good or service and the marginal benefit to the consumer. For example, the introduction of an income tax distorts the decision at the margin between work and leisure and between purchasing services and providing for them within the household. Where appropriate, these costs of distortion should be incorporated into a cost-benefit evaluation.

One note of caution should be made here. Where regional development requires the provision of public goods that are necessarily provided by the state and funded communally, then the cost of taxation is as outlined in the paragraph above. However, if the aim of regional policy is to correct the distortions generated by market failure, then there is no reason for the total tax take to rise. Policy could be carried out through changes in tax rates that would be neutral overall. For example, a labour subsidy in problem regions could be brought about by a reduction in existing taxes on labour paid by employers in problem regions, accompanied by an increase in the same tax for employers in more prosperous regions. The 'distortions' created by the tax system are exactly what the policy maker wants in this case, because the operation of the free market is perceived as being sub-optimal.

Where large projects need appraising, such as the regional development implications of a successful bid for staging the Olympic Games, a very detailed CBA is appropriate. However, the real strength of the CBA approach should be that it can be used to make many relatively small decisions in a decentralised manner. To do this it is important that key rules of thumb can be developed concerning the size of the relevant shadow prices. There are two reasons for this. First, if the CBA evaluation is costly and time consuming it negates its value. Second, in practice these decisions have to be made and understood by officials who are not necessarily economic specialists. A set of easily followed rules should be the foundation for a successful appraisal and evaluation programme.

This involves being able to assign shadow prices to different inputs and outputs for different types of region, and use these shadow prices to replace the actual market prices in an appraisal or evaluation procedure. Once this is done it is possible to develop straightforward rules for appraisal and evaluation. In the next section this is illustrated for perhaps the most straightforward case, where a firm is seeking assistance for a particular project. I will assume that this is discretionary assistance, and the issue therefore is, using cost-benefit principles, how much subsidy should the government be prepared to pay?

13.6 A COST-BENEFIT APPROACH TO DISCRETIONARY REGIONAL AID

The treatment here is schematic. I assume that the government is able to verify fully the true facts concerning the project seeking a subsidy. I assume also that there is complete certainty concerning the outcome and that there is no initial investment, so that the problem of discounting in order to put the flow of costs and benefits on the same footing is not encountered. It is as if the project simply rents all the inputs.

The output of the project generates a sales revenue, R, which is equal to output (Q) times price (p) so that:

$$R = pQ \tag{13.3}$$

The private costs are C, which can be disaggregated to the sum of all input costs. In this case I just take two inputs, capital and labour (K and L) with the corresponding unit costs, k and w, though the extension to any number of inputs is trivial, so that:

$$C = Kk + Lw \tag{13.4}$$

Because the firm needs a minimum grant, G, for the project to go forward, this is defined as:

$$G = C - R > 0 \tag{13.5}$$

It will be useful to express the minimum grant as a proportion of the project's costs. Using expression (13.5), this subsidy rate, g, is therefore

$$g = \frac{G}{C} = 1 - \frac{R}{C} > 0 \tag{13.6}$$

Adopting a cost-benefit approach, the requirement for the project to go ahead through a subsidy would be:

$$p^S Q \geq k^S K + w^S L + t^S G \tag{13.7}$$

where the S superscripts indicate shadow prices and t is the cost of taxation. Dividing both sides of inequality (13.7) by C, multiplying the terms Q, K and L by p/p, k/k and w/w respectively and then simplifying produces:

$$\left[\frac{R}{C}\right]\dot{p}^S \geq \left[\frac{Kk}{C}\right]\dot{k}^S + \left[\frac{Lw}{C}\right]\dot{w}^S + \left[\frac{G}{C}\right]t^S \tag{13.8}$$

where the dot notation represents the shadow price as a proportion of the market price, so that, for example:

$$\dot{p}^S = \frac{p^S}{p} \tag{13.9}$$

Formulating expression (13.8) as an equality and then substituting in equation (13.6) gives the maximum value of the proportionate grant, g_{max}, that would be justified on cost-benefit criteria as:

$$g_{\max} = \frac{\dot{p}^S - \sigma_K \dot{r}^S - \sigma_L \dot{w}^S}{\dot{p}^S + t^S} = \frac{\dot{p}^S - \sigma_K (\dot{r}^S - \dot{w}^S) - \dot{w}^S}{\dot{p}^S + t^S} \qquad (13.10)$$

where σ_K and σ_L are the shares of capital and labour in the project cost, so that $\sigma_K + \sigma_L = 1$.

Equation (13.10) makes clear that for a subsidy to be efficient from a cost-benefit perspective, the term $\dot{p}^S - \sigma_K \dot{k}^S - \sigma_L \dot{w}^S$ must be positive. Recall that $\dot{p}^S$, $\dot{k}^S$ and $\dot{w}^S$ are the ratios of the shadow to market prices and that the share parameters, σ_K and σ_L sum to unity. First, if there is no difference between shadow prices and market prices there should be no subsidy. Essentially this just says that under these circumstances, the market performs efficiently. If a project is not viable on commercial criteria it should not be aided, and if it is viable it does not need to be aided.

A corollary is that if the shadow price of output (inputs) is higher (lower) than the market price, then a grant payment up to some maximum level would be efficient. Partially differentiating the extreme left-hand expression in equation (13.10) gives:

$$\frac{\partial g_{\max}}{\partial \dot{r}^S}, \frac{\partial g_{\max}}{\partial \dot{w}^S}, \frac{\partial g_{\max}}{\partial t^S} < 0$$

$$\frac{\partial g_{\max}}{\partial \dot{p}^S} > 0$$

and:

$$\frac{\partial g_{\max}}{\partial \sigma_K} > 0 \textit{ iff } \dot{w}^S > \dot{r}^S$$

These results are intuitive. For the input prices, the lower the shadow price, the greater is the efficient maximum subsidy. In these cases the market price overstates the true opportunity cost of the input, and under commercial decision making too little of the input will be used and also too little of the output will be produced. The opposite result holds for the shadow price of output. The higher this is, relative to the market price, the greater the maximum grant.

For the shadow price of taxation, the higher the value, the lower the maximum grant. In this case, administration, organisational and compliance costs of the tax and grant-giving process, plus any distortionary costs of taxation, will reduce the efficiency of financial support.

Finally variations in the capital intensity of production will affect the maximum grant, in so far as the ratio of the shadow price to market price varies across capital and labour. If labour has the lower relative shadow price, then the more capital intensive a project is, the lower the maximum grant that it should attract. Of course, if the relative shadow price is lower for capital, capital intensive projects should have the higher maximum grant.

In equation (13.10) the grant is expressed as a proportion of the projects total costs. However, it could also be expressed in relation to the inputs. In particular, equation (13.10) can be recast to give the maximum grant as a proportion of labour costs, and we can then extend this to give a maximum cost per job calculation too.

Expressing the maximum grant measured as a share of labour costs as $\eta_{\max}$ then:

$$\eta_{\max} = \frac{Cg_{\max}}{Lw} = \frac{\dot{p}^S - \sigma_K \dot{r}^S - \sigma_L \dot{w}^S}{(\dot{p}^S + t^S)\sigma_L} = \frac{\dot{p}^S - \sigma_K(\dot{r}^S - \dot{w}^S) - \dot{w}^S}{(\dot{p}^S + t^S)(1 - \sigma_K)} \tag{13.11}$$

Variation in the maximum grant as a proportion of the wage bill is generated by changes in the relative shadow prices and the cost of taxation in qualitatively the same way as for the grant as a share of total costs. The only qualitative difference from the earlier analysis concerns the impact of a change in capital intensity. Whereas previously, variation in the maximum grant as the capital intensity increased depended only on the relative shadow price ratios for capital and labour, now there is an additional effect that is always positive. This is because the subsidy is measured against the labour share base and, *ceteris paribus*, the smaller is that base, the greater the maximum grant rate will be.

If the maximum grant is measured against an input, it is usually simply in terms of the grant cost per job (*cpj*). This is given as:

$$cpj = \frac{G}{L} = w\eta_{\max} = w\left[\frac{\dot{p}^S - \sigma_K \dot{r}^S - \sigma_L \dot{w}^S}{(\dot{p}^S + t^S)\sigma_L}\right] = w\left[\frac{\dot{p}^S - \sigma_K(\dot{r}^S - \dot{w}^S) - \dot{w}^S}{(\dot{p}^S + t^S)(1 - \sigma_K)}\right] \tag{13.12}$$

As in Swales (1997) equation (13.12) shows how the maximum subsidy determined using cost-benefit criteria can be transformed into a maximum cost per job measure.

13.7 THE POLITICAL ECONOMY OF COST-BENEFIT ANALYSIS

The position this chapter has taken up to now is the standard one for economic policy appraisal and evaluation. The problem is set up as one in which a benign government maximises the welfare of the population, as represented by the social welfare function. Policy makers are implicitly assumed to act altruistically and the population to react atomistically.

However, in the last five decades, there has been a growing political economy literature that does not accept these assumptions. To be more specific, this literature generally assumes that policy makers have interests of their own and that they will attempt to further these interests in seeking and retaining office. Further, individuals or groups within the population will attempt to influence government policy for their own purposes. Difficulties for policy evaluation identified by this political economy approach are discussed here in a very informal and non-rigorous manner. However, they are of fundamental importance and require more detailed analysis.

13.7.1 Accountability and Transparency

One of the most powerful arguments put forward in favour of cost-benefit analysis is the transparency of the process. Potential costs and benefits are identified and explicitly weighted to produce a decision-making procedure. Also CBA is, in principle, a holistic approach, where the vector of all elements relevant for the policy decision is reduced to a scalar positive or negative signal. However, these key strengths of the cost-benefit method that so attract economists, namely transparency and comprehensiveness, might also repel the policy maker.

The ability of policy makers to follow their own interests instead of that of the electorate is generally limited in situations where the electorate is more informed. On this score, policy makers might wish low levels of policy assessment in order that their real intentions and effort levels are not revealed.[6] Of course, electoral competition might force policy makers to disclose their intentions, and thus encourage the appraisal of policy. Also it might be in the interests of policy makers to reveal their own efficiency in policy implementation through *ex post* evaluation. However, there might be characteristics of CBA that policy makers find undesirable, especially for the analysis of regional policy.

The first is that the weighting of costs and benefits makes explicit the government's social welfare function. Essentially the weights used in the cost-benefit calculations reveal the extent that policy makers are prepared to trade off policy impacts at the margin. One important issue is the way that changes in income or activity in one part of the country are to be offset against changes in other areas. Whilst it is implicit in its overall policy stance, this is something which the government might not necessarily want to highlight, and certainly not to quantify with any precision. A second major problem is that, in so far as the cost-benefit approach is seen as comprehensive, it reduces the extent of the policy maker's freedom of action. Policy makers may prefer a number of partial assessment methods which allow more room for manoeuvre.

13.7.2 Gaming the System

Changing the prices faced by decision makers to reflect the true opportunity cost of using resources and the social, as well as the private, benefits of outputs, is an idea which is difficult to criticise as such. Further, one would think that it would be an idea that supporters of the market mechanism would embrace. Where CBA is used for *ex ante* appraisal, the advantages of decentralised decision making are retained whilst adjustments are made for known market failures. This argument is even further strengthened if CBA principles are used to set tax and subsidy rates so that agents maximising private net benefits are simultaneously maximising social net benefits.

However, there are two basic questions that can be asked about the use of CBA. The first is: can the market failures that underpin the use of CBA be tackled in a more effective way? For example, imagine that a commodity is produced under conditions of imperfect competition, so that price is greater than marginal cost. From welfare theory, *ceteris paribus*, the production in this industry is too low; output should be higher and the price to the consumer should be lower. A CBA analysis would support the use of a subsidy in the production of the good, which would bring about these changes. However, it might be thought that instead of increasing the price of the good to the producer, it would be more appropriate to attempt to increase competition in the market. Of course the monopolist would prefer a situation where the market imperfections were retained, together with the application of a subsidy to induce the production of the optimal output.

The second question is an extension to the first: might accommodating the market failures through CBA actually entrench these market failures or even exacerbate their impact? This can occur through moral hazard problems or, putting it more colloquially, where economic agents 'game the system'. Take an example from the labour market. It might be that the alternative employment in a particular area is very limited, so that the shadow wage is below the actual wage. This implies that the government should be prepared to subsidise activity in existing industries. However, if this is known, it might be in the interests of both workers and management in existing plants to reduce effort, thereby reducing efficiency and triggering a subsidy. This might not even be a conscious strategy, but occur simply as a result of the efficiency constraints on the plant being weakened. This problem is not restricted to the use of CBA, though a set of known shadow prices which are different to market prices may exacerbate this problem.

13.8 CONCLUSIONS

There are two major strengths to the cost-benefit analysis approach. The first is conceptual. It forces policy makers to make explicit the nature of the market failure that they believe needs to be tackled through regional policy. Moreover, in the requirement to identify shadow prices for inputs and outputs, the nature of these market failures needs to be studied in greater depth. The second is practical. CBA should be adapted to produce workable rules of thumb for directing decentralised decision taking on project appraisal and evaluation in problem regions.

However, CBA principles are not generally used in practice for directing regional economic intervention. This might be because the transparency and comprehensiveness of the technique leaves too little leeway for policy makers. Yet, in adopting other appraisal and evaluation methods, policy makers are turning their back on a consistent and systematic method for shaping regional policy.

NOTES

1. A Pareto optimal outcome is one in which one person cannot be made better off without making some other person worse off. These optimality of market mechanisms depends on other conditions holding (de Graaf, 1957). Some of these are discussed in the text.
2. Public goods score high on the criteria of non-exclusion and non-rival. However, the criteria are not clear cut: the issue is one of degree.
3. Moral hazard problems are where post contractual opportunism occurs.
4. A non-marketed output or input would be given a zero weight because it would have a zero price. No explicit account is taken of uncertainty.
5. The notion of shadow prices comes initially from the linear programming literature where the shadow price of a resource equals the increase in optimal output generated by increasing the supply of the resource by one unit (Dorfman et al., 1958). A similar interpretation can be made in non-linear, market-based models (Dreze and Stern, 1994).
6. Evaluation and appraisal are themselves costly, resource-using, activities, so that from a cost-benefit standpoint there will be an optimal level of these activities. The issues raised here are such that it is unlikely to be in the interests of the policy maker to introduce this optimum level of policy assessment.

REFERENCES

Armstrong H. and Taylor, J. (2000), *Regional Economics and Policy*, 3rd edition, Oxford: Blackwell.

Bergson, A. (1938), 'A Reformulation of Certain Aspects of Welfare Economics', *Quarterly Journal of Economics*, **52**, 310–334.

Dorfman, R., Samuelson, P. and Solow, R. (1958), *Linear Programming and Economic Analysis*, New York: McGraw Hill.

Dreze, J. and Srern, N. (1994), 'Shadow Prices and Markets: Policy Reform, Shadow Prices and Market Prices', in R. Layard and S. Glaister (eds), *Cost Benefit Analysis*, 2nd edition, Cambridge: Cambridge University Press.

Gibbons, S., Green, A., Gregg, P. and Machin, S. (2005), 'Is Britain Pulling Apart? Area Disparities in Employment, Education and Crime', in N. Pearce and W. Paxton (eds), *Social Justice: Building a Fairer Britain*, London: Politicos Publishing.

Graaff, J. de V. (1957), *Theoretical Welfare Economics*, Cambridge: Cambridge University Press.

Hayek (von) L. A. (1937), 'Knowledge and Economics', *Economica*, **4**, 33-54.

HM Treasury (2003), *Green Book: Appraisal and Evaluation in Central Government: Treasury Guidance*, 3rd edition, London: The Stationery Office.

Mas-Collel, A., Whinston, M. D. and Green, J. R. (1995), *Microeconomic Theory*, Oxford: Oxford University Press.

Miller, R. E. and Blair, P. D. (1985), *Input Output Analysis: Foundations and Extensions*, Edgeworth Cliffs, NJ: Prentice Hall.

Persson, T. and Tabellini, G. (2000), *Political Economics: Explaining Economic Policy*, Cambridge MA: The MIT Press.

Porter, M. (1990), *The Competitive Advantage of Nations*, New York: Free Press.

Samuelson, P. A. (1954), 'The Pure Theory of Public Expenditure', *Review of Economics and Statistics*, **36**, 387–389.

Stiglitz J. E. (1996), *Whither Socialism*, Cambridge MA: MIT Press.

Stiglitz, J. E. and Greenwald, B. (2003), *Towards a New Paradigm in Monetary Economics*, Cambridge: Cambridge University Press.

Swales, J. K. (1997), 'A Cost-Benefit Approach to the Evaluation of Regional Selective Assistance', *Fiscal Studies*, **18**, 73–85.

14. Regional Policies in Italy, the Netherlands and the United Kingdom: A Shift in Paradigm?

Ilaria Mariotti

14.1 INTRODUCTION

In all its forms, national regional policy (from now on regional policy) has remained focused on the problem of differential spatial development within national economies. Since World War II, European governments have implemented specific policies to redress regional economic disparities which were based on policy measures in the form of financial incentives to influence location decisions by companies. By promoting core-periphery relocation, governments were effectively providing subsidies to reduce local unemployment and inequalities in income. This regional policy approach is labelled 'traditional' and was undertaken by central governments (top-down approach).

From the mid-1970s, regional policy began to change. The theoretical debate criticised the role of the State in the economy and proposed, as an alternative, liberalisation, deregulation, and privatisation (Viesti and Prota, 2004). The effectiveness of policy incentives to promote economic development was questioned and in the 1980s the traditional approach was partially superseded by a decentralised model for regional development, the so-called 'contemporary' approach. This was based on decentralised intervention (bottom-up approach) and focused on integrated development plans and strategies designed and delivered by a partnership between regional and local players (Bachtler and Yuill, 2001). Core-periphery relocation was not the goal of the new approach which was mainly focused on endogenous development by promoting small and medium- size enterprises (SMEs) and innovation and by encouraging geographical clusters of firms.

However, empirical evidence shows that the boundaries between the traditional and the contemporary approaches are not so well defined and that, especially as far as policy measures are concerned, the legacy of the old framework still survives. Even in contemporary-approach oriented policies, in fact, some of the tools typically labelled as 'traditional', such as incentives promoting core-periphery relocation, are widely used. There is a broad area of overlap which highlights a discrepancy between the ideal situation as planned by theorists and the reality designed by policy makers. The present chapter aims to provide evidence on this issue through the analysis of the approaches to regional policy in Italy, the Netherlands and the United Kingdom from the 1990s onwards. Specifically, particular attention is devoted to the policy measures promoting core-periphery relocation and their importance as a location factor is investigated.

The chapter is organised as follows. Section two presents a review of the approaches to regional policy in the three European countries and the policy measures promoting core-periphery relocation. Section three is dedicated to the empirical analysis of those investments subsidised by three policy measures: the Area contract (AC) in Italy, the IPR premium (IPR) in The Netherlands and the Regional selective assistance (RSA) in the United Kingdom. The last section concludes the chapter.

14.2 APPROACHES TO REGIONAL POLICY AND POLICY MEASURE PROMOTING CORE-PERIPHERY RELOCATION

There are two main regional policy approaches in 1945–2000s: the traditional approach (from 1945 to the 1980s) and the contemporary approach (from 1990 to the 2000s). The traditional approach refers to the view that regional disparity is caused by structural weaknesses in the regional economy, coupled with a fundamental deficiency of investment due to a drain of financial capital from poor to rich regions (see Martin, 1988, 1989; Mackay, 1997). The provision of hard infrastructures together with the assistance in the form of business aid are adopted to reduce economic disparities between regions. Core-periphery relocation is subsidised and cost penalties and permit systems are introduced in the core areas to encourage firms to consider moving to less congested, problem locations in the region (see Pellenbarg et al., 2002 for an overview on firm migration in Europe in 1945 to the 2000s). The mode of organisation is essentially top-down: the design of policy, the delimitation of assisted areas and the delivery of instruments were undertaken by central government (see Yuill et al., 1980; Bachtler and Yuill, 2001; Yuill and McMaster, 2001).

The contemporary approach emphasises the role of endogenous local factors in promoting growth. Industrial sectors and the State stopped being viewed as the engines of development while firms and the market became the key ingredients for promoting economic success. Consequently, peripheral areas should reduce regional unemployment disparities by creating their own growth. Policy measures should work on the causes of lagging in the periphery, allowing depressed areas to exploit their potential resources in order to optimise in this way the contribution of regional resources to the creation of the national wealth. One of the main components of the contemporary approach concerns the reduction of the traditional automatic incentives, in favour of softer measures aiming to improve local development, stimulated in part by the EU structural funds. The promotion of local development stresses the importance of stimulating new firm formation and inducing SMEs to expand and develop the latest technology and agglomerate in geographical clusters. The mode of organisation is bottom-up: design and implementation of integrated development plans and strategies fall within the competence of a partnership of regional and local players.

Within the framework of the traditional approach, in Italy, the Netherlands and the United Kingdom regional incentives were offered to firms mainly in the form of capital grants, as stimulants to locate to the problem areas (periphery). In Italy, the *Intervento Straordinario* development policy offered financial incentives to firms moving from the core areas in the centre-north to the depressed regions in the Mezzogiorno (literally 'land of the midday sun', Southern Italy)[1]. The state-owned firms (*Partecipazioni Statali*) had to place at least 40 percent of their total investment in the Mezzogiorno and 60 per cent of their new plants (see Camagni, 1983; Saraceno, 1977, 1980; King, 1985). They had to act as growth poles for the whole economy in order to stimulate ancillary development (Perroux, 1966).

In the same period, in the Netherlands, financial incentives (i.e. the IPR investment premium) were awarded to firms locating in development nuclei within the designated problem regions and infrastructural investments were implemented (Boersma, 1980; Molle, 1983; Atzema and Wever, 1994; Van Dijk and Pellenbarg, 1996).

In the UK, a variety of policy incentives were introduced to reduce unemployment in the problem areas, named Assisted Areas. These comprised loans and grants to firms (i.e. Regional Development Grant and Regional selective assistance), location controls (Industrial Development Certificate) and labour subsidies (Keeble, 1972, 1976; Yuill, 1980; Armstrong and Taylor, 2000; Wren, 1996a, 1996b).

In the 1980s, the traditional approach to regional policy came under sustained attack in most of the European countries because it did not alleviate regional disparities. As a consequence, since the beginning of the 1980s, the traditional approach was partially superseded by the contemporary approach, which focuses on the creation of appropriate framework conditions for business development ('business climate') and is based on integrated regional development plans and strategies, designed and delivered by partnerships of regional and local actors. Nevertheless, policy measures promoting core-periphery relocation are still in place (Table 14.1).

Table 14.1 Policy measures promoting core-periphery relocation in Italy, the Netherlands and the United Kingdom

Countries	Regional Incentives	Regional Strategies
Italy	Grant-based Law 488, plus automatic fiscal assistance	Negotiated plans (Area contract, Programme contract, Territorial pact)
Netherlands	Grant-based, IPR	
United Kingdom	Grant-based, Regional selective assistance (RSA)	

In Italy, in 1980 to the 2000s, the policy measures adopted are of two main types: a semi-automatic system of incentives (Law 488) and partnership-based instruments or regional strategies (Negotiated Plans) (see Barca, 2001; Del Monte, 1998). While Law 488 is mainly a grant-based measure (Table 14.2), the tools of the Negotiated Plans (Territorial pact, Programme contract and Area contract) are of a different nature (Table 14.3). Their aim is to promote the local context and encourage cooperation among different actors (firms, local institutions, central institutions). Although each has its own characteristics, Law 488, Programme contract (PC) and Area contract (AC) promote the relocation of external investments to the Mezzogiorno.

Firms, independent of size, industry, or nationality, are subsidised by Law 488 to invest in the depressed areas of the country through new plants, expansions or renovation. Large companies (national or foreign) or consortia of SMEs and representatives of the Italian industrial districts are eligible for PC incentives to realise industrial investments, together with R&D centres and projects and the training of personnel (Giunta and Florio, 2002; Viesti, 2002). Finally, new initiatives and expansions (national and foreign) in industry, agro-industry, services and tourism are subsidised by the AC to locate in the crisis areas of the country.

While the analysis on investments in the Mezzogiorno, which have been promoted by Law 488 in 1996–2002, shows a negligible amount of investments coming from the centre-north (5%[2]), the AC and PC tools attracted, on average, a larger volume of 'external' projects (Mariotti, 2005). The investments, promoted by both Law 488 and PC, are mainly of a large or medium size, while those subsidised by the AC are smaller on average. Most of the projects receiving grants in the south are capital-intensive and belong to 'scale-intensive' and 'traditional' sectors, which are the typical sectors of the southern economy. This outcome emphasises and reinforces the north-south dualism because it does not induce a diversification of the southern production basis.

The goal of regional policy in the Netherlands in 1990–2000s is no longer to establish interregional equity but to strengthen the competitiveness of all regions, with a specific effort on reinforcing the economic structure of the peripheral areas (northern provinces, Flevoland, South Limburg, Twente and Overijssel), through regional incentives and strategies, and framework measures (Oosterhaven, 1996). It is intended that the Netherlands fosters and maintains an outstanding business location climate, through reinforcing network infrastructures, and especially those serving the main ports of Amsterdam and Rotterdam. In addition, a programme of investment in high quality industrial estates is also promoted (the Provincial Investment Programme Tendering Scheme, TIPP). The programme-based TIPP encourages provinces to cooperate with municipalities (and other provinces) to develop strategic plans and related investment programmes for their areas.

Table 14.2 Law 488/1992

Law 488/1992 provides discretionary grants and operates in all depressed areas of the country, both in the southern regions (areas under Objective 1) and in the centre-north. Law 488 is characterised by a semi-automatic selection scheme in order to accelerate the process of bank processing and ministerial evaluation so that firms can obtain immediate financing for new investment in eligible areas. The scheme is administrated on a competitive basis in the sense that applications are submitted to an annual deadline and assisted in order of merit until the funds for that tranche of applications are exhausted. The ranking of a project in the application process is determined with respect to a score based on five evaluation criteria: (i) the proportion of own funds invested in the project in relation to the total investment; (ii) the number of jobs created in relation to investment; (iii) the value of assistance sought as a proportion of the maximum rate applicable in the region; (iv) regional priorities regarding project type and sector; (v) an indicator related to the environmental impact of the project. This incentive scheme is targeted at firms independent of their size, which are operating in the industry, services or tourism sectors. There are six areas and they are: Industry, Tourism, Commerce, Handicraft, Environment and Smaller Islands.

Table 14.3 The measures of the negotiated plans

The Programme contract (*Contratto di Programma*) was adopted by the Italian government in the 1980s and it was incorporated in the new policy. With Programme contracts, the central administration promotes the implementation of large-scale industrial development projects in lagging areas in the Mezzogiorno or, since 1992, in some target areas in the centre-north (law 64/1986). The Programme contract, stipulated autonomously (e.g. at the request of the parties concerned) or in the execution of a programme agreement or understanding, is meant specifically for large companies, either national or foreign, or consortia of SMEs (and the representatives of industrial districts). This policy measure provides discretionary grants for the realisation of industrial investments together with research activities, both R&D centres and projects and training of personnel, whether for newly created or pre-existent jobs.

The Area contract (*Contratto d'Area*) is a policy tool which provides discretionary grants for the implementation of new business initiatives and the creation of new jobs in industry, agro-industry, services and tourism, in designated territories facing an employment crisis (Law 662/96). It can be defined as an operative instrument agreed by the national and local governments, businesses organisations and unions, as well as local actors and other parties concerned, for the realisation of measures oriented to speed up the development and the creation of new employment in designated territories facing employment crisis. A series of agreements characterise this instrument. These include: an agreement between social partners on labour flexibility and wage moderation; agreements among public institutions to simplify and speed up procedures; agreements with the banking system for access to credit; and legal protocols for improved public order conditions in the territory. The government guarantees financial incentives to companies in response to agreements made.

The Territorial pact (*Patto Territoriale*) is specifically designed to promote trust in the territory and to stimulate and coordinate investment by private enterprises and local administrations. This instrument aims at territorial development from the bottom up, by means of infrastructure investments combined with incentives for companies that undertake integrated projects in the fields of industry, agro-industry, services and tourism.

Source: Elaboration of Giunta and Florio, 2002; Viesti and Capriati, 2003; SVIMEZ, various years.

Specific regional incentives, the Centralised and the Decentralised IPR premiums, aim to reinforce the economic structures of the northern provinces. Both the IPR premiums are addressed to firms that sell 50% or more of their output outside the region where they are located. The eligible projects for the Centralised IPR are,head offices and laboratories in the manufacturing and service sectors, while basic firms, head offices, pilot plants and laboratories are eligible for the Decentralised IPR. The Centralised IPR subsidises significantly larger projects than the Decentralised measure (Table 14.4).

The analysis of the projects that have been granted by the Decentralised IPR premiums in 1993–2003 shows, however, negligible relocations from the Randstad and from abroad (Mariotti, 2005). Indigenous investments, in the form of expansion projects, dominate the scene. Nevertheless, such projects are more effective than new establishment projects in terms of amount of investment per job created. The investments subsidised by the IPR are mainly in capital- and labour-intensive industry sectors, with knowledge-intensive sectors in the third position. The knowledge-intensive industry registers the highest level of job creation.

Table 14.4 The Centralised and Decentralised IPR

The Centralised IPR (*Decree on Subsidies for Regional Investment Projects*) provides discretionary grants to investments located in the Northern Netherlands, Flevoland, South Limburg and Twente. The subsidy may be granted for investments in: (i) establishment projects in the core zones of the above-mentioned regions, aimed at setting up an industrial company, a basic service company, a head office or a laboratory or a basic tourism business; (ii) strategic expansion projects in the core zones; (iii) projects involving fundamental changes in the regional economy of a core zone (i.e. the employment effect has to be at least 400 jobs – directly and indirectly). The Centralised IPR is addressed to larger projects than the Decentralised IPR and is managed by the Ministry of Economic Affairs in The Hague. .
The Decentralised IPR (*Northern Netherlands Investment Subsidy Scheme)* provides discretionary grants to investments located in the Northern Netherlands and Overijssel. A subsidy may be granted for investments in: (i) establishment projects aimed at setting up a basic company, a head office, a pilot plant or a laboratory in a core zone; (ii) expansion projects concerned with expanding the capacity of a 'basic' company, head office or laboratory; (iii) restructuring projects which effect a fundamental change in the production programme of a basic company. The Decentralised IPR is managed by the Cooperation Northern Netherlands – SNN.

Regional policy in the UK in 1990-2000s focuses on improving competitiveness of an industry in international markets, promoting indigenous development in the disadvantaged regions, stimulating entrepreneurship, small firms and new technology. Among the other business programmes, the regional incentives which subsidise core-periphery relocation are the RSA and EGS (Table 14.5).

Table 14.5 *The regional selective assistance and the enterprise grant scheme*

The RSA (*Regional Selective Assistance*) provides discretionary grants to investments, which induce a net positive impact on Assisted Areas' employment levels (Objective 1 and 2), create and safeguard jobs, help to attract and retain internationally mobile investment and to contribute to the regeneration and competitiveness of the regions. The RSA provides discretionary grants to large capital investment projects and is operated by the UK central government, so it is essentially a national scheme, with the larger cases determined centrally in London. The RSA is aimed at the manufacturing industry as well as at service sector businesses markets beyond the local level (e.g. retail activities, hotels and restaurants do not qualify). Assistance is provided to establish a new project or expand or modernise an existing business, to set up research and development facilities, or to take the next step from development to production. An important change was made to RSA in 1984 when it was no longer available to plants relocated within the UK, and displacement was taken into account in deciding the offer. This was a major change in the orientation of regional policy, which since the early post-war period had been about encouraging firms to relocate in the regions.

The EGS (*Enterprise Grant Scheme)* is a discretionary scheme which substitutes for the RSA, in England, in the case of projects with an investment amount equal or lower than £500 000, and is available only to SMEs. These should be high value added investments providing high productivity levels and a higher number of skilled jobs. The EGS operates in Enterprise Grant Areas, which include the Assisted Areas as well as those areas outside the Assisted Areas designated to include areas with particular social problems.

The selection criteria of the EGS differ from those of the RSA in three important respects: (i) projects do not have to create or safeguard jobs; (ii) the application and appraisal process has been streamlined; (iii) the grant is available in a broader range of areas. In addition, particular attention is placed on the project's contribution to the vocational training. The evaluation of the programme is scheduled every three years.

These are discretionary grant schemes, which contribute to the attraction of investments (both manufacturing and services) in the Assisted Areas. RSA and EGS are not directed to specific sectors; they focus on high-quality, knowledge-based projects which provide skilled jobs. Foreign investments and new projects are eligible for these measures, while plants relocating within the UK, which displace jobs, cannot be granted.

The analysis on the RSA-granted projects in 1991–1995 shows that manufacturing sectors are the main beneficiary of the RSA grant, while most consumer-type services, serving purely the local market, are excluded because of their displacement effects (AEP, 2000 in Mariotti, 2005). UK firms account for one-tenth of the value of the foreign mobile projects. These are larger than national investments. Unfortunately, a lack of data means that we do not have a very clear picture of core-periphery firm relocation. However, we can assume a negligible amount of those types of projects because of the RSA criteria.

The analysis of the policy measures promoting core-periphery relocation demonstrates that they differ mainly in the aim, typology and eligibility of the subsidies. While the Italian and Dutch measures promote the location and relocation of business establishments in the peripheral areas, RSA and EGS are not available to plants relocating within the UK, which lead to job displacement. This represents a major change in the orientation of UK regional policy, which since the early post-war period had been about encouraging firms to relocate in the regions. Regional policy in the UK mainly fosters the attraction of internationally mobile investments, and the data on RSA-granted projects showed that the country succeeded in this goal. The main beneficiary of the RSA grant is industry and the larger investments are foreign mobile projects. In 1990 to the 2000s the Italian Mezzogiorno exhibits a concentration of 'scale-intensive' and 'traditional' sectors, which fails to induce any diversification of the southern production base. The investments subsidised by the IPR are mainly in capital- and labour-intensive industry sectors while the knowledge-intensive sector ranks third, although it registers the highest level of job creation. In addition, in the three countries, policy measures have subsidised more indigenous investments than external investments.

14.3 EMPIRICAL ANALYSIS

In order to investigate the role of policy measures in promoting core-periphery relocation, this section presents a firm-level analysis of a sample of grant recipient firms which invested in the peripheral areas. The difficulties encountered in collecting firm-level data as well as securing a reasonable number of respondents, led to the use of both primary and secondary information. The Italian and the Dutch cases are based on surveys carried out by the author (for a detailed description see Mariotti, 2005), while secondary information has been used for the British case (AEP, 2000). Nevertheless, both primary and secondary data collection refers to direct interviews with managers of randomly selected samples of AC-, IPR- and RSA-assisted firms. The survey in Italy is, however, much bigger than the other two because of the easier access to information in that country as well as the larger number of grant-receiving firms that invested in the Mezzogiorno, which is more extensive than the peripheral areas in the Netherlands and in the UK.

14.3.1 AC-granted Firms

In Italy there are 18 areas, named 'crisis areas', eligible for the AC funds and their additional protocol. Of those, 15 are located in the Mezzogiorno and 3 in the centre-north (Figure 14.1). The focus of the analysis is on the 15 southern areas, including both deindustrialised area and other depressed areas.

The survey was conducted via questionnaire and telephone interviews. The sample included 528 firms which invested in the Mezzogiorno (at 31st May 2003) and were AC grant recipients. The average response was 52%, which gave us a final sample including 254 firms. It is, however, important to emphasise that the sample is partially biased because it is made up of entrepreneurs who have actually invested. Their opinion may therefore sound as an *ex post* justification of their decision. This is a typical response regarding the rationalising of decisions afterwards.

The descriptive analysis showed that external investments account for 27.2%, of which 12.6% are new firms with investors from the central-northern part of Italy, 13.8% are branches of central-northern firms and 0.8% are branches of foreign firms. Among the pull factors promoting investments in the Mezzogiorno, it was confirmed that government incentives were decisive in the location and relocation of firms in this area, but that other so-called pull factors (i.e. availability of industrial areas and labour force) also played a relevant role.

In order to investigate more in-depth commonalities and differences among firms that invested in different areas as well as the role played by pull factors in the location/relocation choice, a factor and cluster analysis was carried out. The factor analysis allowed us to group the variables resulting from the survey into two factors: (i) Factor 1 includes supply-side variables; (ii) Factor 2 represents the demand side of the economy together with accessibility levels of the areas. Factors 1 and 2, together with four characteristics of the firms (investment origin, size, share of regional and national market), were used to run a cluster analysis. Four clusters have been identified; firms belonging to the same cluster are supposed to share certain regularities in behaviour which differ from firms belonging to other clusters.

Figure 14.1 *Geographical distribution of Contratti d'area and Protocolli aggiuntivi*

Cluster 1 is composed by external medium-large sized investments which look at the Mezzogiorno for its market opportunities. These firms need government subsidies to compensate for the southern diseconomies. Centre-northern firms operating in various manufacturing sectors, tend to transfer medium-high value-added activities or new production lines to the southern areas, by making use of the government subsidies. These firms open new branches or new firms in which the whole production process or a new production line is transferred.

The firms belonging to Cluster 2 are local or southern, medium-large sized firms and they operate both in the national and regional market. Supply-side factors are extremely important for them, while demand side and accessibility do not play an important role in relocation. Those firms tend to make capital-intensive investments. Clusters 3 and 4 include local small or very small sized firms. The firms in the third cluster operate mainly at a regional level and consider demand-side and accessibility factors more important than supply-side variables. The firms in Cluster 4 considered both the factors not relevant in the decision of relocate, thus they would have probably invested in the AC areas regardless of incentives.

14.3.2 IPR-granted Firms

This section presents the results of a telephone survey on relocation determinants, which was addressed to a small sample of 20 firms which relocated to the northern provinces (Groningen, Drenthe and Friesland) between 1998 and 2003. The selection of these 20 firms was made on the basis of their identification and their willingness to cooperate. As for the Italian case, it is however important to stress that the sample is partially biased because it comprises entrepreneurs who have all actually relocated.

Table 14.6 Origin of IPR-granted firms that have invested in the northern provinces

Origin	Groningen No. (%)	Drenthe No. (%)	Friesland No. (%)	Total No. (%)
Randstad	4 (44.4%)	2 (22.2%)	3 (33.3%)	9 (100.0%)
Northern provinces	5 (55.5%)	/	4 (44.4%)	9 (100.0%)
Abroad	1 (50.0%)	1 (50.0%)	/	2 (100.0%)
Total	10 (50.0%)	3 (15.0%)	7 (35.0%)	20 (100.0%)

Among the interviewed firms, 30% were subsidised by the Centralised IPR and the rest by the Decentralised IPR. Forty-five per cent come from the Randstad, 45% are 'local' – they were already located in the Northern provinces – and the remaining ones originated from abroad (Table 14.6). The survey showed that for a large part of the firms concerned, the IPR instrument was not decisive. The IPR policy measure as such seems not to foster the development of the northern provinces because it mainly subsidises expansion, capital-intensive and medium-large sized investments. These firms do not induce a higher employment growth and do not lead to differentiation of the production structure of the area which is mainly constituted by standard- and capital-intensive activities. On the other hand, the quality of the local environment is perceived as a relevant determinant to relocate, followed by the good cooperation with the local government and local institutions.

14.3.3 RSA-granted Firms

For the UK case, the results of a survey of 165 firms that were RSA grant recipients for investing in the Assisted Areas in 1991–1995 are presented (AEP, 2000). However, as before, it is worthwhile to note that the evaluations such as the one for the RSA, which are based on face-to-face interviews with senior managers of randomly selected samples of RSA-assisted firms, are subject to biases, and some caution must therefore be exercised in their interpretation.

Table 14.7 Geographical distribution of RSA-granted projects

Location comprising	Number of cases	Percentage of total	Percentage of total gross job outputs achieved	Percentage of grant paid (net of taxes)	Average net grant per project (£)
England	88	53.3	56.6	34.5	389 396
of which:					
East Midlands	3	1.8	5.1	2.7	883 650
Merseyside	8	4.8	1.1	2.6	318 416
North East	16	9.7	10.2	10.5	690 357
North West	11	6.7	12.3	1.9	169 615
South West	7	4.2	3.3	2.6	363 714
West Midlands	11	6.7	12.6	8.2	729 293
Yorkshire & Humberside	9	5.5	6.2	3.4	370 286
East	11	5.7	1.2	1.0	92 978
London	5	3.0	4.0	1.2	242 009
South East	7	4.2	0.7	0.3	46 354
Wales	37	22.4	17.4	24.2	680 068
Scotland	40	24.2	26	41.3	1 128 871
Total	165	100.0	100.0	100.0	622 273

Source: Elaboration of AEP (2000).

Fifty-three percent of the sample projects are located in English Assisted Areas and 22% and 24% in Welsh and Scottish Assisted Areas, respectively

(Table 14.7). Concerning the country of ownership, 28.5% of the projects were foreign-owned companies (Table 14.8). This included a reasonable mix of other EU (15 projects), USA (17 projects) and Far Eastern companies (15 projects).

Table 14.8 Country of ownership

	Number of projects (sample)	% of total
UK	118	71.5
Foreign-owned	47	28.5
Total	165	100.0

Source: Elaboration of AEP (2000).

The results of the survey indicate the effect of the grant on location, especially for foreign-owned firms. The RSA changed the project location in around a fifth of cases, with most firms reporting that without the investment they would otherwise have located outside the UK. Attracting foreign investors is one of the priorities of the regional policy in the UK; foreign companies are more footloose than indigenous firms, which tend to keep their location because they are more embedded in their spatial environment. In addition, the analysis also suggests that the RSA was influential in encouraging larger projects which are more likely to be foreign-owned.

The importance of RSA as a pull factor is however not as important as was shown for the Italian case. The RSA represents the main pull factor for early-stage location of foreign-owned firms, but it is easily overtaken by agglomeration economies in the course of time. Moreover, Jones and Wren (2004) argue that inward investment policy is applied most intensively in the regions where the foreign direct investment (FDI) is already most concentrated in particular activities and where the effect of the policy is expected to be weakest.

14.4 CONCLUSIONS

The description of the approaches to regional policy and the related measures shows that the call for changing the focus of regional policy has been really more in its philosophy than in its practice. The tools promoting core-periphery relocation are of old fashioned type, with capital grants being the favourite incentives. Even the AC and the PC, which are negotiated plans' measures, make use of capital grants to attract business investments in the peripheral areas. The contemporary approach can therefore be considered to be a more balanced and pragmatic approach to regional policy. Although indigenous development remains very much at the heart of modern regional policy, the merit of a certain amount of inward investment as part of a balanced regional development strategy is now widely accepted.

In order to investigate whether capital grant measures constitute one of the principal reasons to invest in the peripheral areas, a firm-level analysis was carried out. This analysis focuses on the initiatives which have been granted by the three policy measures (AC for Italy, IPR for the Netherlands and RSA for the United Kingdom). The outcome of the analysis, although based on slightly different methodology for each country, shows that Government incentives are the most important determinant for the Italian firms investing in the Mezzogiorno, followed by the availability of industrial areas at competitive prices and labour force. As the factor and cluster analyses have shown, while centre-northern firms are particularly inclined to invest in the south because of the availability of incentives and industrial areas, small and very small local firms would probably have located in the areas without incentives. Incentives are necessary to overcome the diseconomies of the southern areas, and centre-northern firms benefit from them in order to be able to transfer medium-high value-added activities or new production lines in the Mezzogiorno, through the opening of new establishments. In the Dutch case, the availability of business sites together with their cost and the quality of living remain the most important factors, while none of the interviewees stated that without incentives he/she would never have invested. The RSA measure in the UK was considered an important pull factor by only a minority of investors, and mainly foreign-owned firms.

Nevertheless, it is important to emphasise that the role achieved by Government incentives in the location and relocation of activities to the Italian Mezzogiorno is probably linked to the following patterns: (i) the size of the Mezzogiorno and (ii) the amount of public grants offered by the AC. In other words, the Mezzogiorno cannot be compared to the Dutch Northern provinces as far as the level of economic development is concerned; similarly, the AC tool cannot be compared to the IPR or the RSA because of the different premium percentage for business investments, which totals 50–80% in the first case, compared to 15–30% and 10–35% in the IPR and RSA, respectively.

However, it cannot be denied how difficult it is to assess the effect of regional policy: the investment and location decisions are based on a rather complex set of variables and it is not entirely clear whether or not firms indeed know exactly the role played by regional policy measures. From the foregoing discussion, it can be concluded that the effectiveness of regional policy measures is at least questionable. Given the doubts about the contribution of regional policy measures to the solution of regional disparities, the question that arises is whether or not regional policy should be terminated, and particularly the old-fashioned type incentives.

A first sign of change has been provided by the most recent Memorandum of Dutch regional policy, in which it was announced that after 2006, the IPR premiums and the special incentives for the Northern Netherlands were to be suppressed, and the focus was to be on reinforcing the areas of the country which are already strong. The new policy, which is named 'spatial economic policy', holds that only development projects of national importance can receive special financial support, and this possibility is now open for places and regions everywhere in the Netherlands (Minez, 2004). As expected, such projects are more easily found in the core areas than in the periphery.

NOTES

1. The Mezzogiorno area includes the following southern regions: Abruzzo, Molise, Campania, Puglia, Basilicata, Calabria, Sicilia and Sardegna.
2. However, this number is partially underestimated because, mainly for fiscal reasons, centre-northern firms sometimes prefer to establish a new firm in the south rather than open a branch plant (Chiri et al., 1998).

REFERENCES

AEP (2000), 'Evaluation of Regional Selective Assistance 1991–1995', London: Arup Economics and Planning.

Armstrong, H. and Taylor, J. (2000), *Regional Economics and Policy*, Third Edition, Oxford: Blackwell Publishers.

Atzema, O. A. L. C. and Wever, E. (1994), *De Nederlandse Industrie, Ontwikkeling, Spreiding en Uitdaging*, Assen: Van Gorcum.

Bachtler, J. and Yuill, D. (2001), 'Policies and strategies for regional development: a shift in paradigm?', Regional and Industrial Policy Research Paper, **46**, August.

Barca, F. (2001), 'New trend in the Mezzogiorno and the change of development policies', in T. Padoa-Schioppa and S. R. Graubard (eds), *Italy: Resilient and Vulnerable: The European Challenge*, Daedalus, Journal of the American Academy of Arts and Sciences, **1**, Spring.

Boersma, L. (1980), 'Regional incentives in the Netherlands', in D. Yuill, K. Allen amd C. Hull, *Regional Policy in the European Community*, London: Croom Helm, pp. 165–86.

Camagni, R. (1983), 'Italy', in L. H. Klaassen and W. T. M. Molle (eds), *Industrial Mobility and Migration in the European Community*, Rotterdam: Gower, pp.146–87.

Chiri, S., Pellegrini, G. and Sapino, C. (1998), 'L'attuazione dell'intervento per lo sviluppo degli investimenti nelle aree depresse', *Rassegna Economica*, **1**, 157–212.

Del Monte, A. (1998), 'La nuova politica per il Mezzogiorno: dalla centralizzazione allo sviluppo locale incentivato', *Economia e Politica Industriale*, **100**, 45-71.

Dijk, J. Van and Pellenbarg, P. H. (1996), 'Regional policy in the Netherlands: problems, actors and strategies', in P. H. Pellenbarg, F. Schuurmans and J. de Vries (eds), *Reisgenoten, Liber Amicorum Prof.dr. W.J. van den Bremen*, Amsterdam/Groningen: Netherlands Geographical Studies, **214**, 211–37.

Giunta, A. and Florio, M. (2002), 'L'esperienza dei contratti di programma: una valutazione a metà percorso', *L'Industria*, **2**, 359–96.

Jones, J. and Wren, C. (2004), 'Do foreign inward investors achieve their job targets?', *Oxford Bulletin of Economics and Statistics*, **66** (4), 483–513.

Keeble, D. (1972), 'Industrial movement and regional development in the United Kingdom', *Town Planning Review*, **43** (1), 3–25.

Keeble, D. (1976), *Industrial Location and Planning in the United Kingdom*, London: Methuen & Co.

King, R. (1985), *Italy*, London: Harper & Row.

Mackay, R. R. (1997), 'Unemployment as exclusion', in P. Lawless, R. Martin and S. Hardy (eds), *Unemployment and Social Exclusion: Landscapes of Labour Inequality*, London: Jessica Kingsley/ Regional Studies Association, pp. 49–68.

Martin, R. (1988), 'The political economy in Britain's north-south divide', *Transactions of the Institute of British Geographers*, **13**, 389–418.

Martin, R. (1989), 'The new economics and politics of regional restructuring: the British experience', in L. Albrechts, F. Moulaert, P. Roberts and E. Swyngedouw (eds), *Regional Policy at the Crossroads*, London: Jessica Kingsley, pp. 27–51.

Mariotti, I. (2005), 'Firm relocation and regional policy. A focus on Italy, the Netherlands and the United Kingdom', *Netherlands Geographical Studies*, **331**, Utrecht/Groningen.

Minez (2004), 'Pieken in the Delta', Den Haag: Ministerie van Economische Zaken.

Molle, W. T. M. (1983), 'The Netherlands', in L. H. Klaassen and W. T. M. Molle (eds), *Industrial Mobility and Migration in the European Community*, Rotterdam: Gower.

Oosterhaven, J. (1996), 'Dutch regional policy gets spatial', *Regional Studies*, **30** (5), 527–32.

Pellenbarg, P. H., Van Wissen L. J. G. and Van Dijk J. (2002), 'Firm migration', in P. McCann (ed), *Industrial Location Economics*, Cheltenham, UK and Lyme, US: Edward Elgar, pp. 110–48.

Perroux, F. (1966), *L'Economia del XX Secolo*, Milan: Etas Kompass.

Saraceno, P. (1977), 'Risultati e nuovi obiettivi dell'intervento straordinario', Rome: SVIMEZ.

Saraceno, P. (1980), 'Cattedrali nel deserto?', *Informazioni Svimez*, **1**, Janauary.

SVIMEZ (2001), 'Rapporto 2001 sull'economia del Mezzogiorno', Bologna: Il Mulino.

SVIMEZ (2002), 'Rapporto 2002 sull'economia del Mezzogiorno', Bologna: Il Mulino.

SVIMEZ (2003), 'Gli investimenti "esterni" nel Mezzogiorno', Rapporto 2003 sull'economia del Mezzogiorno, Bologna: Il Mulino, pp. 672–691.

SVIMEZ (2004), 'Gli strumenti in forma negoziale', I contratti d'area, Rapporto 2004 sull'Economia del Mezzogiorno, Bologna: Il Mulino, pp. 414–420.

SVIMEZ (2005), 'Rapporto 2004 sull'economia del Mezzogiorno', Bologna: Il Mulino.

Viesti, G. (2002), 'Il sud che attrae', Quaderni Formez, Rome: Donzelli.

Viesti, G. and Capriati, M. (2003), 'The impact of national policies on territorial cohesion: the case of Italy', Background Report prepared for the Third EU Report on Cohesion, Mimeo, Bari.

Viesti, G. and Prota, F. (2004), 'Le politiche regionali dell'Unione Europea', Bologna: Il Mulino.

Wren, C. (1996a), *Industrial Subsidies: The UK Experience*, Basingstoke: Macmillan Press.

Wren, C. (1996b), 'Grant equivalent expenditure on industrial subsidies in the post-war United Kingdom', *Oxford Bulletin of Economics and Statistics*, **58** (2), 317–53.

Yuill, D. (1980), 'Regional incentives in the United Kingdom', in D. Yuill, K. Allen and C. Hull, *Regional policy in the European Community*, London: Croom Helm.

Yuill, D. and McMaster, I. (2001), 'Benchmarking regional policy in Europe: a review of national regional policies', paper presented at the Senior Policymaker Conference, Forest Hills, Scotland, 9–12 September 2001.

Yuill, D., Allen, K. and Hull, C. (1980), *Regional Policy in the European Community*, London: Croom Helm.

15. Innovation Policy after the 'Celtic Tiger'

Declan Jordan and Eoin O'Leary[1]

15.1 INTRODUCTION

Ireland, after the 'Celtic Tiger' boom of the 1990s, may be regarded as a mature industrial region. A policy consensus exists that the way forward is to develop Ireland as a knowledge-driven economy. Accordingly the government has significantly increased its investment in science and technology in recent years, the promotion of which is at the heart of Irish enterprise policy. The objective is that 'Ireland, by 2013 will be internationally renowned for the excellence of its research and will be at the forefront in generating and using new knowledge for economic and social progress, within an innovation driven culture' (Department of Enterprise, Trade and Employment, 2006, p.8).

This chapter shows how this consensus reflects the international literature on the importance for promoting innovation of clusters and networks of interacting businesses and third-level institutions. Based on a survey of 180 Irish high-technology businesses, it questions the particular type, if any, of local or regional clusters that might be reasonably expected for the promotion of innovation in Ireland. It suggests that greater emphasis should be placed on policies targeted at those elements of the Irish innovation system that are likely to achieve the greatest return. These include improved transport and communications infrastructure to facilitate local, regional and international interaction, investment in education and training and support for commercialisation of research.

The chapter begins by presenting the policy consensus and relating it to the international literature. It proceeds to present the results of the innovation survey of foreign-owned and indigenous Irish high-technology businesses in the period 2001–2003. The concluding section draws out the policy implications of the results.

15.2 THE IRISH POLICY CONSENSUS

The extraordinary performance of the Irish economy since the 1990s has been inextricably linked to highly successful foreign-owned businesses, in sectors such as pharmaceuticals, electronics and computers (Gallagher et al., 2002). By 2004 the stock of foreign direct investment in Ireland reached €168 billion, 'the highest in the world in per capita terms, after Hong Kong' (Enterprise Strategy Group, 2004, p. 5). These businesses were largely drawn from a buoyant US economy locating in Ireland to gain access to increased intra-EU trade facilitated by the Single European Act. They were attracted to Ireland by low corporation taxes, sound macro-economic policies and the availability of a relatively young well-educated workforce (Kennedy, 2000/01). Yet, at the beginning of the new millennium, full employment, increasing costs and international tax competition make Ireland a less attractive proposition to multinationals in search of a low-cost location.

As a result the consensus has emerged for Ireland to develop as a knowledge-driven economy. This involves increasing public and private sector investment in research and development in order to entice greater levels of innovation by both foreign-owned and indigenous enterprises. There is now a widespread acceptance among policy makers that the factors that contributed to recent strong growth will not be sufficient to ensure sustained growth into the future. This consensus may be traced to a number of key policy statements.

In terms of industrial policy, the Culliton Report (1992) was the first to stress the importance of local or regional clusters around internationally competitive businesses for improved national competitiveness, and in particular for embedding foreign-owned businesses. These prescriptions were strongly influenced by the work of Porter (1990). The National Economic and Social Council subsequently commissioned research on the applicability of Porter's cluster approach to Ireland, which was somewhat critical (see for example, O'Malley and Van Egeraat, 2000).

More recently, the National Competitiveness Council recommended support for clusters and networks (2004, p. 3). Forfas, which is Ireland's national policy and advisory board for enterprise, trade, science, technology and innovation, recommended that the government focus on inter-firm networks as a key building block in the development of the innovation capacity of Irish industry (2004, p. 7). The Enterprise Strategy Group, appointed by the government to prepare an enterprise strategy for growth and employment in Ireland to 2015, also advocated networks involving industry, academic and public sector cooperation to drive the development of knowledge and expertise (2004, p. 53).

Irish regional policy has also advocated clusters. The *National Spatial Strategy* (Department of Environment, Heritage and Local Government, 2002) was formulated as a response to increasing regional divergence, primarily caused by the impressive performance of the Greater Dublin Area.[2] It proposed an urban hierarchy model, consisting of eight gateways and nine hubs. It envisaged, for example, gateways as having 'large clusters of national/international scale enterprises, including those involved in advanced sectors' (2002, p. 40). O'Leary (2003) has argued that this strategy is more concerned with distribution rather than the mainstream policy goals of growth and competitiveness. When it comes to enterprise policy the official view seems to be of Ireland as one region, rather than consisting of a number of sub-regions.[3] This can be seen for example in the Enterprise Strategy Group attaching little importance to the *National Spatial Strategy*, which is mentioned only once and is not seen as a key source of future competitive advantage (2004, pp. 97–8). There is little appetite in Irish regions for developing distinctive policies on industrial development. The Regional Planning Guidelines, which gave effect to the National Spatial Strategy in Ireland's eight regions, were remarkably similar in identifying the same kinds of industry clusters in high-technology knowledge-based sectors.

The continuing enterprise policy consensus on clusters and networks has been influenced, as in other EU countries, by the performance of particular industrial clusters, which have been associated with strong innovation performance, such as Silicon Valley, Emiglia-Romagna in Italy and the science-based cluster in Cambridge, UK (Forfas, 2004). These clusters have been highlighted by the work of authors such as Scott (1988), Porter (1990) and Castells and Hall (1994).

In recent years substantial public funding has also been devoted to research and development. In the lifetime of the *National Development Plan 2000–2006* (Department of Finance, 2000) public spending on science and technology doubled. In addition to establishing funding of research in Third Level institutions, the government also established Science Foundation Ireland, the purpose of which was to undertake and support strategic research of world class status in key areas of scientific endeavour, which would underpin Irish economic development (Forfas, 2005a). Science Foundation Ireland's task has been to establish a world-class research capability in niche areas of ICT and biotechnology, including the underlying scientific disciplines. Public spending on this initiative has increased from €10 million in 2001 to €121 million in 2005 (Forfas, 2005b). The Enterprise Strategy Group endorsed this funding and proposed additional initiatives, such as building enterprise capability, funding collaboration between industry and Irish Third Level Colleges and introducing tax credits for research and development (2004).

The *Strategy for Science, Technology and Innovation to 2013* (Department of Enterprise, Trade and Employment, 2006) has pledged the Irish government to further increases in public spending on science and technology to 2013. Commitments are to be front-loaded to 2007 and 2008 which will witness increases of 40% over existing levels. Among the targets specified are: a doubling of the number of PhD graduates by 2013 and an increase in private sector research and development from 0.9 to 1.7% of GNP over the same period (2006).

In an EU context, the commitment to increased Irish research and development spending is in line with targets established as part of the 'Lisbon Agenda'. At present Irish public spending on research and development is 1.6% of GNP, well below the EU target of 3%. Like its enterprise policy, Ireland's science and technology policy draws on the international high-technology cluster literature already identified. In addition, the *Strategy for Science, Technology and Innovation to 2013* cites the work of Florida and Tingali, who suggest that the driving force of economic growth is 'the ability of nations to attract, retain and develop creative people' (2004, p. 5).

By presenting survey-based evidence on the sources of innovation in Irish high-technology industry, this chapter makes an important contribution to the challenges faced by Irish policy makers. In doing so it contributes to the wider debate on policy development in an international context. The chapter continues by outlining the survey design.

15.3 SURVEY OF INNOVATION IN IRISH HIGH-TECHNOLOGY BUSINESSES

The survey, conducted towards the end of 2004, was targeted at Irish high-technology businesses, which have been crucial to the recent strong performance of the Irish economy. For example, Information and Communications Technologies (ICT) and Chemicals (which includes Pharmaceuticals), between them, accounted for approximately 90% of the growth in industrial output and employment during the second half of the 1990s (Department of Enterprise, Trade and Employment, 2003, p. 67). The particular sectors chosen for the survey are classified as *Chemicals and Pharmaceuticals, ICT* and *Electronic Devices and Engineering.*[4] These sectors are identified by the Enterprise Strategy Group as having future growth opportunities (2004, pp. 41–45).[5] The list of businesses in the selected sectors was constructed from the IDA Ireland database,[6] which relates to foreign-owned businesses, and the Enterprise Ireland (EI) SourceIreland website,[7] for indigenous companies.[8] While most of the businesses included may be classified as manufacturing, some may also provide services. In addition, some businesses, such as those in the software sector, are classified as services. It is estimated that 38% of the population of *ICT* and *Electronic Devices and Engineering* and 73% of *Chemicals and Pharmaceutical* businesses are covered in the resulting database used for the survey.[9] In terms of employment, the survey is representative of the population, although there are a small number of very large businesses in the survey.[10]

A self-administered questionnaire, containing 25 questions on levels of product and process innovation and their sources, was circulated to 857 businesses.[11] Table 15.1 details the response rate achieved by sector and type of business. In the context of a lengthy questionnaire and survey fatigue by businesses (Central Statistics Office, 2001), a total of 184 responses with an overall response rate of 22% is quite satisfactory. As can be seen from Table 15.1, the response rate is relatively evenly spread across sectors and types of business.

Table 15.1 Responses by sector and type of business[a]

	Sample	Responses	Response Rate %
Chemicals and Pharmaceuticals			
Foreign[b]	86	27	31
Indigenous[b]	97	16	17
Total	183	43	24
ICT			
Foreign	129	25	19
Indigenous	222	40	18
Total	351	65	19
Electronics Devices and Engineering			
Foreign	156	34	22
Indigenous	167	41	25
Total	323	75	23
Total	857	184[c]	22

Notes:
a. A detailed account of the population, the survey instrument and the survey findings are available from the authors on request.
b. Foreign-owned and indigenous businesses are from the IDA and EI databases respectively.
c. One respondent was anonymous and could not be classified by sector or by location.

Given the obvious differences in the development of both indigenous and foreign-owned businesses it is important to note the different characteristics of these respondents. The 98 indigenous respondents had an average of 49 employees in 2003, 54% of whom had a third-level degree. The 86 foreign-owned respondents had an average of 182 employees, 29% of whom had third-level education. The average age of indigenous businesses was 14 years compared to 23 for foreign-owned. *ICT* respondents were significantly smaller businesses and a greater proportion of their workforce had third-level education. These differences in age, employment and proportion in third-level are statistically significant at the 95% level. Geographically, respondents are spread throughout the country, with 48% in the Dublin/Mid-East regional authority areas, 22% in the South-West, 10% in the West and the remaining 20% spread between the Border, the Mid-West and the South-East.[12]

Following studies such as Roper (2001), MacPherson (1998) and the Community Innovation Survey conducted by the Department of Trade and Industry (2004) in the UK, product innovation is defined as the introduction of new or improved goods/services to the market over a reference period, which is taken as 2001 to 2003 in the survey. Process innovation, which is less tangible and, as a result, more difficult to measure, is defined as the introduction to the business of a new method of producing or delivering existing goods/services, the reorganisation of support activities, management structures or distribution channels, the introduction of existing goods/services to new markets and the introduction of new sources of supply of materials or other inputs over the same period (Schumpeter, 1934; Kline and Rosenberg, 1986; Gordon and McCann, 2005; Department of Trade and Industry, 2004). Businesses were asked to indicate whether they introduced process innovations continuously, frequently, regularly, rarely or never in the reference period.

These comprehensive definitions of what may be referred to as innovation output, is closely linked to what managers might be expected to observe in their businesses. It avoids some of the thorny issues involved in trying to elicit whether an innovation is new to the market and whether its impact is commercially significant (Kline and Rosenberg, 1986; Gordon and McCann, 2005). Given the breath of this definition, it is not surprising that 80% of respondents were product innovators with 76% introducing process innovations, either regularly, frequently or continuously between 2001 and 2003. Similar results are found in other national and international studies (Forfas, 2005a; Becker and Dietz, 2004).

In order to determine the sources of both product and process innovation, businesses were asked whether they undertook research and development, either formally through dedicated research and development departments, or otherwise. They were then asked their frequency of interaction with other group companies (which relate to other companies in the same organisation and might be especially important for foreign subsidiaries), suppliers, customers, competitors, academic researchers (in universities and Institutes of Technology) and innovation support agencies (such as IDA Ireland and Enterprise Ireland). Interaction may include meetings, networking or other communications that affect innovation. It may range from social or informal, perhaps unintentional, networking to formal or contractual collaboration that might generate new knowledge used for product or process innovation. Frequency of interaction was measured on a scale from continuously, to frequently, regularly, rarely and never. This approach to the study of interaction is slightly more detailed than generally found in the literature, which typically involves asking businesses whether or not they engage in interaction (see for example MacPherson, 1998; Love and Roper, 2001; Freel, 2003).

In order to understand the importance of local and regional sources of innovation, businesses were asked to estimate the one-way driving time from their most important interaction agents for both product and process innovation. Driving times were categorised in intervals of less than half an hour, a half to one hour, one to two hours, two to four hours and greater than four hours. The lower end of this range represents local interaction, with the upper end including interaction with agents outside the state. This method of measuring the importance of geographical proximity follows that of MacPherson (1998). It is preferred to the standard measure, which involves asking businesses whether they are co-located with interaction agents.[13]

15.4 INNOVATION AND THE ROLES OF INTERACTION AND PROXIMITY

Table 15.2 presents the percentage of respondents who undertook research and development by sector and type of business. This shows that 67% of businesses indicated that they undertook research and development between 2001 and 2003. At the 99% confidence level, indigenous firms were more likely than foreign-owned businesses to perform research and development. There was no significant difference in the likelihood of performing research and development across sectors.[14]

Table 15.3 presents the frequency of interaction for product innovation by interaction agents, sector and type of business in terms of percentage of respondents. This shows a striking pattern across the three sectors and for both foreign and indigenous businesses. For a clear majority of businesses, regular, frequent or continuous interaction occurs with other group companies, suppliers and customers. This strong interaction is in stark contrast to the noticeably weaker interaction with competitors, Third Level Colleges and innovation support agencies, as indicated by the majority of businesses which never or rarely interact with these organisations. This difference is significant at the 99% level. There are no statistically significant exceptions to these clear trends.

Table 15.2 Businesses conducting research and development, by sector and type of business (% of respondents)

Chemicals and Pharmaceuticals		58
of which	Foreign	48
	Indigenous	75
ICT		74
of which	Foreign	60
	Indigenous	83
Electronic Devices and Engineering		65
of which	Foreign	50
	Indigenous	78
Total		67
of which	Foreign	52
	Indigenous	80

Table 15.4 presents the time distance between businesses and their most important interaction agents for product innovation. This shows that for those agents with whom interaction is strong, there is a clear tendency for the most important agent to be located more than one hour and usually more that 4 hours driving time from high-technology businesses. Thus, for other group companies, suppliers and customers interaction occurs over relatively long distances, and clearly not locally. This result holds for all sectors and types of businesses and is significant at the 99% level. For competitors, with whom it was seen in Table 15.3 that businesses do not interact strongly, the most important agent is located more than four hours away for a clear majority of businesses. For both Third Level Colleges and innovation support agencies, where interaction is weaker, no clear pattern emerges, with the most important agent being spatially spread across local and international locations. It should be noted that these results relate to the most important interaction agent, from the perspective of the businesses themselves. Thus, it

is possible that interaction between these high-technology businesses and, for example, suppliers occurs locally, but this is not regarded by the businesses as most important for innovation.

Tables 15.5 and 15.6 present the frequency and proximity of interaction for process innovation. From Table 15.5 it is clear that in the vast majority of cases interaction with other group companies, suppliers and customers is, once again, regular, frequent or continuous. Interaction for the purposes of process innovation with competitors, Third Level Colleges and innovation support agencies, occurs rarely or never in the majority of cases. This difference is significant at the 99% level. There are no statistically significant exceptions to these clear trends. A significant exception is indigenous *ICT* businesses, 59% of which never or rarely interact with suppliers.

Table 15.6 indicates that once again for agents with whom interaction is strong in Table 15.5 (i.e. other group companies, suppliers and customers), the location of the most important interaction agent is not local and may be international. This result holds for all sectors and types of businesses and is significant at the 99% level. A significant exception is indigenous *ICT* businesses, 43% of whom interact with suppliers located less than 1 hour from their business. The weak interaction with competitors occurs over long distances. For Third Level Colleges and innovation support agencies the most important source of interaction is again spatially spread, although there is a tendency for the interaction by indigenous *ICT* and *Electronic Devices and Engineering* businesses to be local in the majority of cases.

Tables 15.3 and 15.5 concentrate on the relationship between the frequency of interaction for both product and process innovation and each interaction agent separately. No account is taken of the relative importance of interaction with the different agents for the probability of Irish high-technology businesses introducing new products or processes. Table 15.7 presents the results of estimating logit equations for both product and process innovation. Only the variables that are significant at the 5% level are included. For product innovation, unsurprisingly those businesses that indicate they perform research and development have the largest effect. In terms of interaction agents, frequency of interaction with suppliers, customers and innovation support agencies improves the chances of product innovation. Of these interaction agents, innovation support agencies are most important and customers are more important than suppliers. It is worth noting that interaction with competitors has no significant effect on product innovation. Interestingly, the probability of engaging in product innovation is found to be negatively related to the frequency of interaction with Third Level Colleges. Thus, the more high-technology businesses interact with academia, the lower their chances of introducing new products.

Table 15.3 Frequency of interaction for product innovation by interaction agent, sector and type of business (% of respondents).[a]

	Group	Suppliers	Customers	Competitors	3rd Level	Agencies
Chemicals and Pharmaceuticals						
Foreign						
Never/Rarely[b]	0	11	11	81	67	63
Regularly to Continuously[b]	100	89	89	19	33	37
Indigenous						
Never/Rarely	n.a.[c]	13	13	69	56	44
Regularly to Continuously	n.a.	88	88	31	44	56
ICT						
Foreign						
Never/Rarely	21	24	13	68	71	75
Regularly to Continuously	80	76	88	32	29	25
Indigenous						
Never/Rarely	27	30	8	62	75	48
Regularly to Continuously	73	70	93	38	25	52
Electronic Devices and Engineering						
Foreign						
Never/Rarely	0	15	9	78	72	61
Regularly to Continuously	100	85	91	22	28	39
Indigenous						
Never/Rarely	n.a.	12	10	59	59	51
Regularly to Continuously	n.a.	88	90	41	41	49

Notes:
a. Numbers may not add to 100% due to rounding.
b. Respondents indicated frequency of interaction based on 5 categories as follows: never, rarely, regularly, frequently and continuously. For the purposes of this table the categories are grouped.
c. Not available for confidentiality reasons.

Table 15.4 Time-distance from most important interaction agent for product innovation by sector and type of business (% of respondents).[a]

	Group	Suppliers	Customers	Competitors	3rd Level	Agencies
Chemicals and Pharmaceuticals						
Foreign						
< 1 hour[b]	0	5	14	0	33	20
1 to 4 hours[b]	0	24	18	18	47	60
> 4 hours[b]	100	71	68	82	20	20
Indigenous						
< 1 hour	n.a.[c]	13	20	17	20	36
1 to 4 hours	n.a.	7	40	17	30	55
> 4 hours	n.a.	80	40	67	50	9
ICT						
Foreign						
< 1 hour	5	20	30	27	13	56
1 to 4 hours	5	13	10	0	38	33
> 4 hours	91	67	60	73	50	11
Indigenous						
< 1 hour	25	36	15	20	53	76
1 to 4 hours	0	8	21	20	24	10
> 4 hours	75	56	64	60	24	14
Electronic Devices and Engineering						
Foreign						
< 1 hour	7	12	13	33	40	24
1 to 4 hours	4	46	33	17	27	65
> 4 hours	89	42	54	50	33	12
Indigenous						
< 1 hour	n.a.	25	8	22	48	61
1 to 4 hours	n.a.	31	34	26	36	36
> 4 hours	n.a.	44	58	52	16	4

Notes:

a. Numbers may not add to 100% due to rounding.

b. Respondents indicated one-way driving distance in 5 categories as follows: < ½ hour; ½ to 1 hour; 1 to 2 hours; 2 to 4 hours and greater than 4 hours. For the purposes of this table the categories are grouped.

c. Not available for confidentiality reasons.

Table 15.5 Frequency of interaction for process innovation by interaction agent, sector and type of business (% of respondents).[a]

	Group	Suppliers	Customers	Competitors	3rd Level	Agencies
Chemicals and Pharmaceuticals						
Foreign						
Never/Rarely[b]	11	7	41	89	85	78
Regularly to Continuously[b]	89	93	59	11	15	22
Indigenous						
Never/Rarely	n.a.[c]	19	38	75	63	50
Regularly to Continuously	n.a.	81	63	25	38	50
ICT						
Foreign						
Never/Rarely	4	33	8	84	70	70
Regularly to Continuously	96	67	92	16	30	30
Indigenous						
Never/Rarely	30	59	26	77	85	82
Regularly to Continuously	70	41	74	23	15	18
Electronic Devices and Engineering						
Foreign						
Never/Rarely	13	24	34	93	73	65
Regularly to Continuously	87	76	66	7	27	35
Indigenous						
Never/Rarely	n.a.	33	29	78	83	71
Regularly to Continuously	n.a.	68	71	22	17	29

Notes:

a. Numbers may not add to 100% due to rounding.

b. Respondents indicated frequency of interaction based on 5 categories as follows: never, rarely, regularly, frequently and continuously. For the purposes of this table the categories are grouped.

c. Not available for confidentiality reasons.

Table 15.6 Time-distance from most important interaction agent for process innovation, by sector and type of business (% of respondents).[a]

	Group	Suppliers	Customers	Competitors	3rd Level	Agencies
Chemicals and Pharmaceuticals						
Foreign						
< 1 hour[b]	4	5	11	0	40	17
1 to 4 hours[b]	4	33	17	20	40	58
> 4 hours[b]	92	62	72	80	20	25
Indigenous						
< 1 hour	n.a.[c]	0	15	30	30	27
1 to 4 hours	n.a.	20	54	10	20	55
> 4 hours	n.a.	80	31	60	50	18
ICT						
Foreign						
< 1 hour	5	13	24	33	11	38
1 to 4 hours	5	13	12	22	44	50
> 4 hours	90	73	65	44	44	13
Indigenous						
<1 hour	25	43	15	19	58	79
1 to 4 hours	0	10	26	19	33	7
> 4 hours	75	48	59	63	8	14
Electronic Devices and Engineering						
Foreign						
< 1 hour	4	18	14	18	36	41
1 to 4 hours	3	36	27	9	29	35
> 4 hours	93	46	59	73	36	24
Indigenous						
< 1 hour	n.a.	21	26	28	60	58
1 to 4 hours	n.a.	36	32	22	25	30
> 4 hours	n.a.	42	42	50	15	12

Notes:
a. Numbers may not add to 100% due to rounding.
b. Respondents indicated one-way driving distance in 5 categories as follows: < ½ hour; ½ to 1 hour; 1 to 2 hours; 2 to 4 hours and greater than 4 hours. For the purposes of this table the categories are grouped.
c. Not available for confidentiality reasons.

Table 15.7 Logit estimations of the probability of innovating

	Product Innovation Coefficients	Process Innovation Coefficients
Age	+0.0385	-0.0267
	(0.1770)	(0.0131)
Perform R&D	+1.1788	+2.4232
	(0.6455)	(0.6710)
Frequency of Interaction with:		
Suppliers	+0.4210	+0.5691
	(0.2155)	(0.2119)
Customers	+0.5748	+0.4046
	(0.2200)	(0.1776)
Third Level	-0.7520	-0.6510
	(0.3127)	(0.3122)
Agencies	+0.8368	
	(0.3353)	
N	175	170
Psuedo R^2	0.3431	0.2672

Note: The figures in parentheses are standard errors.

Process innovation is defined as businesses engaging in this activity regularly, frequently or continuously. Once again, the performance of research and development has a positive and sizeable effect on the probability of being a process innovator. Frequency of interaction with suppliers and customers also improves the probability of engaging in process innovation, with suppliers being more important in this case. Interaction with competitors and innovation support agencies has no significant effect on process innovation. As for product innovation, frequency of interaction with Third Level Colleges has a negative effect on the probability of engaging in process innovation. This, together with a similar finding for product innovation, is especially noteworthy. It may perhaps be explained by these businesses only turning to academia as a last resort with complex and, perhaps insoluble, problems. Also the lack of on-going relationships and differences in culture between business and academia may inhibit interaction that is beneficial for these businesses. Clearly, this result is worthy of special attention.

15.5 THE CHALLENGE FOR IRISH POLICY MAKERS

This chapter presents new evidence, from a survey of 184 foreign and indigenous high-technology businesses on the critical issue of the roles of interaction and proximity in promoting innovation in Ireland. High-technology sectors are regarded by policy makers as crucial for future Irish competitiveness. Overall, while care must be taken in generalising from these results, they are remarkably consistent across sectors and for both foreign-owned and indigenous industry. The following is a summary of the results:

- Interaction for the purposes of both product and process innovation is strong between high-technology businesses and other group companies, suppliers and customers, as indicated by interaction being regular, frequent or continuous with these agents for 81% of businesses on average. The probability of engaging in product and process innovation is increased by more frequent interaction with suppliers and customers.
- This strong interaction with other group companies, suppliers and customers occurred over long distances, as indicated by the average driving time from high-technology businesses to the most important of these agents being greater than four hours in 67% of cases. This clearly implies that such interaction does not occur locally or regionally within Ireland, and may be international.
- Interaction for both product and process innovation with competitors is weak. Less than a third of businesses report at least regular interaction and this interaction has no significant effect on the probability of engaging in product or process innovation. When such interaction occurs, it does so over long distances.
- 68% of businesses rarely or never interact with Third Level Colleges and innovation support agencies in the promotion of both product and process innovation. This interaction is spread between local, regional and more distant agents. The greater the frequency with which Irish high-technology businesses interact with Third Level Colleges, the lower their chances of introducing new products and processes.

These results have important implications that merit debate in policy circles. They indicate the absence of any strong interaction for the purpose of promoting innovation between locally- or regionally-based concentrations of suppliers, customers, competitors, Third Level Colleges and agencies and Irish high-technology businesses.

These findings raise questions about the particular type, if any, of local or regional clusters and networks, which might reasonably be expected for the promotion of innovation in Irish high-technology businesses. Given the long-standing industrial policy of building competitive advantage on the back of foreign direct investment by successful high-technology businesses, it is perhaps not surprising that the survey found that interaction is strong and occurs between these businesses and other group companies over long distances. Love and Roper (2001) also find that technology transfer within multinational enterprises in Ireland is relatively high. In addition, the limited size of the domestic market and the overriding importance of international selling, might suggest the finding that the most important interactions by foreign-owned and indigenous businesses with customers are not local or regional, is to be expected. In this respect the finding of a positive association between the frequency of interaction with innovation support agencies and the probability of introducing new products is noteworthy. It may reflect the role, for example, of Enterprise Ireland in assisting the development of links between Irish businesses and international customers.

However, despite repeated efforts devoted to building backward linkages locally and regionally, especially between foreign-owned and indigenous businesses, it is notable that the survey finds interaction between high-technology businesses and suppliers for the purpose of promoting innovation occurs over long distances. The findings of weak interaction with innovation support agencies and, in particular, the absence of any significant relationship between the frequency of interaction with agencies and the probability of engaging in process innovation, are notable. These institutions, as part of their role, facilitate the process of developing backward linkages at the local or regional level.[15] Given the importance of innovation, a failure to forge backward linkages with high-technology businesses locally and regionally is a cause for concern, especially in the context of continued public funding devoted to developing networks and clusters.[16]

It was outlined earlier that Irish policy makers' view of clusters has been influenced by the work of authors such as Scott (1988), Porter (1990) and Castells and Hall (1994). From the perspective of innovation, Gordon and McCann (2005) have argued that these are an idealised type of cluster, which may not always be superior to alternative forms of agglomeration.[17] Thus, for example, in the Irish case more attention may need to be devoted to urbanisation economies, which depend on a diversity of businesses and skills in an urban setting. In the first instance this requires research to be undertaken to investigate the issue.

The idea that businesses interact, whether formally or informally, with competitors in order to promote their own innovation receives little support, with the survey showing that such interaction is weak and with no discernible relationship between this form of interaction and the probability of product and process innovation. In addition, where it occurs, such interaction is not local. The notion of collaboration between competitors has arisen from a number of celebrated examples in places such as Silicon Valley, Emiglia-Romagna and Cambridge (Scott, 1988; Castells and Hall, 1994; Forfas, 2004), where the businesses are small and flexible, thus enabling alliances to form easily. These special cases may not be easily generalised (Gordon and McCann, 2005). In the case of Ireland, the applicability of this concept is open to question, as typically high-technology businesses located in the country are a mix of very large foreign-owned and smaller indigenous businesses, operating in particular international market niches, with few firms competing with each other.

The survey findings are very important given the recent public funding of research and development which is to be substantially increased in the coming years. If high-technology businesses interact only weakly with Third Level Colleges, the question arises as to whether the return to Ireland from this public investment is likely to be satisfactory. Moreover, the finding that interaction with academia is detrimental to the probability of innovation raises serious issues for both policy makers and Third Level Colleges.

As regards the return on public investment in research and development, international evidence suggests that research and development spillovers are significant across international boundaries (see for example, Coe et al., 1997). The experience of the 'Celtic Tiger' testifies to this, where the hosting of established foreign-owned businesses has clearly resulted in the country benefiting from international knowledge spillovers, which may have originated, either directly or indirectly, from research and development activity by Third Level Colleges in other countries. Thus, the benefits of public investment in research and development are not likely to be retained within Ireland.

There is some international evidence of positive spillovers from university research to business innovation (Jaffe, 1989; Audretsch and Stephan, 1996; Anselin et al., 2000). However, the extent to which a region can benefit from research and development in Third Level Colleges depends on factors such as the relevance of the research to businesses, businesses' absorptive capacity, the strength of local knowledge dissemination networks and the integration of public and private knowledge mediating institutions (Roper et al., 2003, p. 114). The survey finding of weak interaction between high-technology businesses and Third Level Colleges casts doubt on the existing strength of Irish regional innovation systems.

These findings may reflect an overemphasis on the funding of basic research. An expert evaluation of Science Foundation Ireland found that although impressive progress had been made towards developing a strong research capability in ICT and biotechnology, there was a need to fund more applied rather than basic research and to focus on niche areas in a country as small as Ireland. It is recommended that more attention be paid to commercialising research (Forfas, 2005b). In addition, evaluations of particular measures geared at strengthening the links between Third Level Colleges and businesses suggest some weaknesses. Innovation Partnerships, a programme administered by Enterprise Ireland, does not involve real partnership, as businesses pay for, but do not participate in projects, which seem to benefit academics more than businesses (Technopolis, 2004). The evident lack of this form of collaboration is supported by a further evaluation, which revealed company concerns regarding intellectual property being a factor hindering interaction with Irish Third Level Colleges, who regard intellectual property as an income stream in a way which, in the light of US examples, seems unrealistic (Evaltec, 2004).

The greater emphasis placed by the Enterprise Strategy Group on the funding of applied research and in-firm research and development (2004, p. 69) indicates some cognisance of the importance of developing connectivity between Third Level Colleges and businesses. Although the return from substantial actual and planned public funding of research and development may be long term, it can be inferred from the results presented here, that the level and nature of the existing interaction between these businesses and Third Level Colleges may seriously compromise the achievement of satisfactory future returns.

Overall, the survey results suggest a limited role for geographical proximity in regard to innovation by Irish high-technology business. This may be partly due both to the distinctive development of Ireland's internationally competitive industry, with the dominance by foreign-owned businesses, as well as to the small size of the country. However, it may also be attributable to Ireland's undeveloped regional innovation systems, which currently seem to have little to offer these businesses in pursuit of enhanced innovation performance. Given the agreed imperative of developing alternative urban centres to the greater Dublin area, the failure to decentralise power regionally may undermine the ability of Ireland's other major urban centres[18] to use 'joined-up' government in order to address these deficiencies.

In order to develop stronger innovation systems, Irish regions require realistic policies targeted at facilitating interaction between businesses and regionally based suppliers, Third Level Colleges and support agencies. A clear implication of the survey results is the need to also facilitate national and, more crucially, international interactions by Irish-based business. Thus, policies to address the transport and communications infrastructure deficits, which policy makers often fail to connect to innovation, may have a crucial role in improving the innovation performance of Irish regions by facilitating interaction over long distances. It should not be taken for granted that policy is to be focused on high-technology clusters as the presence of urbanisation economies may warrant support for a diverse range of industries and services. The new emphasis on the funding of PhD's in Ireland's new 'fourth level' may very well yield a disappointing return. It may also result in the more basic educational and training needs of the Irish workforce, which were important during the 'Celtic Tiger' boom, being undermined.

Finally, more research should be undertaken on these important issues. Larger and more detailed surveys and case-study research are required on how high-technology businesses interact with agents and, perhaps more importantly, how Third Level Colleges, innovation support agencies and locally-based suppliers, interact with these businesses. In addition research is required to determine the relative importance of industry clusters and urbanisation economies in Irish urban areas. The overriding objective of further research must be to identify realistic policies to improve local and regional interaction in Irish regions. Clearly public funding of research and development in Third Level Colleges and of clusters or networks to support high-technology business should only be committed if the return is justified, following a detailed analysis of the economic costs and benefits.

NOTES

1. The authors wish to thank Eleanor Doyle and Bernadette Power of the Department of Economics, UCC and Brendan Whelan of the ESRI for helpful comments. They also wish to acknowledge the kind support of Enterprise Ireland who funded the survey. The views expressed are those of the authors.
2. Consisting of Dublin and the counties of Meath, Kildare and Wicklow (Mid-East).
3. Ireland is divided into two NUTS 2 and eight NUTS 3 regions primarily for the purpose of qualifying for EU Structural Funds. Paradoxically, the success of the economy has increasingly resulted in Irish regions being ineligible for maximum funding.
4. *Electronics Devices and Engineering* includes Medical Devices.
5. The Enterprise Strategy Group also identified Food and Consumer Goods, neither of which is considered to be high-technology, according to the OECD classification (OECD, 2004). The Enterprise Strategy Group additionally identified Internationally Traded Services as having growth opportunities. A sizeable number of businesses were identified ranging across a variety of sub- sectors such as Financial Services, Education Services and Creative Industries (2004, xii). Investigation of these businesses was considered to be outside the scope of this study.
6. www.idaireland.com
7. www.enterprise-ireland.com/sourceirelandsearch
8. This involved identifying and removing businesses double-counted on the databases and removing businesses that, for the purposes of this study, were inappropriately classified.
9. The population of businesses was constructed using the Census of Industrial Production (CIP), 2002 (CSO, 2004) for manufacturing local units and the National Software Directorate (www.nsd.ie) for software.
10. In each sector mean employment reported in the CIP, 2002 (CSO, 2004) is similar to the 5% trimmed mean for respondent employment.

11. The survey was addressed to establishments, with respondents being requested only to consider the activity at the particular location of their business.
12. These are NUTS 3 regions.
13. Thus, in the standard measure, interacting businesses in east Cork (South-West) and west Waterford (South-East) are not co-located, even though they are geographically close.
14. This and subsequent tests are based on the Pearson Chi-Squared test.
15. For example, Enterprise Ireland have been responsible for the National Linkage Programme.
16. For example, the Enterprise Strategy Group has proposed that €20 million per annum over five years is devoted to building networks (2004, p. 73).
17. Parr (2002) gives a clear account of the different kinds of agglomeration economies.
18. These are Cork in the South-West, Limerick in the Mid-West, Galway in the West and Waterford in the South-East.

REFERENCES

Anselin, L., Varga, A. and Acs, Z. J. (2000), 'Geographic and sectoral characteristics of academic knowledge spillovers', *Papers in Regional Science*, **79** (4), 435–43.

Audretsch, D. B. and Stephan, P. E. (1996), 'Company-scientist locational links: the case of biotechnology', *American Economic Review*, **86**, 641–52.

Becker, W. and Dietz, J. (2004), 'R&D co-operation and innovation activities of firms – evidence for the German manufacturing industry', *Research Policy*, **33**, 209–23.

Castells, M. and Hall, P. (1994), *Technopolies of the World: The Making of 21st Century Industrial Complexes*, New York: Routledge.

Coe, D., Helpman, E. and Hoffmaister, A. (1997), 'North-south R&D spillovers', *The Economic Journal*, **107**, January, 134–49.

Central Statistics Office (CSO) (2004), *Census of Industrial Production, 2002*, Dublin: Central Statistics Office, www.cso.ie.

Central Statistics Office (CSO) (2001), 'Characteristics of Sample Surveys in Ireland', paper presented at Conference of European Statisticians, United Nations Economic and Social Council, Geneva.

Culliton, J. (1992), *A Time for Change: Industrial Policy for the 1990s*, Report of the Industrial Policy Review Group, Dublin.

Department of Enterprise, Trade and Employment (2006), 'Strategy for Science, Technology and Innovation to 2013', Dublin: Stationery Office.

Department of Enterprise, Trade and Employment (2003), 'Review of Industrial Performance and Policy 2003', Dublin: Department of Enterprise, Trade and Employment.

Department of Environment, Heritage and Local Government (2002), 'National Spatial Strategy for Ireland 2002–2020: People, Places and Potential', Dublin: Stationery Office.

Department of Finance (2000), 'National Development Plan 2000–2006', Dublin: Stationery Office.

Department of Trade and Industry (2004), 'Detailed Results from the Third UK Community Innovation Survey (CIS3)', London: Department of Trade and Industry, http://www.dti.gov.uk/iese/cis.htm.

Enterprise Strategy Group (2004), 'Ahead of the Curve: Ireland's Place in the Global Economy', Dublin: Forfas, www.forfas.ie.

Evaltec (2004), 'Evaluation of Agency Supports for R&D Performed in the Business Sector', Dublin, http://www.forfas.ie/publications/forfas041025/webopt/forfas_rdevaluation_041025.pdf.

Florida, R. and Tingali, I. (2004), 'Europe in the Creative Age', February 2004, http://www.creativeclass.org/acrobat/Europe_in_the_Creative_Age_ 2004.pdf.

Forfas (2005a), 'Making Technological Knowledge Work: A Study of the Absorptive Capacity of Irish SMEs', Dublin, www.forfas.ie.

Forfas (2005b), 'Report of an International Evaluation Panel – Science Foundation Ireland The First Years 2001–2005', Dublin: Forfas, December, http://www.forfas.ie/publications/forfas051215/webopt/forfas051215_sfi_evaluation_report_webopt.pdf.

Forfas (2004), 'Innovation Networks', Dublin: Forfas, www.forfas.ie.

Freel, M. (2003), 'Sectoral patterns of small firm innovation, networking and proximity', *Research Policy*, **32**, 751–70.

Gallagher, L., Doyle, E. and O'Leary, E. (2002), 'Creating the Celtic Tiger and Sustaining Economic Growth: A Business Perspective', *Quarterly Economic Commentary*, Dublin: Economic and Social Research Institute, Spring, pp. 63–81.

Gordon, I. and McCann, P. (2005), 'Innovation, agglomeration and regional development', *Journal of Economic Geography*, **5** (5), 532-43.

Jaffe, A. (1989), 'Real effects of academic research', *American Economic Review*, **79** (5), 957–70.

Kennedy, K. (2000/01), 'Symposium on economic growth in Ireland: Where has it come, where is it going? reflections on the process of Irish economic growth', *Journal of the Statistical and Social Inquiry Society of Ireland*, **XXX**, pp. 123–39.

Kline, J. and Rosenberg, N. (1986), 'An Overview of Innovation', in R. Landau and N. Rosenberg (eds), *The Positive Sum Strategy: Harnessing Technology for Economic Growth*, Washington: National Academy Press.

Love, J. and Roper, S. (2001), 'Location and network effects on innovation success: evidence for UK, German and Irish manufacturing plants', *Research Policy*, **30**, 643–61.

MacPherson, A. D. (1998), 'Academic-industry linkages and small firm innovation: evidence from the scientific instruments sector', *Entrepreneurship and Regional Development*, **10**, 261–75.

National Competitiveness Council (2004), 'NCC Statement on Innovation', Submission by the National Competitiveness Council to the Enterprise Strategy Group, National Competitiveness Council, February.

OECD (2004), *OECD Science, Technology and Industry Scoreboard 2004*, Paris: OECD.

O'Leary, E. (2003), 'A Critical Evaluation of Irish Regional Policy', in E. O'Leary (ed.), *Irish Regional Development: A New Agenda*, Dublin: The Liffey Press, pp. 15–37.

O'Malley, E. and Van Egeraat, C. (2000), 'Industry clusters and Irish indigenous manufacturing: Limits of the Porter view', *Economic and Social Review*, **31** (1), 55–79.

Parr, J. (2002), 'Missing elements in the analysis of agglomeration economies', *International Regional Science Review*, **25** (2), 151–68.

Porter, M. (1990), *The Competitive Advantage of Nations*, London: Macmillan.

Roper, S., Hewitt-Dundas, N. and Love, J. (2003), 'Research and Development Centres in Less Favoured Regions: Towards an Ex Ante Impact Assessment', in E. O'Leary (ed.), *Irish Regional Development: A New Agenda*, Dublin: The Liffey Press, pp. 97–124.

Roper, S. (2001), 'Innovation, networks and plant location: Some evidence for Ireland', *Regional Studies*, **35** (3), 215–28.

Schumpeter, J. (1934), *The Theory of Economic Development*, Cambridge, US: Harvard University Press.
Scott, A. (1988), *New Industrial Spaces*, Pion.
Technopolis (2004), 'An Evaluation of the RTDI for Collaboration Programme', June, http://www.forfas.ie/publications/show/pub88.html.

16. Grants and the Location of Foreign Direct Investment: Evidence from the UK Regions

Colin Wren and Jonathan Jones

16.1 INTRODUCTION

Reflecting its phenomenal growth, the attraction of foreign direct investment (FDI) is an important component of the economic development strategies of many industrialised countries at both national and regional levels.[1] It is no less true of the United Kingdom, which relative to its GDP has received the greatest FDI inflow of any nation over the 1990s, and which has sought to actively encourage inward FDI (see Jones and Wren, 2006). In principle, the support to attract FDI can take many forms (e.g. infrastructure, factory or site provision, corporate tax policy, deregulated labour market conditions, the exchange rate, advice, etc), but an important element of many inward FDI policies is the provision of direct financial assistance to industry in the form of grants (see Mariotti, chapter 14 in this book and UNCTAD, 2003). In the UK, grant support for FDI has been substantial, amounting to more than £2 billion since the early 1990, and it has been important component of policy in many other industrialised countries, e.g. France (Crozet et al., 2004).

While the grants may just result in a zero-sum game between nations and regions, the issue that this chapter addresses is whether these grants are any use in determining the location of FDI plants. At the outset, three points should be stressed. The first is that it is perhaps strange that financial inducements for inward investment should be permissible in European Union countries at all, given that state aids are presumed to be prejudicial to competition by the European Commission and ruled unlawful.[2] However, exemptions make the grants available in certain geographical areas for the purpose of economic development where either the standard of living is abnormally low or there is serious underemployment (Article 87(3)(a)) or it facilitates certain activities and in areas without adversely affecting trading conditions (Article 87(3)(c)). In practice, this means that the grants tend to be available in the relatively underdeveloped areas and regions of the European Union, which in the UK are known as Assisted Areas.[3] Second, the grants work in an asymmetrical fashion, as while they are available to inward investment they are usually ruled illegal when applied to prevent outward investment, often in an *ad hoc* way. Finally, the fact that the UK has attracted a relatively large amount of FDI to its regions does not necessarily mean that the grants are an important factor in location, as it could just be that the grants just reflect the pattern of FDI location rather than determine it.

The purpose of the chapter is two-fold. The first is to review the large literature on the locational determinants of FDI in order to examine the effect of grants. This draws widely to consider the relative effects of the classical and agglomeration factors, but focusing on three recent UK studies of FDI grant assistance, each using a different methodology. In the UK context the FDI grants are made under the Regional Selective Assistance scheme, which operated from 1972, although in England it has recently been superseded (see Wren, 2005). The second purpose of the chapter is to examine the UK regional pattern of FDI and the operation of the grants, focusing on North-East England. In the micro-econometrics literature the agglomeration factors dominate the classical location terms, including grants, but our contention is that these studies may be intrinsically biased against the classical factors, and that as a result the grant effect may be misinterpreted. Indeed, the areas that are newly designated for regional grants rapidly accumulate FDI, and in a very narrow range of activities, so that the key task for future research is not only to be clearer about the effect of the grants, but about the relationships that underlie the agglomeration terms used in these studies.

16.2 FACTORS DETERMINING FDI LOCATION

The regional science and economics literatures are replete with empirical studies on the factors determining the location of internationally mobile industrial investment. In the literature a useful distinction is made between the *classical factors* and *agglomeration factors* of location. The former are the traditional location factors that directly affect the firm's costs (e.g. size of market or labour costs), while the latter includes those factors that are not firm specific, but which arise from advantages from a location close to other firms, e.g. Marshallian economies. The distinction between these factors is useful, although not always clearcut in empirical work, as some studies use industrialisation or infrastructure to proxy agglomeration, but which impact on firms' costs and could be regarded as classical factors. Likewise, terms to measure the classical factors (e.g. size of market) could be correlated with the agglomeration economies, so that again there is a problem of interpretation.

Nevertheless, the issue of whether it is classical or agglomeration factors that are important in determining location of FDI is a crucial concern. It is of interest here, as if it is the classical factors that are important then it suggests that FDI policy (including grants) should operate on the traditional location determinants in order to attract FDI. However, if the agglomeration factors are crucial, then not only is inward investment potentially self-perpetuating, with FDI attracting further FDI, but the grants are ineffective. Of course, this is ultimately an empirical matter, and to which attention now turns. Initially, the relative importance of the classical and agglomeration factors is explored, and then the location effect of the grants is investigated, based on an in-depth examination of UK evidence. A substantial literature has accumulated over the years, which, as we see, can appear inconsistent. However, it is argued that the contradictory results that are obtained, including on grants, can partly be explained by the different methodologies that are adopted.

16.2.1 Classical and Agglomeration Factors

The well-known seminal investigation on the locational determinants of FDI is Scaperlanda and Mauer (1969), who examine US investment in Europe. Later studies of the European Union (EU) include Pain (1997) and Barrell and Pain (1999a), while studies on specific EU Member States are Mariotti and Piscitello (1995) on Italy, Billington (1999) on the UK and Guimaraes et al. (2000) on Portugal. These all feature below, as well as some state-level studies for the US, including Coughlin et al. (1991), Head et al. (1995) and Hines (1996). Other research is carried out across countries in order to gain a broader perspective, including Kravis and Lipsey (1982), Wheeler and Mody (1992) and Braunerhjelm and Svensson (1996), while Wei et al. (1999) examine FDI in Chinese regions. The studies use single-choice models, but Mucchielli and Puech (2004) obtain interesting results from estimating a 'hierarchical model', in which the investor first chooses a country before the decision is made where to locate within that country.[4]

Classical explanations

Overall, the empirical studies provide evidence on a range of classical factors potentially affecting the location of FDI, although it is not always consistent. The main factors seem to be the market size, labour market conditions and fiscal regime. The findings of these studies are now briefly reviewed, while the issue of grants is examined in more detail below.

The *market size* is measured by GDP. In cross-country studies, Kravis and Lipsey, Wheeler and Mody and Braunerhjelm and Svensson all find it has a positive effect on US multinational FDI, while Coughlin et al. find similar evidence for the US. Some studies find the growth rate is more important (e.g. Billington, Barrell and Pain and Wei et al.), although Scaperlanda and Mauer fail to find any effect at all. As regards *labour market conditions* there are several measures. Wheeler and Mody measure the *cost of labour* by the average hourly wage rate and find an adverse effect on FDI, although these costs can be offset by higher productivity, as in Friedman et al. (1992). An alternative approach is to use the unit labour costs, and Coughlin et al., Barrell and Pain and Wei et al. all find that it negatively affects FDI location. A further measure of labour market conditions is the *rate of unemployment*. Higher unemployment may imply lower wage costs, not least as individuals are prepared to work for a lower wage rate. Billington, Coughlin et al. and Friedman et al. (1992) find that the unemployment rate has a positive effect on FDI location, although Taylor (1993) argues that too high a rate may be a disincentive, as it signals that an area is depressed.

Several studies find that *taxation* has a negative effect on FDI, including Billington for the UK. Both Coughlin et al. and Hines find that higher tax rates in US states a negative impact on FDI levels, while Hines also finds that firms that cannot claim tax credits from their home country reduce their level of FDI within high tax areas of the US. These pecuniary effects would seem to indicate a likely positive effect for grants, although neither Glickman and Woodward (1989) nor Wheeler and Mody fail to find an effect of taxation on FDI location. Blonigen (1997) and Gorg and Wakelin (2001) also fail to find a significant relationship between the *exchange rate* and FDI location.

Agglomeration economies

Whether measured by an area's infrastructure, the level of industrialisation or even the amount of existing FDI, Wheeler and Mody, Billington and Wei et al. all find a significant positive relationship between these and FDI location, which they attribute to agglomeration economies. Wei et al. also use the population density and find a positive relationship with FDI, but of course it could just as easily measure the market size or even superior transport links, which are classical factors. Coughlin et al. (1991) use the density of manufacturing activity as a proxy for agglomeration in their study of the US, and also find a positive influence of this on location.

In an effort to get at the nature of agglomeration economies, a distinction is sometimes made between the economies that occur between industries (Jacob externalities) and those that occur between firms in the same industry (known as MAR externalities).[5] The Jacob externalities are economy wide, occurring between firms in different industries, and are sometimes known as urbanisation economies, while the MAR externalities arise from economies in the form of labour market pooling, effects associated with demand and supply linkages or from technology spillovers. Most studies are conducted across industries, focusing on the Jacob externalities, although some recent FDI studies have firm-level data for a single industry and are able to examine MAR externalities, which could be of greater significance.

Braunerhjelm and Svensson (1996) measure agglomeration by the number of employees in each manufacturing industry relative to total employment. They find a positive effect on FDI location, especially in industries that are characterised by a high degree of technology, and it is in these industries that knowledge spillovers appear strongest (see Döring and Schnellenbach, 2006). Similarly, Barrell and Pain (1999b) find that agglomeration economies are present in technology-related industries, but use both the relative production scale and the relative size of the research base within an industry to measure the agglomeration economy. A recent analysis at the firm level by Devereux et al. (2007) suggests that firms tend to locate near to other firms in the same industry. This study includes terms for both MAR externalities, proxied by the number of plants in the same industry (that are possibly foreign owned), and Jacob externalities, as measured by a Herfindahl index between plants across industries, but it is the former that proves significant.

Agglomeration versus classical location factors

Like many other studies, the Devereux et al. study includes both classical and agglomeration terms. It finds that the inclusion of the agglomeration terms causes the classical terms to become significant, which suggests that the two kinds of variable should be considered together in order to examine which is more important. This is also the case for Head et al. (1995). They argue that it is not possible to list all the potential classical terms, so that at the US state level they include a single term to capture *all* of the classical variables that potentially influence a Japanese investor's location decision.[6] Since Japanese investment was relatively new in the 1980s then the existing pattern of US establishments is used to proxy the locational attractiveness of an area for an industry. The agglomeration terms include industry-specific variables, which are measured by the number of US and Japanese plants in an industry, as well as terms for the effects occurring across states. They find that the Japanese investors prefer a location where there already exists a concentration of previous Japanese FDI in the same industry, and that these agglomeration effects cross over into nearby states. Overall, Head et al. conclude that no state-specific effects are significant in determining the location of Japanese FDI in the US, so that the classical variables appear to have no effect.

Another study to use both classical and agglomeration terms is Guimaraies et al. (2000). The level of disaggregation is the Portuguese Concelho. This is a smaller geographical unit than the US state and it is believed to be better able to capture agglomeration. Four kinds of agglomeration economy are identified: industry-specific, manufacturing, foreign-specific and business-service effects. In their case the classical variables are the costs of land and labour, but these are each insignificant, whereas the agglomeration terms are each significant except for the foreign-specific effect, which is measured by the share of employment in foreign-owned firms in a Concelho. Mucchielli and Puech (2004) examine the classical and agglomeration location variables at both national and regional levels. They find that the classical variables are important to the choice at the national level only, and that the agglomeration effects are present at both levels. This raises the possibility that the grants, which are regional in nature, may in fact work at the national level.

Overall, the studies find that agglomeration economies play an important role in the attraction of FDI and that they dominate the classical factors. This appears conclusive, but caution should be exercised for two reasons. First, it may be difficult to distinguish empirically between the different factors, so that measures of industrialisation may just capture the market size, and hence classical effects. Second, it is likely that there is relatively little variation in the classical variables at the country level (i.e. taxes tend not to be spatially differentiated), but much more variation in the agglomeration terms, which is supported by Mucchielli and Puech (2004). Nevertheless, if the evidence is taken at face value it suggests that the agglomeration terms work for mature FDI when there is a build-up of foreign-owned plants in a particular locality, whereas the classical location factors may at best work for early-stage FDI. This could carry over to the grants, and this evidence is now considered

16.3 STUDIES ON THE EFFECTS OF GRANTS

Our discussion on the effect of grants on FDI location focuses on the UK, for which there is a long history of financial inducements for FDI and several in-depth studies. In particular, this review examines three studies of industrial location: Taylor (1993), Devereux et al. (2007) and AEP (2000). These are useful as these embody different methodologies and offer important clues on the effect of the grants on FDI location. Of course, a significant issue is the endogeneity problem that was highlighted in the introduction. In practice, the grants tend to be of two types: 'automatic', at predetermined grant rates to projects satisfying published eligibility criteria; and 'discretionary', where the grant rate is variable, possibly bargained. In the former case the *amount* of grant disbursed depends on the amount of FDI, and this simultaneity bias needs to be allowed for in applied work, while for discretionary assistance even the grant *rate* is endogenous. It is shown below that each of the three studies of interest is potentially able to circumvent the endogeneity problem, either directly or indirectly, but varying with the choice of methodology.

16.3.1 The Taylor Study

Taylor (1993) examines the geography of Japanese manufacturing FDI in the UK using county data for 1984–91. Japanese firms were beginning to increase their investment overseas at this time, with the number of Japanese establishments in the UK increasing in number from 40 to 219 over this period. A Poisson distribution was used to model the number of Japanese firms setting-up production in the UK. Explanatory variables were included for the labour market (earnings to measure labour costs and unemployment for availability), infrastructure (population density) and the agglomeration effect of supplier networks (employment in manufacturing and in services). To capture the effect of grant assistance, dummy variables were included for the different tiers of Assisted Area, representing different intensities of aid (these are Development and Intermediate Areas).[7] It has the advantage that the policy terms are exogenous, at least over the study period, although over the longer run, if the investment improves prosperity, then the designation of Assisted Areas may not be independent of the amount of FDI. In fact, major changes were made to UK regional policy instruments in 1988, although the Assisted Areas map was fixed until August 1993 (see Wren, 1996).

Reflecting the above policy changes, the analysis of FDI location was conducted for two sub-periods over 1984–91, representing pre- and post-1988 policy changes. The agglomeration terms were significant in both periods, but the labour costs, population density and infrastructure terms (which in this study is included as a classical variable) were always insignificant. The unemployment term was significant in the second period (perhaps reflecting a severe worsening of economic conditions), but with a positive coefficient when entered in levels and negative when entered as a rate. Taylor interprets it as firms preferring to locate where there is a plentiful supply of labour, but where the unemployment rate is not too high so that an area is depressed. As regards the Assisted Area terms these are significant in both periods, and they provide evidence for the effect of the FDI grants even with the inclusion of the agglomeration terms, which has proved problematic elsewhere.

16.3.2 The Devereux, Griffith and Simpson Study

The Devereux et al. (2007) study is interesting, as this seeks to address the endogeneity problem directly. This study uses firm-level data to examine the location of new plants over 1986–92, taking data for both UK and foreign-owned firms. It looks at the effect of grants, matching firm-level data with a dataset on grant offers, based on the postcode location and 4-digit industry.[8] A two-stage approach is taken. In the first stage, the grant rate is predicted for each firm based on its size, industry, region and ownership structure using a Tobit estimation that takes account of those observations where the grant offer is zero. In the second stage, a location model is estimated, including firm characteristics, the predicted grant rate and industry, region and time fixed effects. To capture the agglomeration effects, industry-specific, region-specific and region-industry variables are included.

The first iteration of the model included the dummy variables only, and showed that the firms were less likely to locate in an Assisted Area, but this is because nearly 50 percent of the plants chose to enter non-Assisted Areas in the South of England. When the (predicted) grant term was included then it revealed that the Assisted Areas were still a less attractive destination for FDI, but that higher grant rates lead to a greater likelihood of a firm setting up production in these areas. However, when the agglomeration terms are introduced, the effect of the grant remained significant but very small. A 1 percent increase in the regional grant increases the probability of location by only between 0.04 and 0.13 of 1 percent, and when calculated at the mean a £100 000 increase in the grant increases in the probability of location in an area by only 0.0001! This offers little support for grants as a location factor, which is in stark contrast to Taylor. The Taylor study includes agglomeration terms, but in the form of Jacob externalities, whereas these are rejected by Devereux et al. in favour of MAR externalities.

16.3.3 The Arup Economics and Planning Study

The AEP (2000) study uses questionnaire survey techniques to determine the effect of the UK regional grants on the employment of grant-aided firms, including their location. It is examined in Wren (2005), making comparison with the previous evaluations of the regional grants that were carried out in 1990 and 1993 using the same methodology. This is known as the 'industrial survey approach', and it involves probing managers of assisted plants about the counter-factual position in order to determine the policy effect (i.e. it asks what would have occurred in the absence of assistance, including where they might otherwise have located). The survey approach is subjective rather than objective in nature and it suffers from potential biases (Bartels et al., 1982). However, it has the advantage of providing useful, possibly supplementary, information on the way in which policy works (Smith, 2004), and as such it can help open the 'black-box' of evaluation (Wren, 2007).

The key issue in the survey approach is the 'additionality' of assistance, which is the extent to which the support alters the nature of investment, either in its scale, timing or location. The results for the most recent evaluation of the regional grants are shown in Table 16.1. Overall, it reveals that 19 percent of projects were changed in location (first column). This is similar to the result for all three of the evaluations taken together (i.e. 18 percent from a total of 526 respondents), but increasing from 11 percent for the first evaluation over 1985–88. The grants have increasingly been taken up by FDI, so that by value half the grant assistance went to foreign-owned plants over the 1990s. The AEP study does not disaggregate the results by the nationality of ownership, but it gives a breakdown by the investment scale of projects, which is interesting as the foreign-owned plants implement larger scales on average.[9] This is also shown in Table 16.1 (other columns). Overall, for investment scales of less than £100 000, less than 10 percent of projects are changed in location, whereas for investments above this scale it is in excess of 40 percent of projects, and 50 percent for scales greater than £1 million. Given that many FDI projects are in fact re-investments (see below), which may be less mobile, it suggests a powerful effect of the grants in altering FDI location. Further, of the larger projects reporting a change in location two-thirds say they would otherwise have located outside the UK, suggesting that the grants work at an international level, so that they seem to determine the location choice between countries rather than just between UK regions.

Table 16.1 The 'additionality' of the grants

Project Scale	All	< £25k	£25k–£100k	£100k–£1m	> £1m
Response (%):					
Gone Ahead Unchanged	19.5	17.0	31.3	18.2	22.7
Gone Ahead, but:					
later in time	25.1	45.0	27.6	12.2	7.1
on a smaller scale	14.4	20.9	9.3	9.8	0.0
elsewhere outside UK	13.7	2.1	1.4	29.8	33.3
elsewhere inside UK	5.3	0.0	11.8	11.7	15.8
Abandoned Altogether	22.0	15.0	18.6	18.3	21.1

Notes: The table reports the percentage of firms in each category responding to questions about what would have occurred to the grant-assisted project in the absence of the grant, given the alternatives shown. Adapted from survey results of 165 grant recipients over 1991–95. It excludes a very small number of cases where there was a combination of effects.

Source: Arup Economics and Planning, 2000, and authors' own calculations.

16.3.4 Some Difficulties

Using different methodologies, the three studies offer conflicting views for the effect of the grants on FDI location, so a key issue is to reconcile these results. However, rather than come to firm conclusions at this stage on what is the likely effect of the grants on FDI location, it is useful to first consider other evidence on the location of inward investment. Again, this is for the UK, and it involves considering the distribution of FDI both for the regions as a whole and in more detail for a single English region.

All the same, it can be noted that all three of the above approaches suffer from difficulties. The Taylor study includes terms for Jacob externalities, but Devereux *et al.* (2007), find that the MAR externalities are more important, and it is these that cause the grant term to be insignificant. Second, for plants that are grant aided the Devereux *et al.* study calculates the predicted grant rate, but for larger projects these grant rates are determined centrally in London, so there may not be much variation between these. The study also operates at the UK national level, whereas the survey-based results on the grants suggest that the FDI choice is between the UK and elsewhere. Finally, the survey evidence suffers from a number of biases, of which one is strategic bias. In their study, AEP (2000) find evidence that the respondents may reply in order to ensure the continuity of the grants. In particular, 'some companies have come to realise that RSA [the grants] will be assessed largely on the grounds of additionality and that the best way they can help secure its continuation is to claim maximum additionality' (AEP, 2000, para. 7.2.2).

16.4 EVIDENCE FROM THE UK REGIONS

16.4.1 The Regional Distribution of Investment

The above three studies offer contradictory evidence for the attractiveness of the Assisted Areas to FDI. Devereux et al. (2007) find that the Assisted Areas are relatively unattractive to foreign-owned start-up plants, but which diminishes with the grant rate, while Taylor (1993) and AEP (2000) find that the Assisted Areas are attractive to new investment. Of course, the policy terms in the Taylor study could just pick up agglomeration effects (e.g. the missing MAR terms), while the survey approach is subject to strategic bias. However, there is evidence for the UK that the designation of a region, or a part thereof, for Assisted Area status leads to a sustained increase in inward investment. There are several pieces of evidence to support this view.

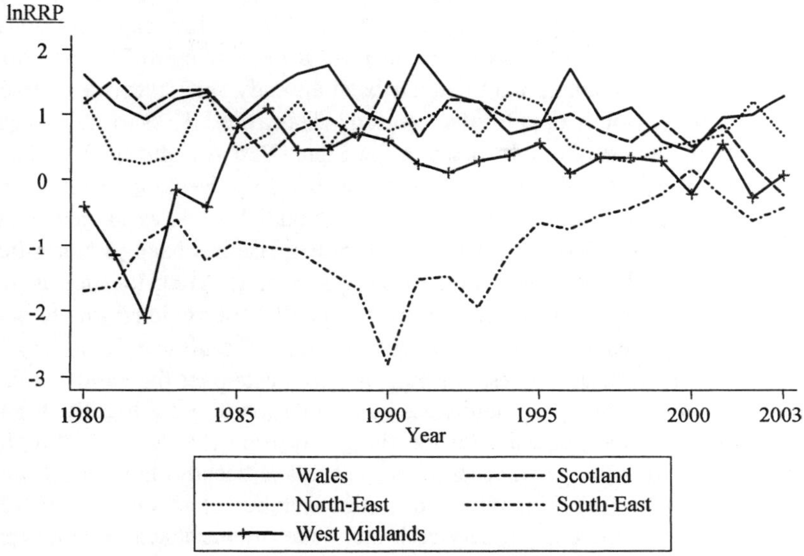

Note: The natural logarithm of the index of Relative Regional Performance (ln RRP), measured according to the number of jobs in projects.

Source: Jones and Wren (2006).

Figure 16.1 Regional location quotients for FDI

First, for selected regions, Figure 16.1 calculates location quotients for the regional share of employment in foreign-ownership relative to the share of UK employment in foreign hands.[10] It is for the years 1980 to 2003, and includes the three regions and countries that have been the most-intensively designated for the grants over this period (and continue to be so), namely Scotland, Wales and North-East England. A location quotient greater than zero implies that an area has been relatively more successful in attracting FDI. Figure 16.1 shows that the three regions have among the highest quotients of all the regions over the period, and which generally exceed zero.

The second piece of evidence, also apparent from Figure 16.1, is that the West Midlands region has substantially increased its share of FDI since 1984 when it was first designated for regional grants. From 1981 its share of UK manufacturing employment in foreign ownership nearly doubled by 1997 (from 7.9 to 15.3 percent), by which time it had the largest regional share of foreign-owned manufacturing employment in the UK. A similar increase in

FDI is evident for Scotland, Wales and North-East England after they were first designated for large-scale regional grants in the early 1960s. It is not shown in Figure 16.1, but it is apparent from Hill and Munday (1992). The grants have been taken-up mainly to manufacturing (owing to EU conditions on capital requirements), and there has been a steady shift in this investment away from the South East to other regions. Thus, the UK share of foreign-owned manufacturing employment in the South-East region fell from 51 to 28 percent over 1963–90, and by 1997 it was just 13 percent.

Not surprisingly, all regions outside the South East have gained in the share of foreign-owned jobs, but the biggest increase has been in North-East England and in Wales. The power of the grants in affecting location is also evident in relation to UK-owned firms. Originally, the regional grants were available for plants relocating within the UK, and Ashcroft and Taylor (1979) find that policy caused 500 establishments to relocate over the 1960s, raising the employment of the Assisted Areas by about 100 000 jobs. Finally, Figure 16.1 shows that the trend in FDI locating away from the South-East region has reversed over the past decade or so, but this is because there has been a shift in the nature of FDI, with a declining number of manufacturing FDI projects, and an increase in service sector investments that has more than compensated for this. However, most service FDI occurs in London, and in the form of mergers and acquisitions, so that there are limited new jobs.

Agglomeration

One striking aspect of the regional pattern of FDI is its concentration in a few activities within regions, and no doubt underpinning the success of the MAR externality terms in the studies of FDI location. Brand et al. (2000) examine the five largest manufacturing activities in the West Midlands, Wales and Scotland combined for the years 1994 and 1995.[11] They find that the foreign-owned sector's share of output in these five activities is greatest in Scotland, at 52 percent, but that Wales and the West Midlands have 39 and 23 percent respectively. This partly reflects the nature of FDI that arises from different countries.[12] Further, a disaggregation of the five activities shows its location also varies between locations. In Scotland FDI accounts for 80 percent of output in electronics and electrical engineering, in the West Midlands it is 60% of transport equipment, and in Wales the most important sectors are electronics (38 percent), metals (28 percent) and transport (21 percent). As we now see, there is a similar concentration of FDI in a few activities in the North-East of England.

16.4.2 Grants for FDI in North-East England

Further insights into the nature of FDI and role of grants can be obtained by examining these for a single region. This is the North-East England, which relative to its size and by the number of jobs has received more FDI than any other British region. It was also subject to some of the earliest regional policy efforts dating back to the 1930s. These insights are based on a study of FDI reported in Jones and Wren (2006), where further information can be found. This study matched details of regional investment grants with project-level data on foreign-owned investments, where information on the latter is based on the returns made by the Regional Development Agency to the national inward investment agency (and forming part of the data reported in Figure 16.1). It is believed to be a comprehensive account of FDI in the region since the mid-1980s, and it has advantages over similar datasets. For example, the UK production 'census' samples a proportion of smaller companies, which can be as little as 1 in 5 for firms with less than 100 employees (see Griffith, 1999), but in the North-East region three-quarters of foreign-owned start-up plants have less than 100 employees at entry (Jones and Wren, 2006). A good match was obtained for the grants, including smaller unpublished cases, which were supplied by the former Department of Trade and Industry.[13] The matching was helped by the fact that both the FDI and the grant data were recorded on a historical basis, so that plant details generally correspond.

In total, the dataset identifies 550 FDI projects in North-East England over 1985–98, carried out by 337 plants. Of the 550 projects, 164 are start-ups, 93 are acquisitions (with some mergers) and 24 are joint ventures.[14] The other 269 projects are reinvestments, which the development agency describes as 'expansions', indicating that they represent a 'significant' increase in plant scale. The plants tend to be located towards the populated Eastern seaboard of the region, which is in the Assisted Areas where the grants are available. A summary of the grant data is given in Table 16.2, with a disaggregation by the four project types. The data give the total offered amount of grant on which at least a payment is made, with details recorded at this time. Not all of the grant may be taken up, but these are the data used in other UK studies, while they at least reveal whether a plant has been assisted or not. The grants are discretionary in nature and given towards job creation investment projects in the Assisted Areas (see Wren, 2005).[15]

Pattern of grants

Table 16.2 shows that a third of the projects received a grant, i.e. 182 of the 550 projects. The project type most likely to be assisted is a reinvestment, with 41 percent of these receiving assistance against 30 percent of start-ups. This might be expected given that the primary purpose of the grants is job

Table 16.2 Grant offers to FDI projects

	All projects	SU	AC	JV	RE
Number of projects	550	164	93	24	269
Total Assisted Projects	182	49	16	6	111
% Grant Assisted	33%	30%	17%	25%	41%
Grant amount					
Total	£283.2m	£117.1m	£23.5m	£11.5m	£131.1m
Mean	£1.56m	£2.39m	£1.47m	£1.92m	£1.18m
Coefficient of Variation	3.22	2.87	1.86	1.72	4.33
Grant distribution	No. (Cum. %)				
≤£0.1m	39 (1.1)	10	4	1	24
>£0.1m - £0.5m	63 (9.2)	16	4	1	42
>£0.5m - £1m	33 (15.9)	7	2	2	22
>£1m - £2m	25 (27.8)	8	3	1	13
>£2m - £5m	14 (43.0)	4	2	0	8
>£5m - £10m	4 (52.9)	2	0	1	1
>£10m	4 (100.0)	2	1	0	1
Median grant	£0.42m	£0.47m	£0.50m	£0.75m	£0.38m
Gini Coefficient	0.77	0.81	0.72	0.79	0.76
Grant rate					
Mean	17.3%	17.2%	12.0%	14.7%	18.2%
Coefficient of Variation	1.57	0.94	0.67	0.75	1.82

Notes: Grant offers of Regional Selective Assistance to projects on which a first payment is made, at 1995 prices. The grant rate is calculated where the investment is known. The cumulative percentage refers to the grant amount, while the Gini coefficient is for the cum. % of grant expenditure relative to the cum. % of grant offers. SU = start-up; AC = acquisition; JV = joint venture; and RE = reinvestment project.

creation. The 182 projects attracted £283.2m in offered grant over 1985–98, at an average of £1.56 million, but with higher mean grants going to start-ups compared to reinvestments. The Gini coefficient for the start-ups shows that this grant expenditure is relatively concentrated in these projects, so that the median grant is £470 000 against £380 000 for the reinvestment projects. The grant rate can be calculated as the ratio of the offered grant amount to the proposed investment scale. It reveals that the mean grant rate is around 17 percent, but again there is less variation in the grant rate for start-up projects.

The analysis suggests that the grants tend to favour the re-investments, but that the larger grants go to start-up projects, reflecting their larger scale. The characteristics of 337 foreign-owned *plants* are shown in Table 16.3. The 'grant quotient' shows the propensity of a given plant type to be grant-assisted at some time over 1985-98 relative to all plants.

The plant details are recorded according to the date of the first observed project over this period, but this is of little significance, as plant details tend not to change. Of the 337 plants, 72 (21 percent) commenced in foreign ownership prior to 1985, and these are referred to as 'Mature plants'. The initial investment is not observed for Mature plants nor any other project prior to 1985, but Table 16.3 shows that they are more likely to be assisted. Overall, about half the plants are start-up plants (so that many of the re-investments projects are in fact undertaken by these), and these are much more likely to be assisted compared to the acquired plants.

The remainder of Table 16.3 shows that the foreign-owned plants are likely to be located in the metropolitan areas (61 percent), but that they are less likely to be assisted in these areas (0.85). This is because plants outside the metropolitan areas are more likely to arrive by start-up and to undertake reinvestment, and which is possibly a reason for their initial location in the non-built-up area. In terms of the supranational origin, plants from the Far East are more likely to be assisted, as these tend to be start-ups and hence to reinvest, while those from North America are more likely to be acquisitions and in services. The final part of Table 16.3 shows that the 'grant quotient' for activities outside of manufacturing is 0.21, which is just eight grant awards over 1985–98, so that virtually all the assistance has been taken-up by manufacturing. Within this sector the grants are concentrated in four activities, reflecting the agglomerative nature of FDI by activity, which as previously noted is apparent for other UK regions (see Brand et al., 2000).[16] The lower 'grant quotient' for chemicals is because it is mostly in the form of acquisitions, while for the other three manufacturing activities the quotients are close to unity. Overall, the FDI is concentrated in certain activities in the manufacturing sector, and (except for chemicals) the grants just seem to reflect this pattern of investment.

Table 16.3 The plant-level distribution of grants

	Plants		'Grant Quotient'
	No.	(%)	
Plant type[a]			
SU	164	(49)	0.91
AC	79	(23)	0.80
JV	22	(7)	0.92
Mature plant	72	(21)	1.44
Plant location			
Metropolitan[b]	204	(61)	0.85
Non-metropolitan	133	(39)	1.22
Plant origin			
Western Europe	133	(39)	1.09
North America	128	(38)	0.84
Far East	67	(20)	1.22
Rest of the World	9	(3)	0.41
Plant activity[c]			
Non-manufacturing	95	(28)	0.21
Manufacturing	242	(72)	1.31
of which:			
Chemicals	38	(16)	0.71
Machinery	26	(11)	1.07
Communications	26	(11)	0.97
Transport	29	(12)	1.11
Other	95	(28)	0.21

Note: The 'grant quotient' shows the proportion of plants in a category receiving support from the agency relative to the proportion of all 337 foreign-owned plants receiving a grant. a = Mature plants commenced prior to 1985, but the other plants from this time onwards up to 1998, where SU = start-up, AC = acquisition and JV = joint-venture plant. b = Location in the metropolitan county of Tyne and Wear or Teesside. c = Activity groupings are the NACE 2-digit classification. For the four manufacturing activities shown these are NACE codes 24, 29, 32 and 34 respectively.

Multivariate analysis

The above analysis suggests that the pattern of FDI and grants vary according to a number of characteristics. To get a handle on which of these is important a multivariate analysis is undertaken according to the variables in Table 16.3. The parsimonious result from a logit regression for the probability that a plant is grant assisted is given in Table 16.4. The right-hand side terms are the plant characteristics, which are mainly included in a binary form, and comprising up to four categories in each case in order to ease interpretation (e.g. for the region of origin these are Western Europe, North America and the Far East, while the baseline case is the Rest of the World). The table shows that the grants are more likely to go to plants in manufacturing, which have larger projects, and to younger plants, which includes start-ups. A term was included for whether the plant survived up to the year 2000 (about one-fifth did not), but this was insignificant, suggesting that the grant bodies were unable to discriminate in favour of these plants.

The other two significant terms in Table 16.4 must be treated with caution, as a plant is observed only if it implements a project over 1985–98, which is a form of data truncation. It shows that plants implementing more than one project over this period (multiple investors) are more likely to be aided. This is not surprising, as it is picking up the reinvestment projects, with up to one third of start-up and acquisition plants commencing after 1985 each receiving more than one grant offer by 1998 (see Wren and Jones, 2009). Table 16.4 shows that plants commencing in foreign ownership prior to 1985 (Mature plants) are more likely to get a grant, but since only plants that subsequently invest are observed then it is not representative of all plants commencing prior to this time. Of interest from Table 16.4 is that the location and country of origin are insignificant, which suggests it is the intrinsic characteristics of a project (scale, type and activity) that are important for the grant.[17]

Table 16.4 The probability of a grant

Dependent variable: offered grant = 1 (yes) or 0 (no)		
Variable description	Estimated coefficient	z-value
Constant term	- 1.029	3.41
Non-manufacturing activity	- 1.916	4.97
Number of project jobs	0.003	2.92
Age since plant establishment (years)	- 0.028	3.17
Multiple investor	1.545	4.98
Mature plant	1.275	3.45
n = 301, pseudo R^2 = 0.29		

Note: Parsimonious results from plant-level logit regression of grant offer. Variables are included in binary form, except jobs and age. Appendix Table 2 of Jones and Wren (2006) gives the variables initially included. Number of observations is less than 337 due to some missing observations on age and the number of jobs.

16.5 OVERVIEW AND CONCLUDING REMARKS

This chapter has ranged widely in its review of the evidence on the effects of government grants for inward foreign investment, which in the UK represent the main financial inducement for FDI. The chapter has spanned the literature on the determinants of FDI location, the UK evidence on grants, the pattern of FDI in the UK regions and it has considered the operation of the grants in a single English region. In relation to the role of grants in determining FDI location it remains to consider what can be made of all this.

On the positive side, all three studies considered in detail suggest that the grants affect FDI location, although they differ markedly in the magnitude of the effect. The industrial survey approach finds a strong effect on location, with half the large projects changed in location and often at an international scale, while the econometrics research finds a significant but extremely small effect. Given the potential strategic bias associated with the survey approach it may well overstate the policy effect, while given that the grants may affect the international location choice, and not just between the regions of Britain, the econometric study may underestimate it. The difficulty is that between these two extremes the grants may either be highly effective or ineffective in determining FDI location.

In the location studies the agglomerative nature of FDI features strongly, but this is not surprising, because as we have seen that FDI concentrates in certain activities, although not all regions specialise in the same activities. It potentially poses a problem for evaluation because the disbursement of grants is correlated with the pattern of investment, especially for start-up plants in manufacturing. However, when the agglomeration terms are included in the regression equations not only do the grant terms become insignificant, but they overwhelm all the other classical location factors. There continues to be debates about the precise nature of the agglomeration economies and in many location studies they are not modelled explicitly, but for several reasons this causes us to be sceptical about the interpretation of these terms.

First, an agglomeration economy is an externality, and its non-market and transient nature not only explains why it is poorly specified in empirical work, but it could mean that it is picking up other effects. For example, following the establishment of the Nissan car plant in North-East England in 1984 this region built-up a substantial number of foreign-owned plants in transport (see Table 16.3), but these form a supply-chain for Nissan, which sources its components 'just-in-time'. Thus, rather than an agglomeration this should really be described as an industrial complex.[18] It is not a matter of semantics, as when agglomeration economies are measured by the stock of FDI it could simply pick up supply linkages, while when it is measured by infrastructure it may just pick up classical location factors. Second, acquisitions have been an important form of FDI over the last decade, but the studies generally do not distinguish between the different types of investment, and it is to be expected that FDI will just reflect the existing industrial structure, which may just arise from historically important classical location factors.

This scepticism about the overwhelming importance of the agglomeration terms is reinforced by the accumulation of FDI in the UK regions that have been designated for grants, especially in manufacturing where the assistance is targeted. In particular, after their designation for regional assistance some regions have done much better in attracting FDI, which is not only true of Scotland, Wales and North-East England in the 1960s, but is the case for the West Midlands after it received Assisted Area status in the 1980s. A possible explanation for this is that the grants determine the location of early-stage FDI, but after which time FDI builds up in the regions where it has gained a foothold in certain activities. However, this build-up could arise for any of a number of possible reasons, including: industrial linkages, e.g. a supply chain; the acquisition of an existing plant that is in a geographically-concentrated activity; a signalling effect, whereby early movers demonstrate that a region is 'a good place to do business'; or it could be an agglomeration economy. This explanation is consistent with the area-based study of Taylor (1993) on the emergence of Japanese inward FDI in the UK in the 1980s and with the micro-econometric studies. However, it is at odds with the survey evidence, where strong location effects are found for the grants even in a mature stage of inward investment. Probably the truth lies somewhere between the results that are indicated by the different kinds of study, and for the reasons outlined above, but what is clear is a need to reconcile the contradictory findings of the different methodological approaches. For this, a much closer modelling of the relations between firms will doubtless prove fruitful.

NOTES

1. The level of FDI in the global economy has risen spectacularly, with world FDI inflows rising by 120-fold from $13 billion in 1970 to $1492 billion in 2000.
2. Article 87 of the European Treaty (formerly Article 92 of the Treaty of Rome) makes state aid unlawful if it distorts (or threatens to distort) competition or impacts on trade between Member States. Guidelines and Frameworks are published setting out the types and levels of support that the Commission is prepared to approve. In fact, the European Commission approves 98 percent of applications for state aid, mainly for distributional or social reasons (see Besley and Seabright, 1999), although clearly it also acts as a deterrent.
3. Regional policy evaluation methodologies and the results are discussed in Taylor (2002).
4. The approach has also been used to test the location decisions of firms within regions and sub-regions of individual countries, e.g. Hansen (1987) and Crozet et al. (2004).
5. The MAR economies are named in honour of Marshall, Arrow and Romer. They arise from the localisation economies of Marshall, and subsequent work on knowledge transfer by Arrow (i.e. 'learning-by-doing') and Romer's growth theory (see Wren, 2008).

6. In fact, they argue that it will lead to an omitted variable problem, and that the significance of the agglomeration variables could arise from these omitted terms.
7. There were seven sub-regional areas where the number of manufacturing employees in Japanese plants was greater than 10 percent of the area's manufacturing employment. Four of these were in Wales and another two in North-East England.
8. The matching was only partially successful, as 2320 grant offers (on which a first payment was made) were successfully matched for Great Britain, but representing only 45 percent of total offers by number or value. Further, only 3 percent of foreign-owned start-ups in the Assisted Areas received an offer of RSA, which compares with 30 percent found by Jones and Wren (2006) for the North-East of England (see below).
9. The foreign-owned plants received half the regional grants over the 1990s, but represent only 10 percent of offers (Wren, 2005, Figure 2).
10. This is the Relative Regional Performance (see Hill and Munday, 1992), which is presented in logarithmic form to aid the clarity of presentation. The figure is for both manufacturing and services, although manufacturing is of greater importance for the Assisted Areas.
11. These are rubber and plastics, metals, engineering, electronics and electrical equipment, and transport equipment.
12. Hood and Taggart (1997) identify 409 German-owned manufacturing companies in the UK at 1994. Of responding firms, 50 percent are in mechanical engineering, chemicals or vehicle-related activities.
13. Only 13 grant cases could not be matched over 1985–98, representing £4.2 million in grant offers, which is just 2 percent of the total value of offers in the region.
14. Standard definitions apply. Start-ups are sometimes known as 'greenfield' investments. Acquisitions are projects where a foreign firm takes more than a 50 percent stake in a UK-owned company, which includes mergers. A joint venture is where a foreign-owned firm has at least a 50 percent stake in a venture, possibly with another foreign-owned firm.
15. The regional grants sought to encourage 'sound projects, which will improve employment opportunities in the Assisted Areas' (House of Commons, 2004).
16. Manufacturing investment accounts for 86 percent of total known investment by the 337 plants in the dataset, and for manufacturing the four activities combined account for 89 percent of known investment and 68 percent of jobs. The investment figures may be overstated, as these data are more likely to be known for assisted plants, but the job data is more or less comprehensive.
17. A similar result is obtained for the project job scales in Jones and Wren (2004), where the scale is independent of where the project comes from, when it arrives and where it goes to.
18. McCann and Sheppard (2003) formally distinguish an agglomeration from an industrial complex and a social network (the latter is usually referred to as a 'cluster'). They argue that an agglomeration involves firms in competition co-locating in an area, but that the nature of relations between the firms is transient; for otherwise it is an industrial complex.

REFERENCES

AEP (2000), *Evaluation of Regional Selective Assistance 1991–1995,* London: Arup Economics and Planning.

Ashcroft, B. and Taylor, J. (1979), 'The Effect of Regional Policy on the Movement of Manufacturing Industry in Great Britain', in D. Maclennan and J. Parr (eds), *Regional Policy: Past Experience and New Directions*, Oxford: Martin Robertson.

Barrell, R. and Pain, N. (1999a), 'Domestic Institutions, Agglomerations and Foreign Direct Investment in Europe', *European Economic Review*, **43**, 925–34.

Barrell, R. and Pain, N. (1999b), 'Foreign Direct Investment, Innovation and Economic Growth within Europe', in C. Driver, and P. Temple (eds), *Investment Growth and Employment: Perspectives for Policy*, London: Routledge, 199–220.

Bartels, C. P. A., Nicol, W. R. and van Duijn, J. J. (1982), 'Estimating the impact of regional policy: A review of applied research methods', *Regional Science and Urban Economics*, **12**, 3–41.

Besley, T. and Seabright, P. (1999), 'State aids: Making EU policy properly reflect geography and subsidiarity', *Economic Policy*, April, 15–53.

Billington, N. (1999), 'The location of foreign direct investment: An empirical analysis', *Applied Economics*, **31**, 65–76.

Blonigen, B. A. (1997), 'Firm specific assets and the link between exchange rates and foreign direct investment', *American Economic Review*, **87**, 447–65.

Brand, S., Hill, S. and Munday, M. (2000), 'Assessing the impacts of foreign manufacturing on regional economies: The cases of Wales, Scotland and the West Midlands', *Regional Studies*, **34** (4), 343–55.

Braunerhjelm, P. and Svensson, R. (1996), 'Host country characteristics and agglomeration in foreign direct investment', *Applied Economics*, **28**, 833–40.

Coughlin, C., Terza, J. and Arromdee, V. (1991), 'State characteristics and the location of FDI within the United States', *Review of Economics and Statistics*, **73**, 675–83.

Crozet, M., Mayer, T. and Mucchielli, J-L. (2004), 'How do firms agglomerate? a study of FDI in France', *Regional Science and Urban Economics*, **34**, 27–54.

Devereux, M. P., Griffith, R. and Simpson, H. (2007), 'Firm location decisions, regional grants and agglomeration economies', *Journal of Public Economics*, **91**, 413–35.

Döring, T. and Schnellenbach, J. (2006), 'What do we know about geographical knowledge spillovers and regional growth? A survey of the literature', *Regional Studies*, **40** (3), 375–95.

Friedman, J., Gerlowski, D. and Silberman, J. (1992), 'What attracts foreign mulinational corporations? Evidence from branch plant location in the United States', *Journal of Regional Science*, **32** (4), 403–18.

Glickman, N. J. and Woodward, D. P. (1989), *The New Competitors: How Foreign Investors are Changing the US Economy*, New York: Basic Books.

Gorg, H. and Wakelin, K. (2001), *The Impact of Exchange Rate Variability on US Direct Investment*, Leverhulme Centre for Research on Globalisation and Economic Policy: University of Nottingham.

Griffith, R. (1999), 'Using the ARD establishment level data to look at foreign ownership and productivity in the United Kingdom', *Economic Journal*, 416–41.

Guimaraes, P., Figueiredo, O. and Woodward, D. (2000), 'Agglomeration and the location of foreign direct investment in Portugal', *Journal of Urban Economics*, **47**, 115–35.

Hansen, R. E. (1987), 'Industrial location choice in São Paulo, Brazil: A nested logit model', *Regional Science and Urban Economics*, **17** (1), 89–108.

Head, K., Ries, J. and Swenson, D. (1995), 'Agglomeration benefits and location choice: Evidence from Japanese manufacturing investments in the United States', *Journal of International Economics*, **38**, 223–47.

Hill, S. and Munday, M. (1992), 'The UK regional distribution of foreign direct investment: Analysis and determinants', *Regional Studies*, **26** (6), 535–44.

Hines, J. R. (1996), 'Altered states: Taxes and the location of foreign direct investment in America', *American Economic Review*, **86** (5), 1076–94.

Hood, N. and Taggart, J. H. (1997), 'German foreign direct investment in the UK and Ireland: Survey evidence', *Regional Studies*, **31** (2), 139–50.

House of Commons (2004), *Annual Report on the 1982 Industrial Development Act*, House of Commons, HC852, Session 2002–03, London.
Jones, J. and Wren, C. (2004), 'Inward Foreign direct investment and employment: A project-based analysis in North-East England', *Journal of Economic Geography*, **4** (5), 517–44.
Jones, J. and Wren, C. (2006), *Foreign Direct Investment and the Regional Economy*, Ashgate, Basingstoke.
Kravis, I. and Lipsey, R. E. (1982), 'The location of overseas production and production for export by US multinational firms', *Journal of International Economics*, **12**, 201–23.
Mariotti, S. and Piscitello, L. (1995), 'Information costs and location of FDIs within the host country: Empirical evidence from Italy', *Journal of International Business Studies*, **26**, 815–39.
McCann, P. and Sheppard, S. (2003), 'The rise and fall again of industrial location theory', *Regional Studies*, **37** (6&7), 649–63.
Mucchielli, J-L. and Puech, F. (2004), 'Globalisation, Agglomeration and FDI Location: The Case of French Firms in Europe', in J-L. Mucchielli and T. Mayer (eds), *Multinational Firms' Location and the New Economic Geography*, Cheltenham, UK and Northampton, MA, USA: Edward Elgar.
Pain, N. (1997), 'Continental drift: European integration and the location of UK foreign direct investment', *The Manchester School Supplement*, **65**, 94–117.
Scaperlanda, A. E. and Mauer, L. J. (1969), 'The determinants of US direct investment in the EEC', *American Economic Review*, **59**, 558–68.
Smith, J. (2004), 'Evaluating Economic Development Policies: Theory and Practice', in *Evaluating Local Economic and Employment Development: How to Assess What Works Among Programmes and Policies*, Paris: Local Economic and Employment Development, Organization of Economic Cooperation and Development.
Taylor, J. (1993), 'An analysis of the factors determining the geographical distribution of Japanese manufacturing investment in the UK, 1948–91', *Urban Studies*, **30** (7), 1209–24.
Taylor, J. (2002), 'The Evaluation of UK Regional Policy: How Much Progress has been Made?', in B. Johansson, C. Karlsson and R. Stough (eds), *Regional Polices and Comparative Advantage*, Cheltenham, UK and Northampton, MA, USA: Edward Elgar.
UNCTAD (2003), *World Investment Report 2003: FDI Policies for Development: National and International Perspectives*, New York: United Nations.
Wei, Y., Liu, X., Parker, D. and Vaidya, K. (1999), 'The regional distribution of foreign direct investment in China', *Regional Studies*, **33** (9), 857–67.
Wheeler, D. and Mody, A. (1992), 'International investment location decisions: The case of US firms', *Journal of International Economics*, **33**, 56–76.
Wren, C. (1996), *Industrial Subsidies: The UK Experience*, Macmillan Press.
Wren, C. (2005), 'Regional grants: Are they worth it?', *Fiscal Studies*, **26** (2), 245–75.
Wren, C. (2007), 'Reconciling practice with theory in the micro-evaluation of regional policy', *International Review of Applied Economics*, **21** (3), 321–37.
Wren, C. (2008), 'Onward the Spatial: Is the Game Up for the 'Region' and Regional Economics?', ESRC Urban and Regional Economics Seminar Group conference, University of Glasgow, January.
Wren, C. and Jones, J. (2009), 'Re-investment and the survival of foreign-owned plants', *Regional Science and Urban Economics*, forthcoming.

17. Cluster Policy Implementation and Evaluation in Slovenia: Lessons From a Transition Economy

Anja Cotič Svetina, Marko Jaklič and Hugo Zagorsek

17.1 INTRODUCTION

In the last two decades increasing attention has been paid to innovation as a key driver of competitiveness (Porter and Stern, 2002), and hundreds of innovation policy programmes are being developed and implemented within EU and other countries (Andersson et al., 2004). Within innovation policy, programmes and measures to promote clusters have gained particular popularity (Nauwelaers and Reid, 2002). However there is a great variety in cluster policy design, and this reflects the strong context-dependency of the approach used with regard to the country's institutional structure, collaboration practices, and even traditions of social dialogue. Policy initiatives range from implementing cluster studies, setting up platforms for dialogue, encouraging cooperation between firms and organisations, establishing collective infrastructures, and providing subsidies for collaborative projects. Despite these differences in national approaches the main driving force behind cluster policies is the aim of stimulating collective learning, thereby leading to innovation and the improved competitiveness of participating firms.

The European Commission launched the Innovation Trend Chart project. This projects maps innovation performance and analyses innovation policies in EU countries in order to improve policy coordination and policy learning across Europe. The purpose of the project is not simply to identify best practices, as the uniqueness of regional success stories makes the drawing of lessons difficult if not impossible (Hospers and Beugelsdijk, 2002). The aim is to encourage a soft version of benchmarking which is based on a 'learning-by-interacting' process (Lundvall and Tomlinson, 2000), where an exchange of experiences takes place between policy makers, who then investigate their own policy practices with regard to the policies used abroad (Nauwelaers and Reid, 2002).

The aim of this chapter is to present the experience of implementing cluster policy in a transition economy, and to discuss the reasons why the cluster programme stopped in 2005. To do this we will present a case study of the cluster policy implementation in Slovenia in the last ten years following a four-stage policy life cycle as introduced by Hogwood (1987). We will describe how the Slovene government followed a policy life cycle approach and why it failed to develop a more long-term programme. We argue that the evaluation phase is crucial to the process of successful policy management. Because of some methodological shortcomings in initiating the cluster policy in 2000 (too much top-down approach) and government changes in 2004, Slovenia to a large extent followed a traditional linear policy model in which the project evaluation was at the end phase of the process, with the emphasis primarily on quantitative (short-term) effects, and where the central role of the government is demonstrated. The alternative could be to have had the assessment(s) as a continuous, structured and integrated learning process in a four-stage framework where different stakeholders (not only the government) play their respective roles.

In the first part we will present how the concept of clustering was introduced in Slovenia, then we explain the mapping and needs analysis phase, followed by the policy design and implementation. Throughout the process of implementation, both internal and external evaluations took place, and we will summarise the main results and implications of those studies. At the end, we will present our view on why the coordinated policy support for clustering stopped in Slovenia, and what could be the consequences of this for the further revitalisation of the Slovene economy, its competitiveness and growth.

17.2 CLUSTER POLICY LIFE CYCLE

Policy making is increasingly recognised as a set of routinised stages which feed back into themselves. The concept of policy life cycle has been addressed for more than two decades by several authors such as Nakamura and Smallwood (1980), Chelimsky (1985), Guba (1984), Jones (1982), Rist (1989, 1990, 1993) and others. If we adopt the policy life cycle model described by Hogwood (1987), then policy can be described as a process of four distinct phases (Raines, 2001):

- Analytical phase: an initial analysis of a problem, a determination of the needs and an audit of the existing policies that address those needs;
- Developmental phase: the development of a policy response to those needs and a definition of key policy objectives, resource views and policy tool investigations;
- Implementation phase: preparation of procedures, selection of the process, budgets and the timetable for the implementation of the programme;
- Evaluation phase: an assessment of the performance of the programme, and an evaluation of the effectiveness and efficiency of the policy after an appropriate operational period.

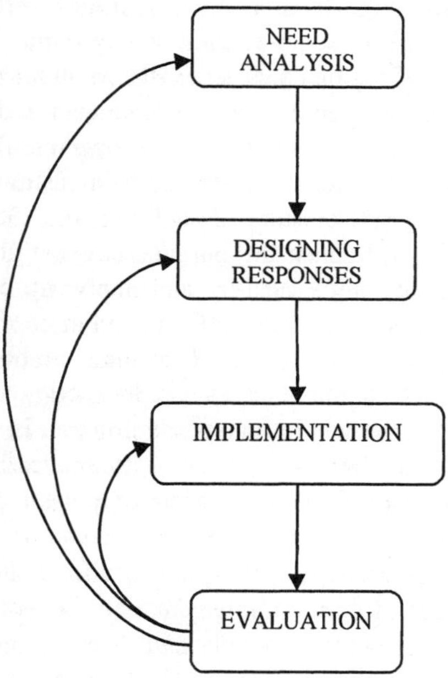

Source: Adapted from Hogwood (1987) and Raines (2001).

Figure 17.1 The policy life cycle model

The four-stage process presented above can be perceived as a circle whereby each stage is not just a linear continuation of the previous phase, but it links back into iterative cycles of future policy development (see Figure 17.1). In the case of clusters, this constant adaptation process of the policy is crucial, because the policy targets an extremely complex phenomenon; namely, a network of firms, R&D and other institutions, all of which are strongly embedded in their environment.

Throughout Europe policy measures to promote innovation are being implemented, often using the cluster approach. The European Trend Chart on Innovation reveals that cluster policies differ considerably among European countries (Boekholt and McKibbin, 2003). The differences are mainly in content (strengthening the triple helix, improving R&D cooperation, encouraging cooperation in general, etc.), approach (bottom-up or top-down), scale (from hands-on to hands-off support), positioning (when aimed at strengthening the existing clusters they are part of an industrial policy, but usually they emerge as part of science and technology policy aimed at encouraging the emergence of new clusters) and the level of implementation (local, regional or national). The diversity of measures reflects the differences in economic development, socio-cultural background, political priorities, existing competences of regions and demonstrates the necessity to adapt the policy life cycle to each specific case (Nauwelaers and Reid, 2002). The WEID research group[1] argues that there is no 'one-size-fits-all' solution and effective policies must respect the differences in institutional contexts and structural conditions across clusters (Tsagdis et al., 2005). Those specific characteristics strongly influence the policy life cycle and demand tailored approaches to designing, implementing and analysing cluster policies. All four phases of the process should be implemented in order to achieve a long-term success of the programme. The analysis made across EU and candidate countries revealed that in many cases cluster policy measures were not implemented following the whole four-phase process. In most cases it is the evaluation phase that has not been systematically undertaken. A comprehensive overview of the evaluation of cluster policies in the EU (Boekholt and McKibbin, 2003) revealed that the evaluation practices varied considerably across countries, reflecting the specific context of the policy as well as different stages of policy implementation. Several countries, such as Luxembourg, the Netherlands, Slovenia and selected regions in Spain and Austria (ibid.), Germany, Finland and the UK have conducted evaluations of clusters[2], aimed at assessing the achievement of the goals set for the cluster performance and the possibilities for future development. Other than these cases, the evaluation of cluster policies was limited. The reason for the limited number of evaluations undertaken is on one hand, due to the youth of the programmes,[3] and, on the other hand, due to a limited understanding of the role that evaluation can play in the policy life cycle.

The field of policy evaluation has greatly matured in the last 20 years, although there are still many theoretical and methodological disputes[4] that make the evaluation work very challenging (Mathison, 2005). In the 1970s, when the pioneering work on the impact of regional policies was done, the evaluations of policy impacts were mainly based on the analysis of the economic data over a period of time using econometric techniques and input–output models (Ashcroft, 1982; Diamond and Spence, 1983). Due to the limitations of these traditional methodologies, more comprehensive evaluation approaches such as cost-benefit evaluations were developed in the 1980s and 1990s (Angeles Diez, 1999). Even these approaches faced methodological challenges related to the accuracy with which they can estimate net effects of the policy, and to the problem of evaluating intangible benefits (Raines, 2002). Consequently, new approaches were developed which introduced more qualitative approaches. The new concept of evaluation is now understood as a dynamic process, open to the participation of the economic and social actors, in which emphasis is laid on the mutual capacity for learning (Guba and Lincoln, 1989).

The increased importance[5] of evaluation is reflected in several theoretical and empirical research papers dealing with this topic and several evaluation techniques and tools recently developed within EU. These include, for example, the MEANS collection developed by the EC (European Commission, 1999) or The Guide for Evaluating Socio-economic Development prepared by the Tavistock Institute (2003). The role of the evaluation is crucial in testing policy approaches and mechanics, providing for a constant assessment of a programme's performance and its adaptation. The evaluation can help in: sharing the knowledge and experience gained through the cluster policy life cycle; upgrading policy learning to the international level, and; in developing a basis for transnational learning. In this respect there is a great need for presenting successful case studies, developing comparative studies, and benchmarking best practices at the international level. In the rest of the chapter, we will present a case study of industrial regeneration undertaken via the implementation of cluster policy.

17.3 CLUSTER POLICY IMPLEMENTATION IN SLOVENIA

In the recent years Slovenia has often been cited as a good practice of how to comprehensively introduce and develop clusters in a transition economy (Sölvell et al., 2003).

> Slovenia is a particularly interesting case that can shed light on the applicability of general conceptions about cluster initiatives to transition economies … the country has made cluster initiatives an important and much publicised element of its economic policy strategy. Its overall success and its significant experience make

Slovenia the prime candidate from which to learn about the success drivers of cluster initiatives in transition economies (Cogan, 2003, p. 71).

At the end of 1990s, when the Ministry of the Economy started an initiative to promote the concept of new inter-organisational structures among companies, such as clusters, technology networks, technological parks and incubators, Slovenia was still in the last place among the EU candidate countries in terms of the number and development of industrial clusters (World Economic Forum, 2002). Yet, there were many obstacles that prevented the dynamic and organic cluster formation in Slovenia. One of these obstacles was associated with the 'integrated self-management', when from 1976 onwards, Yugoslav companies were reorganised into basic organisations of associated labour and then integrated into large conglomerate companies (Hocevar and Jaklič, 1999). This is still influencing Slovenian companies in they are still cautious regarding the formation of networks as they often understand it as a new form of 'forced' integration. With increasing market pressure and R&D and technology demands, the need for the creation of different kinds of networks arose. In the early 1990s, the first ideas for industrial regeneration via clustering were developed (Petrin, 1991). As well as the integrated socialist self-management heritage, some other factors influenced the relatively late processes of clustering, namely: the small size of the economy and a consequential lack of a critical mass of firms in related industries; low levels of trust among actors in the economy; and the relative closure of Slovenia for foreign investments, which prevented the creation of mixed (international) clusters.

17.3.1 Needs Analysis

In order to systematically approach cluster development and to identify possibilities for clustering, in 1999 the Ministry of Economy started a research project called 'Encouraging Company Linkages, Specialisation in Production Chains and the Joint Development of International Markets under a Cluster System' (Dermastia and Kriznic, 2002). Its aim was to identify the potential clusters in Slovenia, and especially those that were of strategic importance to the economy in terms of the number of companies, the level of employment, the existing competitive advantages, and the development potential for domestic and foreign market shares. The research, which was designed as a mapping study, included some 1700 companies (Dermastia, 2002). For the purpose of the research, clusters in Slovenia were defined as 'product/service networks of co-dependent but independent companies, which should include producers of end products or services, specialist suppliers of parts and components, producers of complementary products, suppliers of machinery and hardware, service providers, companies with complementary skills, know-how and technologies, government organisations, research and advisory institutes and customers' (Dermastia, 2002, p. 8). Despite this wide definition of clusters, the primary conclusion of the survey was that there are no real clusters in Slovenia. The results showed that cooperation and networking between companies and other institutions was weak, while the identified networks did not have a clear geographical (regional) dimension and were dispersed throughout Slovenia. Another characteristic was that the identified networks did not fulfil the criterion of a critical mass of companies that promises to develop into a cluster. In addition, the development of infrastructure for cluster development was only in the initial stages and the research pointed to the absence of support structures and specific knowledge that could encourage collaboration and networking. However, ten industries were identified with the potential for cluster development.[6]

17.3.2 Policy Design

On the basis of research results the Ministry of the Economy designed a cluster development programme that aimed at promoting the concept of clusters, acquiring experience in cluster promotion, and strengthening the most prosperous clusters. The programme was planned for the period between 2000 and 2003 and was comprised of three measures: the first measure aimed at encouraging cooperation and networking between companies and between companies and R&D institutions; the second measure aimed at strengthening the skills, expertise and know-how of actors that will promote the development of clusters; and the third measure was the formation of clusters in practice. In contrast to generally accepted definitions of clusters – as networks of SMEs – the cluster policy programme in Slovenia was not only limited to SMEs. The criteria for applying for governmental support was that applications should include at least three partners in the value chain that cooperate with at least three companies and some support institutions in a specific area.

The process of policy design in Slovenia was also influenced by foreign experience. The transnational learning started through cooperation with a Dutch expert group that participated in the second developmental phase. They helped in designing a cluster programme and setting the implementation structure for the programme. After the first international cooperation, in 2001 the Ministry of Economy was invited to participate in an OECD LEED project called 'Clusters in Transition Economies' (OECD, 2005) which enabled the exchange of experiences and knowledge on clustering in transition economies. The transnational learning was further encouraged by cooperation with the Swedish Agency for Innovation System

Vinnova, which enabled Slovenian policy makers to get deeper insights into the structures needed for effective policy implementation, as well as knowledge and skills on policy dialogue facilitation. The Ministry of Economy entered in cooperation with The Competitiveness Institute which enabled them to start dialogue and policy learning on a global level. These forms of cross-national cooperation helped the Slovene government to develop and implement a sound measure for cluster promotion

17.3.3 Implementation

In order to gain knowledge and experience in cluster development, to educate participants about the cluster concept, and to promote the idea among other actors in the economy, the Ministry of Economy launched a pilot project. In 2000 they issued a call for proposals, inviting prospective clusters to apply to receive government assistance for developing and implementing cluster strategies. Out of six applicants three pilot clusters were selected: the automotive sector, the toolmakers sector, and the transportation and logistics sector. The main selection criteria was geographical concentration of companies, a reputation of companies in the community, access to international markets and the ability and willingness to support cluster development. At that point it was important to select such groupings of firms and institutions that would successfully endorse the programme and implement the programme's goals (Rebernik et al., 2004).

The Ministry intended to develop cluster development as a bottom-up initiative, leaving the decisions about the areas (industries) of cluster development, and the organisational and managerial issues to the participating actors. In reality, the programme turned out to be a top-down approach, which will be discussed in the following sections. The government would act as a catalyst for the clustering process, providing the co-financing of start-up costs for selected clusters. The financial scheme aimed at encouraging private initiatives, as government provided 40 percent of the start-up costs and participating companies provided another 60 percent. The financing scheme followed a three-phase cluster development process. This suggests that cluster initiatives begin with the initiation phase when actors develop a common vision and develop an action plan for its implementation; in the early growth phase they implement the action plan and develop the platforms necessary for the final phase of dynamic growth (Rebernik et al., 2004). The financing was initially provided for the first year, but could be extended for two more years on the basis of the detailed action plan for further cluster development (Table 17.1).

Table 17.1 *Governmental co-financing budget for cluster development programme in 2001–2004*

Year	Governmental co-financing (mio EUR)
2001	0.6
2002	1.3
2003	2.6
2004	4.0

Source: Evaluation Report (Deloitte, 2004).

After the pilot project the programme continued and two calls for proposals were issued in 2002 and 2003. The interest in clustering increased and spread to other industries. In the second call, eight new clusters were selected out of 15 applicants and in the third call 14 clusters out of 30 applicants were selected to receive governmental co-financing. By 2004, 17 clusters had been supported in different phases of their development. The clustering initiative attracted more than 300 SMEs, 70 large firms, 80 knowledge institutions and almost 30 other support institutions, involving more than 66 000 employees.

Through the years the programme became quite extensive and promoted more and more clusters. In order to get the feedback information on programme implementation and to gain inputs for further policy intervention, the government decided to carry out the evaluation phase.

17.3.4 Evaluation

The first evaluation in 2002 was based on three pilot projects and brought an interesting insight into the process of cluster development. The usefulness of the first research encouraged the government to continue the evaluation process and in 2004 an extensive research including more than fifteen clusters was completed. The research design of both of the evaluation studies, as well as their results, are described in the rest of this chapter.

First, the *pilot project evaluation* was designed as a mid-term formative evaluation that would show how the process of clustering evolved, and identify potential problems and analyse the business opportunities of clusters. The aim of the study was to collect field experience and use it when designing future cluster policy measures. It was undertaken at the end of 2002, that is, over the two years the pilot clusters had operated. Only three clusters were studied and at that time it was too early to measure the quantitative effects of clustering. Consequently the methodology was based on a qualitative research design based on semi-structured interviews with 32 representatives of participating companies, and workshops with two groups of participants (managers of participating companies). Questions were oriented towards ascertaining the participants' satisfaction with cluster development, identifying areas in which joint projects were realised, the level of trust between cluster members, the functioning of the cluster office, the internationalisation strategies, the communication within the cluster and what kind of policies they anticipated from the state in order to promote the cluster's further development.

The analysis of three pilot clusters revealed that the process of clustering is a long-term process, especially in a transition economy where there is low level of trust in policy intervention (Sölvell et al., 2003, p. 79) and resistance of forced integration is known from the previous system. The research showed that cluster members needed time to get to know each other, to obtain the needed level of knowledge, and to develop collective infrastructure and, most importantly, to build trust before they became involved in more demanding projects. At that time they were mainly focused on encouraging communication among the actors and they still lacked the recognition of an important project that would attract the interest of all the players, bind the cluster actors closer together, and encourage the culture of cooperation and fair play. Although the support for clustering was high, however, all three clusters had the problem of only a low level of commitment of top management, especially from large firms.

The results of the study were presented to the government in a report (Jaklič, 2003), in which several policy implications were also suggested, based on the results of the study. The evaluation report proposed that future policy interventions should be adapted to the specific needs of individual clusters, and that negotiations should be introduced into the selection process in order to overcome the bureaucratic 'take it or leave it' approach. It would enable a selection of truly prosperous projects within a cluster and prevent the selection of 'well-written' proposals that have low prospects for realisation (ibid.).

One of the main conclusions of the study was related to the continuity of the programme. In transition economies there are often doubts regarding the longevity of the programmes and a fear that a programme might be abandoned by a change of the government (Sölvell et al., 2003, p. 79). This was also the case in Slovenia, where the country lacked comprehensive cluster policies which included the harmonisation of approaches taken by all governmental institutions. The evaluation proposed the formation of a microeconomic strategic council in which key government representatives (including the Prime Minister) would participate. At this council, they would discuss key strategic guidelines for cluster promotion and negotiate the harmonisation of operations between the ministries and other institutions that could play a vital role in promoting regional clusters, including Regional Development Agencies, the Small Business Promotion Centre, municipalities, technology parks, incubators, and so on (Jaklič, 2003). In Slovenia, cluster policy was often understood merely as the promotion of cooperation between companies. However cluster policies should also include a wide variety of other initiatives, such as for example, infrastructure development, education and training, research and development, the development of supporting services, human resource policies, and internationalisation policies, all of which have to be comprehensively introduced if we are to aim at quality cluster development.

The study revealed that none of the pilot clusters had the configuration of equal members in the cluster and the process of clustering was mainly driven by large firms, while SMEs only entered the process as followers. As larger companies had the main role and the power in project selection within clusters, small companies were not properly incorporated into the process and there was therefore a problem of trust within clusters. This was identified as one of the main reasons for the relatively slow development of Slovene clusters.

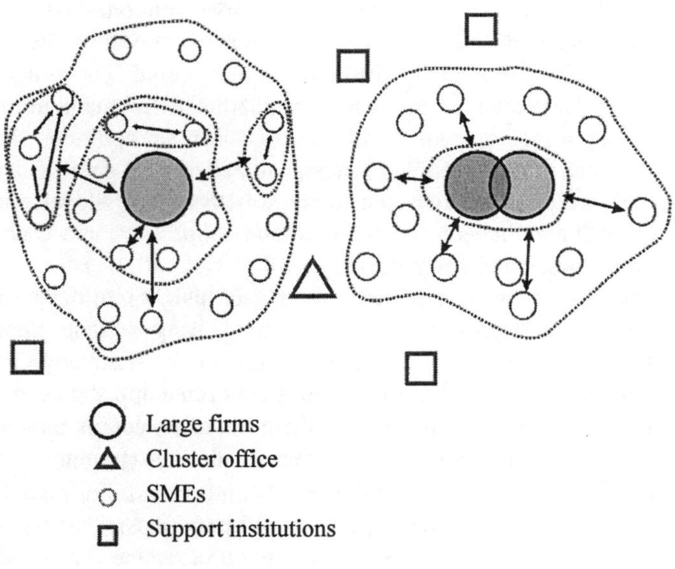

Source: (Jaklič et al., 2003).

Figure 17.2 Slovene model of clustering with dynamic concentric circles

As a possible solution to these problems, the evaluators proposed a model of clustering based on 'dynamic concentric circles' (Jaklič et al., 2003) as presented in Figure 17.2. In this clustering model core firms organise a network of smaller circles of SMEs. Small companies can then form sub-clusters on the basis of their specialisation and geographic concentration.

The picture presents two alternatives of Slovene clustering. The left part of the picture presents the concept, where large companies start to outsource on one side, to existing smaller suppliers, and on the other side to newly created spin-offs. A network of companies then sometimes creates sub-clusters within themselves and grows around the lead firm, which presents a basis for cluster creation. The second alternative presented on the right side of the picture is also based on lead companies and includes mergers, acquisitions or other capital relations. A strong lead centre can use its economic influence to organise a network of suppliers and support institutions around itself. The two presented alternatives do not exclude each other and can coincide within one cluster. Within those networks smaller companies could also achieve competitive capabilities and opportunities for more independent operations and get involved in other (also international) networks. By organising clusters in this way the problem of size and power differences among cluster members, and consequently the low level of trust among cluster actors, could be solved. We believe that the proposed model can only be developed through the companies' initiative and could not be forced on them by any policy intervention. However, the government could encourage larger companies to release a larger number of commercial activities in terms of the formation of spin-off companies. Additionally, start-up activities should be actively promoted within existing clusters. We believe that spin-off and start-up processes would lead to a positive restructuring of Slovene clusters, introducing both simultaneous competition and cooperation and enable them to grow and become more competitive.

The evaluation of the pilot project gave a clear insight into the programme development and suggested the necessary changes that should be implemented when designing further policy measures. The study confirmed the need to improve the vertical connections between companies and also the institutional environment via the development of technological networks, and to enhance the flow of information between the universities and companies. The evaluation process emphasised the establishment of strong links between the Ministry of Economy, Faculty of Economics and cluster actors, and this continued throughout the implementation of the programme. The cooperation took place formally through meetings, conferences and focus groups, as well as informally through casual meetings of different stakeholders. This cooperation encouraged mutual learning that resulted in successful calls for new clusters. However not all the stakeholders were systematically and actively involved in the evaluation phase. The evaluation did not cover other initiatives like the development of small local networks launched by the Small Business Promotion Centre. There was also an insufficient communication of evaluation results between the various public institutions that aim at stimulating regional development. However, the evaluation study was used by the Ministry of Economy for further policy intervention and the encouraging results of the pilot project have had an effective impact on the promotion of cluster initiatives. This was then reflected in an increasing interest in clustering among firms from several industries.

Since the first evaluation was of such significance for policy learning, in 2004 the government decided to order an external evaluation of all measures promoting entrepreneurship and competitiveness between 2001 and 2003 (Deloitte, 2004). Within this project an extensive evaluation of cluster programmes was done. The *mid-term evaluation* analysis included 16 clusters[7] that were supported by the government between 2001 and 2003.

At that point cluster development was already well established and it was important to design a study that would collect qualitative as well as quantitative data. While developing the methodology for the study an extensive overview of current practices in evaluation from other European countries was also prepared. The main conclusion was that the majority of evaluations still follow traditional approaches and are not open to all the possible stakeholders, such as companies, service centres, research institutions, and government administrators. These organisations are mainly taken as a source of data and are not actively incorporated in the evaluation process (Angeles Diez, 2002). The desktop review confirmed that there is no single evaluation model for tackling the evaluation of these new regional policies that might be able to serve as methodological recipe. Each situation requires a unique and specific evaluation design. The evaluation of the cluster policy in Slovenia was conducted by an independent research institute[8] that designed a comprehensive and unique methodological tool. The main goal of the study was to develop policy implications on the basis of collected data that will serve for the future design of cluster policies. The evaluation was mainly focused on assessing the policy initiatives. However, data for the evaluation of individual clusters was also collected as these two concepts are highly interrelated. There is therefore a need to define the link between the performance of individual cluster actors, the performance of the cluster, and the impacts on the wider environment (Raines, 2002).

The evaluation design included the collection of secondary data (national statistics data bases, cluster reports, articles), focus groups with different stakeholders (representatives of the Ministries, Regional Development Agencies, cluster managers, firm representatives and academia), in-depth as well as structured interviews with cluster managers, and structured interviews with representatives of the companies. The structured interviews were based on two questionnaires,[9] one for the firms and the other for cluster managers. The questionnaires were very detailed in order to acquire a thorough insight into cluster development. The design of the questionnaires followed the recommendations made by the EC (European Commission, 1999; Tavistock Institute, 2003) although the questions were developed specifically for the Slovene study.

The results of the study together with policy implications were presented in an extensive report (Jaklič et al., 2004) and were also presented to the government representatives. The main conclusions of the study are discussed in the following sections.

The call attracted groups of firms and other institutions that have previously collaborated, and this increased the chances for successful cluster development. The results demonstrated that the initiative was successful in stimulating cooperation that would otherwise not have happened. Three-quarters of clusters agreed that the governmental initiative was crucial for cluster formation, and almost all the companies also plan to actively participate in the cluster after the termination of governmental co-financing. The participants were already able to identify the benefits of clustering, mainly in terms of improved communication, increased knowledge transfer and also some quantifiable improvements in terms of increased sales, value-added and exports. However, the majority of firms expect the major benefits of clustering to occur over the long run and they estimate the benefits of clustering to outweigh the costs from about six years onwards. This shows that cluster initiatives should be designed as long-term interventions and should not be terminated simply due to a lack of short-term results.

Among the main obstacles for clustering is still an insufficient level of trust among members. However the levels of trust seem to be consistently increasing and this is reflected in an increasing number of joint projects, an increasing number of cluster actors and the improved transfer of information. Other obstacles identified by cluster actors are a lack of financial and human resources, and insufficient knowledge and skills in network management. The last two could be tackled by policy intervention through the co-financing of additional resources and also the active education and training in network organisation and management. On the other hand trust building requires more time and the active participation of firms. Experience showed that the successful completion of initial projects noticeably increased the level of trust, and enabled cluster actors to engage in more demanding projects.

The evaluation confirmed the results of the first study that identified a lack of harmonisation between the ministries and other institutions which should be actively involved in regional development. The process of clustering in Slovenia is too often understood as the sole responsibility of the Ministry of

Economy, which does not actively involve other public and private institutions. Additionally, cluster policies should also include many measures as well as the formation of cooperative networks. In the initial phase of clustering, firms were mainly searching for areas where they can achieve synergetic effects through collaboration, such as, for example, in the development of infrastructure, R&D, education and promotion activities. Later on, the cooperation should be upgraded by competition, and this did not happen in the Slovene clusters. This is partly due to a non-holistic approach to cluster policy design. The programmes were mainly developed for strengthening cooperation, but lacked the dynamic aspect of clustering. The evaluation study identified the need to open Slovene clusters to domestic and foreign competition, and enhance the spin-off and start-up processes which would bring a necessary dynamism into the cluster development.

Both evaluation studies revealed the main advantages as well as the pitfalls of the cluster policy implementation in Slovenia and, as such, should valuably contribute to the types of policy learning which in practice did not happen. If the continuity in policy programmes is to be achieved, the experience gained through the evaluation process and the results of evaluation studies should serve as an input to the further policy life cycle process and the basis for policy learning. In our concluding remarks, we will explain why this continuity is crucial in innovation programmes and why it is so difficult to achieve.

17.4 CONCLUDING REMARKS

The new generation of innovation policies in the literature is proposed as being strongly embedded in the socio-economic framework, since they are tackling complex interactions between regional actors engaged in stimulating collective learning and capacity building. The policy life cycle of those policies is a complex and interactive process with continuous feedback loops, and no linear causal relationship between resources, results, effects and regional impact (Angeles Diez, 2001) Due to these specific characteristics, effective cluster policies must respect the differences in institutional contexts and the structural conditions across clusters, and such an approach should be implemented at all phases of a policy's life cycle. In order to capture the full dynamics of clusters and their environment the literature therefore suggests the use of a participatory and interactive approach where participating stakeholders are involved in all phases of policy life cycle (Acs et al., 2003; Angeles Diez, 2001; Nauwelaers and Wintjes, 2002).

In order to constantly improve and adapt policy intervention to the changing environment, there is a necessity to introduce a learning dimension into the policy-making process (Lundvall and Borrás, 1997). This can be achieved by policy learning, which is a process of systematic adaptation of different policy initiatives in accordance with the changing situation and the needs of the cluster (Tsagdis et al., 2005). There are three central frameworks that support and stimulate the process of policy learning: enhanced communication, the systematic use of evaluation studies, and adaptive benchmarking (ibid.). In Slovenia the first two frameworks were implemented, but not exploited to their full extent. Their role will be discussed below.

In a cluster policy life cycle there is an enhanced need for an on-going and two-way communication between stakeholders, which enables the development of open climate for a continuous self-reflection. This encourages the search for an improvement in collective action and consequently an improvement in the cluster policy implementation. In Slovenia, the communication and exchange of opinions between policy makers was weak. As we have seen, the cluster policy cycle has not yet been open to all of the potential actors, and there have been insufficient communication flows between private and public institutions (Bartlett and Bukvic, 2005). This resulted in an inability of public and private actors to reach strategic agreements on further economic development. Consequently policy interventions aiming at cluster development were not harmonised and the various initiatives were scattered between different Ministries and other institutions.

The evaluation of cluster programmes in Slovenia was implemented using Hogwood's four-phase process. The linear connections between the phases were well established, although the feedback loops deriving from the evaluation phase were not developed to the full extent. The results of the first pilot evaluation study were used for further policy development, while the results of the second evaluation were often misinterpreted and not used to develop further policy intervention. Both evaluations have assessed the cluster programme as one of the most successful measures of industrial policy for increasing cooperation among business enterprises and between the business sector and public research institutions (Deloitte, 2004). Even foreign authors have presented the Slovene case as one of the best practices in cluster policy implementation (Sölvell et al., 2003; European Commission, 2004; Tsagdis et al., 2005). However the new government and the media selectively presented the results to the public, often taking them out of the context, with a tendency to present only the problems of the existing cluster policy rather than the benefits.

In 2005 the new government unexpectedly decided to discontinue the cluster programme. They implemented the existing contracts with clusters, but did not launch any new calls. The main argument for terminating the cluster programme was a lack of tangible and measurable benefits of clustering, although it is clear that cluster policies take time to generate the results from intervention (Landabaso, 2000). New innovation and regional policies should be designed as long-term programmes that are not dependent on the government changes. The effects of clustering cannot be captured by only measuring quantitative effects, such as number of new employees or the increase in sales. All of these characteristics make the evaluation of the results very sensitive to interpretation and require a deep understanding of the role played by innovation policies in establishing institutional collaboration.

The problem of weak communication between stakeholders and the inappropriate use of the evaluation studies suggests that the preconditions for cluster policy learning are still weak in Slovenia. The main problem lies in the perception of the role that evaluations plays in the policy life cycle. As in many other countries, evaluation is perceived only as the final phase in the process and not, of itself, as a tool for implementing policy learning. The discontinuity in cluster programme does not mean that existing clusters will discontinue their collaborations. Some of them managed to gain financial resources from EU Structural Funds, or applied to open calls from the Ministry of Economy, Chamber of Commerce and other institutions but many of the existing clusters are still in the initiation phase where external co-financing and institutional support is crucial. In our opinion the main problem arising from the termination of the cluster programme is actually not related to the financial aspect of cluster development, as prosperous clusters will always find the necessary resources. We see the main problem of this decision in terms of the discontinuity of government intervention. This gives a signal that even the programmes which were demonstrated to bring positive effects on the competitiveness of firms could easily be terminated due to a change of the government. Consequently, companies have little trust in policy interventions.

The Slovene experience demonstrates that despite following a theoretically constructed sequence of phases, the implementation of a cluster policy is not an easy task. The most difficult aspect to establish is the policy-learning process, a lack of which prevents policy makers and other stakeholders from repeating their mistakes. This is a necessary policy feature if we are to foster a constant improvement in our policy interventions. In order to experience the long-term benefits of innovation policies, the government and other stakeholders have to engage in a learning process. It is important to learn from previous experience and to continue with well established innovation programmes, while at the same time constantly adapting them to the changing conditions and requirements of the business environment.

NOTES

1. EU-funded project entitled WEST-EAST-ID (WEID) 'Industrial Districts' Relocation Processes: Identifying Policies in the Perspective of EU Enlargement', 2001–04 (Contract no. HPSE-CT2001-00098). More information can be found on http://www.west-east-id.net.
2. It is important to distinguish between evaluation of cluster development and evaluation of cluster policy. The first has mainly to do with the efficiency of the policy (formative evaluation) and the latter with the effectiveness of the policy (summative evaluation).
3. Some countries such as Sweden, France and Portugal have innovation systems and cluster programmes in their early phase and it was too early for evaluation.
4. For more details see 'Policy Evaluation: Linking Theory to Practice' (Rist, 1995).
5. Increased importance of evaluation has been shaped by four groups of pressures (Bachtler, 2001): political need to determine economic effectiveness, need to account for public spending and demonstrate value added, necessity to improve implementation and delivery systems and to test assumptions regarding the relationship between policy and spatial growth.
6. The following industries: electric-optical, automotive, household appliances, construction, transport, information technology, furniture, textiles, tourism and pharmaceuticals.
7. Data was collected from 16 cluster offices and from a sample of firms participating in selected clusters (84 firms or 21% of the whole population).
8. The Institute for Cooperation and Competition operates within the Faculty of Economics, University of Ljubljana.
9. Can be obtained by the authors of the study on request.

REFERENCES

Acs, Z. J., De la Mothe, J. and Paquet, G. (2003), 'Regional Innovation', in Acs Z. J. (ed.), *Regional Innovation, Knowledge and Global Change*, London: Thomson Learning, 27–49.

Andersson, T., Schwaag, S., Sörvik, J. and Hansson, E. (2004), 'The Cluster Policies Whitebook', Malmö: IKED.

Angeles Diez, M. (1999), 'How to Improve Evaluation Practice? From Traditional to Participatory Evaluation', Regional Studies Association International Conference: Regional Potentials in an Integrating Europe, Bilbao.

Angeles Diez, M. (2001), 'The evaluation of regional innovation and cluster policies: Towards a participatory approach', *European Planning Studies*, **9** (7), 907–23.

Angeles Diez, M. (2002), 'Evaluating new regional policies: Reviewing the theory and practice', *Evaluation*, **8**, 285–305.

Ashcroft, B. (1982), 'The measurement of the impact of regional policies in Europe: A survey and critique', *Regional Studies*, **16** (4), 187–305.

Bachtler, J. (2001), 'Quod erat demonstrandum? The Evaluation of Regional Policy', in H. F. Eckey, D. Hecht, M. Junkernheinrich, H. Karl, N. Werbeck and R. Wink (eds), *Ordungspolitik als konstruktive Antwort auf wirtschaftspolitische Herausforderungen*, Stuttgart: Lucius & Lucius Verlag.

Bartlett, W. and Bukvic, V. (2005), 'The Promotion of Innovation in Slovenia through Knowledge Transfer from Higher Education Institutions to SMEs', UCL Centre for the Study of Economic and Social Change in Europe.

Boekholt, P. and McKibbin, S. (Technopolis Group) (2003), 'European Trend Chart on Innovation, European Commission', Enterprise Directorate-General, www.cordis.lu/trendchart.

Chelimsky, E. (1985), 'Old Patterns and New Directions in Program Evaluation', in E. Chelimsky (ed.), *Program Evaluation: Patterns and Directions*, Washington DC: American Society for Public Administration.

Cogan, A. (2003), 'Cluster Initiatives in a Transition Economy: The Case of Slovenia', in Ö. Sölvell, G. Lindqvist and C. Ketels (eds), *The Cluster Initiative Greenbook*, Ivory Tower AB, 71–9.

Deloitte (2004), 'Zakljucno porocilo: Evalvacija razpisov podrocja za spodbujanje podjetnistva in konkurencnosti v letih 2001–2003J', Prasnikar; Jaklič, M.; Ferligoj A. and T. Skerlj (eds).

Dermastia, M. (2002), 'Country Case Study on Local Clusters in Slovenia', Ljubljana: Ministry of Economy.

Dermastia, M. and Kriznic, A. (2002), 'Promotion of Company Links and Specialisation in Production Chains and the Joint Development of International Markets under the Cluster System', Ljubljana: ITEO.

Diamond, D. R. and Spence, N. A. (1983), *A Methodological Review and the Scottish Example*, Aldershot: Gower.

European Commission (1999), 'MEANS Collection. Evaluating Socio-economic Programmes: Principal Evaluation Techniques and Tools', Luxemburg: Commission of the European Communities.

European Commission (2004), 'Policy Brief', http://europa.eu.int/information_society/activities/gothenburg_conference/doc/pdf/brief_slovenia.pdf.

Guba, E. G. (1984), 'The effect of definitions of policy on the nature and outcomes of policy analysis', *Educational leadership*, **42** (2), 63–4.

Guba, E. and Lincoln Y. (1989), *Fourth Generation of Evaluation*, New Delhi: SAGE Publications.

Hocevar, M. and Jaklič, M. (1999), 'Trust and contractual relations between companies in a transitional economy: The case of Slovenia', *Economic Analysis (Abingdon)*, **3** (1), 1–12.

Hogwood, B. (1987), *From Crisis to Complacency: Shaping Public Policy in Britain*, Oxford: Oxford University Press.

Hospers, G.-J. and Beugelsdijk, S. (2002), 'Regional cluster policies: Learning by comparing', *Kyklos*, **55** (3), 381–402.

Jaklič, M. (2003), 'Cluster Policy Development in Slovenia: Developing Clusters Bottom-up or Top-down', in Organization Analysis Informing Social and Global Development, European Group for Organizational Studies (EGOS), Copenhagen.

Jaklič, M., Cotič Svetina, A., Hribovsek, B., Kokalj, S., Vidmar, S. and Virant, V. (2003), 'Comparison of Clusters in Slovenia, Austria and Italy', in J. Prasnikar (ed.), *SMEs and Economic Growth*, Casnik Finance, 183–309.

Jaklič, M., Cotič Svetina A. and Zagorsek H. (2004), 'Evalvacija ukrepov spodbujanja razvoja grozdov v Sloveniji v obdobju 2001–2003', 170, Ljubljana: Ministry of Economy, http://www.mg.gov.si/fileadmin/mg.gov.si/pageuploads/razpisi/analize_razpisov/evalvacija_ukrepov_spodbujanja_grozdenja_261104.pdf.

Jones, C. O. (1982), *The United States Congress*, Homewood IL: Dorsey Press.

Landabaso, M. (2000), 'EU Policy on Innovation and Regional Development', in F. E. A. Boekema (ed.), *Knowledge, Innovation and Economic Growth: The Theory and Practice of Learning Region*, Cheltenham, UK and Northampton, MA, USA: Edward Elgar.

Lundvall, B.-A. and Borrás, S. (1997), 'The Globalising Learning Economy: Implications for Innovation Policy', Brussels: European Commission.

Lundvall, B.-A. and Tomlinson, M. (2000), 'Policy benchmarking: a Tool of democracy or a tool of authoritarianism?', *International Journal of Benchmarking*, **4** (14), 497–516.

Mathison, S. (2005), *Encyclopedia of Evaluation*, London: Sage Publications.

MEANS (1999), 'Evaluating Socio-economic Programmes: Principal Evaluation Techniques and Tools', Bruxelles: Commission of the European Communities.

Nakamura, R. T. and Smallwood, F. (1980), *The Politics of Policy Implementation*, New York: St. Martin's Press.

Nauwelaers, C. and Reid, A. (2002), 'Learning Innovation Policy in a Market-based Context: Process, Issues and Challenges for EU Candidate- countries', *Journal of International Relations and Development*, **5** (4), 357–79.

Nauwelaers, C. and Wintjes, R. (2002), 'Innovating SMEs and Regions: The Need for Policy Intelligence in Interactive Policies', *Technology Analysis & Strategic Management*, **14** (2), 201–15.

OECD (2005), 'Business Clusters: Promoting Enterprise in Central and Eastern Europe', Paris: OECD.

Petrin, T. (1991), 'Medpodjetniska mre a povezav in prestrukturiranje podjetij pohistvene industrije notranjske regije in ljubljanskega obmocja', RCEF, Faculty of Economics, University of Ljubljana.

Porter, M. and Stern, S. (2002), 'National Innovative Capacity', in K. Schwab, M. Porter and J. Sachs (eds), *The Global Competitiveness Report 2001–2002*, Oxford: Oxford University Press.

Raines, P. (2001), 'The Cluster Approach and the Dynamics of Regional Policy-Making', European Policy Research Centre, University of Strathclyde, UK.

Raines, P. (2002), 'The Challenge of Evaluation Cluster Behaviour in Economic Development Policy', Evaluation and EU Regional Policy: New Questions and Challenges, International RSA Conference.

Rebernik, M., Dermastia, M. and Kroslin, T. (2004), 'Cluster Initiatives, a Firm Fundation and Entrepreneurship in Slovenia', in B. Böhm, H. Frisch and M. Steiner (eds), *Slovenia and Austria: Bilateral economic effects of Slovenian EU accession, The Slovenian EU accession and its bilateral economic effects on Austria and Slovenia*, Graz: Schriftenreihe des Institutes für Technologie- und Regionalpolitik der Joanneum Research, Bd 3.

Rist, R. (1989), 'Management accountability: The signals sent by auditing and evaluation', *Journal of Public Policy*, **9** (3), 355–69.

Rist, R. (1990), 'Program Evaluation and the Management of Government: Patterns and Prospects Across Eight Nations', New Brunswick, NJ: Transaction Books.

Rist, R. (1993), 'Program Evaluation in the United States General Accounting Office: Reflections on Question Formulation and Utilisation', in R. E. A. Conner (ed.), *Advancing Public Policy Evaluation: Learning from International Experiences*, Amsterdam: Elssevier.

Rist, R. (1995), *Policy Evaluation: Linking Theory to Practice*, Aldershot, UK and Brookfield, US: Edward Elgar.

Sölvell, Ö., Lindqvist, G. and Ketels, C. (2003), 'The Cluster Initiative Greenbook', Stockholm: Ivory Tower AB, http://www.cluster-research.org/greenbook.htm.

Tavistock Institute (2003), 'The Guide. The Evaluation of Socio-Economic Development', Tavistock Institute, http://www.evalsed.info/.

Tsagdis, D., Borras, S., Ahedo, M., Rohl, U., Sarcina, R. and Giannini, D. (2005), 'Policies Supporting Clusters' Development and Integration in the Enlarged EU', in S. Borras (ed.), WEID Policy Options Report, West-East ID.

World Economic Forum (2002), *The Global Competitiveness Report*, Oxford, New York: Oxford University Press.

18. Lagging Regions and Policy Options: The Case of Greece in the European Context

Yannis Psycharis and George Petrakos

18.1 INTRODUCTION: SPATIAL STRUCTURE AND REGIONAL INEQUALITIES IN GREECE

Greece has a number of peculiarities that affect its spatial structure and the level and type of its regional imbalances. At the macro-geographical level, it faces a range of constraints arising from its peripheral position and its limited access to major economic agglomerations and markets. In addition, it is characterised by unique border conditions, as it has no common borders with any other EU country and, until recently, had only limited economic (or any other) relations with its neighbours (Albania, FYROM, Bulgaria and Turkey). As a result, the country has been deprived of economic relations with neighbouring countries of similar levels of development and has been restricted to relations with distant and usually more advanced partners. This has affected the type and quality of cross-border infrastructure, as well as the development prospects of border regions.

Internally, Greece is characterised by a highly fragmented physical and economic space, due to the existence of hundreds of inhabited islands and the limitations imposed by its mountainous territory. This fragmentation increases transportation and accessibility costs, requires major investment in infrastructure and inhibits the internal integration of the economy. In addition, physical and political geography have clearly affected development prospects. Limited accessibility to internal and external markets has created a productive structure dominated by small inward-looking firms serving local markets and having limited capacity to adapt and compete in national or international markets.

Partly because of these conditions, the economy has a relatively low level of development and a weak production structure. Greece has the lowest Gross

Domestic Product (GDP) per capita in EU-15 and the highest dependency on the primary sector. In addition, the share of industry in GDP has been in continuous decline (and is the lowest in EU-15) while the productive base is characterised by traditional sectors, small size and low capital and technology intensity. This picture continued even after EU membership in 1981, as the economy experienced strong competitive pressures, which were partly responsible for its poor performance relative to the EU average in the 1980s and early 1990s. Until recently the entire country was defined as an Objective 1 region. GDP per capita was 67 percent of the EU-15 average in 2000. On the other hand, progress has been made in a number of ways: Greece has grown faster than the EU average over the last decade and met the Maastricht criteria and became a member of the EMU.

External asymmetries in the EU context are closely linked to internal disparities at the regional level. Greece is divided into 13 NUTS II regions (Perifereia) and 51 NUTS III regions (Nomos). The metropolitan region of Attica, with 3.7 million inhabitants, contains over one-third of the national population (of just under 11 million). The region of Thessaloniki (belonging in the NUTS II region of Central Macedonia) is second with about one million people, followed by 7 NUTS III regions with populations between 200 000 and 300 000. The rest of the spatial system comprises small NUTS III regions with populations in the range of 20 000 to 200 000. The dominance of Athens (and Thessaloniki) has been consolidated – if not increased – over the last two decades, through the fast growth of their suburbs and satellite cities or regions (Petrakos et al., 1999).

Most studies, including European Commission reports, consider that regional inequalities in Greece are limited compared to other EU-15 countries. These studies make this assertion on the basis of GDP per capita figures where regional disparities are modest (see Table 18.1). Compared to the EU-15 average of GDP per head in PPS, the range in Greece is from 79 percent (South Aegean) to 46 percent (Epirus). Attica, the Athens region, is ranked only third, due to the fact that a significant part of the industrial activities of the metropolitan region have spread to surrounding 'satellite' regions – either to avoid regulations and environmental restrictions or in search of investment subsidies not available in Attica. For the year 2000, the weighted coefficient of variation (CV) was 0.16 and the max/min ratio 1.70, which are modest or low figures by EU-15 standards (Petrakos and Rodriguez-Pose, 2003).

If, however, regional inequalities are measured by a Composite Index of Development (CID), which incorporates 20 indicators of income, production, economic structure, human resources, geography and infrastructure, then regional disparities are significantly higher (see Table 18.1) (Petrakos and Psycharis, 2003). On this ranking, Attica is by far the most developed region with an index value equal to 71.5, followed by South Aegean (56.6), Crete (40.4) and Central Macedonia (40.2), while the least developed regions are West Greece (23.4), Epirus (23.5) and West Macedonia (26.0). The coefficient of variation (CV) increases to 0.58, while the max/min ratio increases to 3.06.

Moreover, the evolution of regional inequalities in Greece is not encouraging. The GDP per capita CV figures have tended to increase in the recent years. Some reports indicate that inequalities in Greece tend to follow the economic cycle, decreasing during periods of low growth and increasing in periods of high growth (Petrakos and Saratsis, 2000). This suggests a degree of tension between the goals of external and internal convergence. On the other hand, the CV figures based on the CID remain stable, but at a relatively high level.

Overall, the metropolitan regions of Attica and Central Macedonia have had a satisfactory performance over time, based on agglomeration economies, while the island regions (especially South Aegean and Crete) have also performed well, despite physical isolation, because of the growing importance of tourism. Mountainous regions, border regions or regions with poor infrastructure have, in contrast, performed poorly; in some cases they are diverging from both national and EU averages (Ioannidis and Petrakos, 2000).

In summary, Greece has significant internal and external spatial disparities that reflect its peripheral position with respect to EU-15 core markets, a lack of adjacency to the 'single European market', a highly fragmented economic and physical space and an unbalanced distribution of regional population. Although GDP per capita based estimates of regional inequality are modest, more advanced indicators of development produce a picture of significant inequalities that, over time, have remained high.

The next section addresses the regional structure of public administration in Greece. The third section then focuses on regional allocation of public investment. The fourth section deals with the allocation of regional operational programmes. The fifth section discusses the effectiveness of the incentive policy to sustain regional development. The sixth section provides an overview for the regional allocation of foreign direct investments in Greece. The last section summarises the regional policy in Greece during the recent period.

Table 18.1 *Regional population, GDP per capita and Composite Index of Development (CID) for the NUTS II regions of Greece*

Regional level	Regional population	Regional share of population	Regional share of GDP		GDP per capita in PPS		GDP per capita in PPS EU-15=100	Composite Index of Development (CID)	
Regions (NUTS II)	*2001*	*2001*	*2000*		*2000*			*Max = 100*	
European Union EU-15					22 576		100		
Greece	10 964 020	100.00			15 098		67		
Attica	3 761 810	34.31	36.96	*1*	17 046	*3*	76	71.5	*1*
Central Macedonia	1 874 214	17.09	17.18	*2*	15 139	*5*	67	40.2	*4*
Continental Greece	605 329	5.52	7.28	*3*	17 474	*2*	77	31.7	*7*
Thessaly	753 888	6.88	6.46	*4*	13 832	*8*	61	28.6	*10*
Peloponnese	638 942	5.83	5.48	*5*	13 010	*9*	58	28.7	*8*
West Greece	740 506	6.75	5.29	*6*	11 379	*12*	50	23.4	*13*
Crete	601 131	5.48	5.25	*7*	14 781	*6*	66	40.4	*3*
E.Macedonia - Thrace	611 067	5.57	4.34	*8*	12 250	*11*	54	28.7	*9*
South Aegean	302 686	2.76	3.05	*9*	17 790	*1*	79	56.6	*2*
W.Macedonia	301 522	2.75	2.91	*10*	15 226	*4*	67	26.0	*11*
Epirus	353 820	3.23	2.47	*11*	10 474	*13*	46	23.5	*12*
North Aegean	206 121	1.88	1.67	*12*	14 563	*7*	65	34.0	*6*
Ionian Islands	212 984	1.94	1.67	*13*	12 956	*10*	57	39.6	*5*
max/min					*1.70*			*3.06*	
CV					*0.16*			*0.584*	

Source: Petrakos and Psycharis (2003)

18.2 THE REGIONAL STRUCTURE OF PUBLIC ADMINISTRATION IN GREECE

During the 1990s a new multi-level administrative system was established in Greece. The top tier consisted of the central public administration at ministerial level, followed by a decentralised regional tier and two tiers of local self-government, one at the prefectural level and one at the municipal and communal level (see Table 18.2). The evolution of the Greek public administration reflects many reforms since the mid 1970s.

Table 18.2 Tiers of local government and decentralised structures in Greece

Municipality and Commune – Demos 1st tier of self-government (local *NUTS V* tier) 1033 *Demoi*	• Fully elected • Underwent drastic amalgamation in 1999 in a process of municipalisation (5 825 previously existing municipalities and communes (438 and 5387 respectively) were reduced into 1033 municipalities and communes (900 and 133 respectively)
Prefecture – Nomos 2nd tier of self-government (prefectural *NUTS III* tier) 51 *Nomoi* grouped into 50 *Self-government organisations*	• Led by a single government-appointed Prefect (Nomarchis) until 1982 • Increased participation with co-opted members in the Prefectural Council established in 1982. Prefect remains centrally appointed • Fully elected tier of self-government since 1994
Region – Perifereia 3rd tier of decentralisation (regional *NUTS II* tier) 13 *Perifereies – Regions*	• Non-existent until 1986 • Legislated for in 1986, but not fully functional until 1997 • Since 1997 fully functional with a centrally-appointed General Secretary (Periferiarchis) and co-opted members participating in the Regional Council

Source: Georgantas and Psycharis (1999).

During the early post-dictator period, after 1974, there were two sub-national elements to the administration: on the one hand, the prefecture level which was an extension of central decision-making and, on the other hand, 'first-level' local self-government. This consisted of very many, mainly small, communes and municipalities and the subsequent over-fragmentation created serious problems. Although this level of local self-government enjoyed a very high degree of local legitimisation and was therefore significant politically, it could not intervene effectively and thus influence regional growth trajectories. In this early period, the pressing political priority was to restore constitutional legitimacy; no serious attempts were made to reform the spatial model of public administration (Georgantas and Psycharis, 1999).

With full EU membership in 1981, along with the emergence and gradual establishment of the PASOK socialist party in 1981, Greek public administration underwent structural change. Over the 1982–88 period, the spatial model of public administration gradually developed, combining elements of the old decentralised prefectural system with some new forms of quasi-representational politics. Law 1235/82 retained the government-appointed prefect as a central feature of the decentralised system. However, the prefect was surrounded in the new scheme by indirectly-elected representatives of local interest groups (chambers of commerce, trade union organisations, etc.), which gave the system a distinct meso-corporatist flavour. At that time, there was no form of regional public administration. Law 1622/86 created the 13 Greek regions but they did not become fully functional until 1997.

Under the combined institutional pressure of European funding programmes after 1988, and their formal requirements for mature and effective mechanisms of regional management, the Greek spatial model of public administration came under extreme strain. The prevailing administrative model could not fulfil even the minimum operational specifications for effectively absorbing EU funding. This became all the more visible with the first Community Support Framework programme (1989–93), which completely changed the country's fiscal profile. Despite the potential for taking advantage of EU funds, Greek governments found themselves in the difficult position of not having the required institutional capacity.

The 1994–99 period, which coincided with the implementation of Community Support Framework II, saw some significant attempts to create an institutional framework, which would satisfy at least some of the specifications required. In the first place, the prefectural level was transformed from an administrative unit of decentralisation, in origin and character, into a fully elected second level of local government (Laws 2218/94 and 2240/94). Second, after eight years of hesitation, regional decentralisation was finally enacted (via Law 2503/97), with the establishment of a unitary institutional body with substantive responsibilities in the fields of planning and development. The presence of a centrally-appointed Regional Secretary, framed by a corporatist structure of interest group representation almost identical to that which surrounded the non-elected prefect, suggests strongly that the previous spatial model of allocating powers was being reproduced at a higher level.

A third feature of the period was the total reconstruction of the first level of local administration. Up until 1998, there were 457 municipalities and 5318 communes in Greece. Subsequently, in a process of drastic amalgamation/municipalisation, following Act 2539/97, these were reduced into 900 municipalities and only 133 communes. This cut down the total number of local government organisations to less than one-fifth of the previous number.

18.3 THE ADMINISTRATION AND REGIONAL ALLOCATION OF PUBLIC INVESTMENT

The Public Investment Program (PIP) is part of the Greek annual budget. It forms a very important element of the budget and, of course, is approved by Parliament. It is the main mechanism for providing the economy with infrastructure and it also encompasses structural funding from the European Union.

From 1988 onwards, the PIP was divided into two main parts. The first included projects co-financed by the Structural Funds and the Cohesion Fund. In the current period, Community Support Framework III and all projects financed by the Cohesion Fund and Community Initiatives are implemented through the PIP. The second part encompasses projects that are financed only from national resources. For the year 2003, the first part represents 65 percent and the second 35 percent of the total budget.

Control of the implementation of projects is shared among the national (ministerial), regional, prefectural and local levels. During the 1980–97 period, the decision-making process for the spatial allocation of funding was under the absolute control of the central and decentralised tiers of the State. At the central level, ministries were responsible for the realisation of almost 80 percent of the programme, a small part of which was allocated to local self-government. The remaining 20 percent was transferred to the prefectures, which were responsible for the allocated amount in line with their own decision-making processes and priorities. At this time, the organisation of the State was based on three distinct tiers. First, there was the central level, based upon the Ministries. Second, there were decentralised structures, based on the prefectures. Third, there was the first tier of self-government, based on municipalities and communes; this was the only elected decentralised level. As mentioned earlier, it enjoyed a high degree of political significance but very limited institutional capacity and authority.

The 13 NUTS II regions were created by Law 1622/86. However, until 1997, regions were not able to participate in the allocation of public investment funding. There are now two tiers of administration at the central (national) and decentralised (regional) levels and two tiers of self-government (prefectural and municipal). The national programme is the dominant one taking up about 70 percent of the total budget. The regional programme accounts for a further 10 percent, while the two levels of self-government (prefectural, municipal) receive about 15 percent and 5 percent of the total public investment budget respectively.

Each one of the four PIPs has a regional dimension, as funds are eventually allocated to different sectors and regions. An analysis of the regional allocation of total public investment in Greece for the period 1988–93 is presented in Table 18.3. The table shows that those regions with major urban centres have received the vast bulk of national public capital investment. The first two regions, Attica and Central Macedonia, with figures well above the average, host the two large metropolitan areas of Greece. Western Greece, in third place, contains the third major urban centre, Patra. Continental Greece, which is in fifth position, is a region adjacent to Attica. It can be argued that here the investment pattern is influenced by the proximity to Athens. The only exception to this pattern is East Macedonia/Thrace, which is ranked fourth. However, this is a rather special border region, to which all Greek administrations have paid special attention for political and economic reasons.

A totally different picture appears if spending per inhabitant is considered. Table 18.3 shows that the North Aegean is the region with the highest public spending per head. Epirus, the least developed region in EU-15, is second, while East Macedonia/Thrace is third. As a general rule, the regional distribution of public investment tends to favour the less developed regions. However, there are some exceptions, notably the Peloponnese and Thessaly. Despite their low levels of development, public investment is below average. Conversely, Continental Greece and the North Aegean have both high levels of economic development and higher percentages of infrastructure spending per inhabitant. In short, although the regional allocation of public investment is inversely related to the level of regional development, indicating a redistributive policy, at the same time some less developed regions appear relatively neglected in terms of resource allocation.

Table 18.3 The regional allocation of public investment 1988-98

	Spatial Unit	Gross Domestic Product (GDP) per capita				Public investments per capita			
		As a percentage of the country average				As a percentage of the country average		Breakdown per spatial unit	
	Regions (NUTS II)	*1988*		*1998*		*1988–1998*		*1988–1998*	
	Greece	100		100		100		100	
1	Continental	123	*1*	120	*2*	116	*8*	7.0	*5*
2	South Aegean	118	*2*	121	*1*	144	*4*	3.6	*10*
3	West	108	*3*	100	*5*	121	*7*	3.5	*11*
4	Attica	105	*4*	111	*3*	83	*11*	27.7	*1*
5	Central	100	*5*	102	*4*	78	*13*	13.1	*2*
6	Peloponnese	100	*6*	84	*10*	80	*12*	5.1	*9*
7	Crete	99	*7*	99	*6*	129	*6*	6.8	*6*
8	Ionian Islands	94	*8*	90	*8*	141	*5*	2.7	*13*
9	Thessaly	93	*9*	89	*9*	88	*10*	6.3	*7*
10	East	90	*10*	82	*11*	146	*3*	7.9	*4*
11	West Greece	83	*11*	76	*12*	115	*9*	8.0	*3*
12	North Aegean	77	*12*	96	*7*	166	*1*	3.0	*12*
13	Epirus	75	*13*	69	*13*	155	*2*	5.3	*8*

Source: Psycharis (2006).

18.4 THE REGIONAL PROGRAMMES IN GREECE

Community Support Frameworks (CSFs) provide the framework for Greek national and regional economic development. Regional programmes constitute an important constituent for regional policy in Greece.

From the inception of the Mediterranean Integrated Programmes in 1986 to the end of CSF III in 2006, 45 EU regional development programmes will have been carried through, at a total cost of €22.6 billion (in 1999 prices). The annual public expenditure for these programmes has increased over time: from €463.6 million in 1986 to an expected €1173.9 million in 2006 (in 1999 prices). In recent years, these resources have accounted for between 1.2 percent and 5.9 percent of regional GDP (not including Attica) (Konsolas et al., 2002). CSF III is divided into seven priority axes and 24 operational programmes; the priority axes are presented in Table 18.4.

Table 18.4 Priority axes of CSF III for Greece, 2000-2006

Priority Axes	Share (%)
1. Development of human resources and employment promotion	10.7
2. Enhancing competitiveness for sustainable development	14.5
3. Transport	28.8
4. Rural development and fisheries	9.0
5. Quality of life	4.3
6. Information society	6.8
7. Regional development	25.8
Technical assistance	0.2
TOTAL	100.0

Source: Psycharis (2004).

Table 18.5 The regional operational programmes of CSF III

	Spatial Unit	Gross Domestic Product (GDP)		Regional Operational Programmes			
		As % national average		Per head		Regional share (%)	
	Regions (NUTS II)	*2000*		*2000–2006*		*2000–2006*	
	Greece	100		100		100	
1	South Aegean	118	*1*	215	*2*	5.6	*10*
2	Continental Greece	116	*2*	127	*7*	8.0	*5*
3	Attica	113	*3*	44	*13*	14.5	*1*
4	West Macedonia	101	*4*	185	*4*	5.3	*11*
5	Central Macedonia	100	*5*	78	*12*	13.3	*2*
6	Crete	98	*6*	125	*8*	6.7	*7*
7	North Aegean	96	*7*	290	*1*	5.0	*12*
8	Thessaly	92	*8*	121	*9*	8.5	*4*
9	Peloponnese	86	*9*	101	*10*	6.4	*8*
10	Ionia Islands	86	*10*	176	*5*	3.4	*13*
11	East Macedonia,	81	*11*	191	*3*	10.2	*3*
12	West Greece	75	*12*	99	*11*	6.9	*6*
13	Epirus	69	*13*	174	*6*	6.2	*9*

Source: Psycharis (2004).

The bulk of the resources, almost three-quarters of the total, relate to the 11 Sectoral Operational Programmes, which belong to the first six axes; the

13 Regional Operational Programmes (one for each region) have been allocated just over one quarter of the total budget. The allocation of these regional funds by region is set out in Table 18.5. The two metropolitan regions of the country receive the largest share of funds. In per capita terms, the allocation pattern is less clear. Regions with relatively high levels of development (like the South Aegean) have received greater than average support, while regions with relatively low levels of development (like West Greece) have been allocated lower than average support. The lack of a clear inverse relationship between levels of support and levels of development under the regional programmes (axis 7) must cast doubt on the regional component of sectoral programmes. Overall, it remains unclear to what extent (if at all) Greek CSFs have contributed to reductions in regional inequality.

18.5 INCENTIVES FOR REGIONAL ECONOMIC DEVELOPMENT

Incentives to promote regional development have a long tradition in Greece. In this section only the current incentive package is presented.

In December 2004, the Greek Government (Ministry of Finance) adopted the new Development Law 3299/2004, entitled 'Private investment incentives towards economic development and regional convergence', which replaced the previous Development Law 2601/1998. The main purpose of the new law is to change the investment profile of the country and the policy approach in a more positive direction through the improvement of the competitiveness and the attractiveness of Greece as an investment location. The main aims of the law refer to the balanced growth and development of the Greek economy combined with regional convergence, focusing on improved competitiveness, employment growth, the promotion of lagging regions, support for specific high value added sectors, technological change and innovation, and environmental protection and energy saving. Nevertheless, as can be seen in the tables which follow, the main focus of the law seems to be more on the support of strategic innovative sectors rather than the development of specific lagging regions.

Investment plans are aided with respect to the coverage they provide in four main respects:

- Innovation and technology promotion through fixed capital investment and the encouragement of research and development.
- Production of new products with high added value, contributing to economic growth.

- Improvement of the trade and export orientation and internationalisation of the economy.
- Promotion of regional convergence, giving priority to lagging or less favoured regions.

For the implementation of incentive policy, the Greek territory is divided into four main areas – Regions D, C, B, and A, with regions classed as D having the lowest levels of economic development and regions classified as A having the highest levels of economic development. Incentive tools include direct subsidies and leasing subsidies, tax exemptions and employment creation subsidies. These incentive types are generally available as alternatives; only in specific cases can a combination of these different incentive types be offered. The incentives provided by the law vary depending on the region, the activity and the type of incentive chosen (see Tables 18.6, 18.7 and 18.8).

Categories 1 and 2 include financial projects in the primary and secondary sectors. More dynamic and technologically advanced projects are found in Categories 3 and 4, while Category 5 includes multi-annual and very high budget business plans.

The main differences between the new law and the previous one can be summarised as: an increase in the level of subsidies; a decrease in the initial investment costs required for eligibility, the absence of discrimination between new and previously existing enterprises; and the separation of the award of incentives from mandatory conditions relating to employment creation.

The new legislation puts much more emphasis on the promotion of competitiveness and entrepreneurship and on the adoption of new technologies and innovative products, as well as on the promotion of exports and the internationalisation of the economy. In particular, the new law provides a stimulus for the technological modernisation of the productive base of established and new enterprises. Enterprises which focus on new technologies, innovative products and exports have priority in receiving subsidies under the new Development Law. The emphasis is put on the promotion of technology and innovation, aiming at the incorporation of new technologies in fixed capital formation and the promotion of R&D.

Despite its intentions, it is uncertain, however, whether the new law will favour regional convergence. Although the whole system of incentives has been modernised, the main effort has been placed on the competitiveness of the economy through investment in large, technologically-advanced projects, which usually locate in the developed areas. For such projects, there is no regional differentiation in the incentive package. This might favour the competitiveness of the economy but, at the same time, it might lead to a higher concentration of economic activity in the most prosperous regions. As a result, the new law seems to be closer to an improvement in industrial policy rather than regional policy. It may succeed in stimulating investment activities and contributing to national growth, but it may be questioned as to whether it will contribute to the reduction of regional inequalities.

Table 18.6 Direct subsidies and leasing subsidies

Investment category	Region A (%)	Region B (%)	Region C (%)	Region D1 (%)	Region D2 (%)	Region D3 (%)
Category 1	0	18	30	35	40	40
Category 2	0	13	25	35	35	35
Category 3	40	40	40	40	40	40
Category 4	30	30	35	35	40	40
Category 5	35	35	35	35	35	35

Table 18.7 Tax exemptions

Investment category	Region A (%)	Region B (%)	Region C (%)	Region D1 (%)	Region D2 (%)	Region D3 (%)
Category 1	0	50	100	100	100	100
Category 2	0	50	100	100	100	100
Category 3	100	100	100	100	100	100
Category 4	100	100	100	100	100	100
Category 5	100	100	100	100	100	100

Table 18.8 Employment creation subsidies

Investment category	Region A (%)	Region B (%)	Region C (%)	Region D1 (%)	Region D2 (%)	Region D3 (%)
Category 1	0.0	18.4	35.1	40.0	45.5	48.1
Category 2	0.0	18.4	33.2	40.0	45.5	45.5
Category 3	40.0	40.0	40.0	40.0	40.0	40.0
Category 4	35.0	35.0	40.0	40.0	45.5	48.1
Category 5	35.0	35.0	35.0	35.0	35.0	35.0

Source: Law 3299/2004 and authors' elaborations.

18.6 REGIONAL ALLOCATION OF FOREIGN DIRECT INVESTMENT IN GREECE

Policies to draw foreign direct investment (FDI) are in general a significant element of regional development strategy. FDI increases income and employment and, in many cases, it is also an instrument for transferring technology and expertise. Through spillover effects, inward FDI can potentially have a considerable impact on the productivity and competitiveness of local enterprises in the region concerned.

The following Table 18.9 shows the allocation of FDI expenditure activity by region in Greece. Foreign investments in Greece that were driven to Greek regions through the Hellenic Centre for Investments are mainly located in Eastern Macedonia – Thrace, in Central Greece and Attica regions. A significant percentage of FDI capital is also invested in the Central Macedonia region.

The conclusion which can be drawn is that FDI tends to locate in regions which have a relatively high level of economic development. This can be explained by considering that these regions have a consistently higher rate of infrastructure, business activity and market potential associated with their higher GDP levels. More specifically, these regions present several general economic and social characteristics which make them more attractive than the rival regions as location for FDI location.

Table 18.9 Foreign direct investments undertaken via ELKE in 1996–2002, per host region

Region	Percentage of FDI via ELKE
Eastern Macedonia – Thrace	20.8
Central Greece	20.1
Attica	18.8
Central Macedonia	12.8
Thesally	9.4
Western Greece	6.0
Peloponnese	5.4
South Aegean	3.4
Crete	1.3
Western Macedonia	0.7
Ionian islands	0.7
Epirus	0.7

Source: ELKE (2003).

Central Greece is the region with the highest GDP per capita in Greece. It is known for the high quality of its agricultural production and for its good infrastructure. The region has a good road and railway network linking it with the populous region of Attica and its easy access to Attica and the capital of Athens is therefore one of its main advantages.

The Attica region represents 34 percent of the country's population and contributes 36 percent to the GDP and is known for its superior infrastructure. The region has a new international airport that provides worldwide connections. There is also a good road network in all local areas and national highways linking it with northern and southern Greece. Moreover, the port of Piraeus is one of the world's busiest ports and has a complete maritime infrastructure.

The only region which attracts FDI investment projects without being among the top regions in terms of GDP per capita is Eastern Macedonia – Thrace. This region is located in the north-eastern part of Greece having borders with Bulgaria and Turkey. The region provides easy access to Bulgaria and Turkey through broad road and railway networks. Foreign investors try to exploit these characteristics, as well as the incentives provided by the corresponding Law, which results in a significant amount of foreign investment capital located in this region.

Although the benefits linked with FDI tend to be more important for the less favoured regions, the comparative advantages for multinationals of investing in such regions are not always sufficient to draw them to locate there. A great deal depends on the capacity of the potential investor to take advantage of particular regional characteristics, such as low cost factors of production, natural resources, proximity to markets, as well as the overall economic and business environment of the region in question. In these cases it often seems difficult for national or regional authorities to influence the investment decisions providing investment location incentives through an administrative pattern, such as an Incentive Law.

18.7 CONCLUSIONS

The preceding analysis reveals a number of distinct aspects of regional policy in Greece during the recent period. First, despite significant steps towards decentralisation at the regional level and self-government at the prefectural and municipal level, Greece remains a highly centralised State in terms of the allocation of power and resources at the various administrative levels.

Second, the regional allocation of public investment in Greece in general follows the distribution of population and economic activities; the largest share of funds (almost one-third) is directed to the metropolitan region of Attica. In per capita terms, however, the allocation has a mildly redistributive character, although some less advanced regions still fare poorly.

Third, the regional differentiation of investment incentives has diminished over time, as new incentives provide more uniform subsidy rates to both advanced and less advanced regions.

Overall, regional inequalities in Greece have proved to be long-lasting and difficult to deal with. Taking into consideration the fact that regional inequalities in Greece have remained significant over the last 25 years, it may be concluded that regional policy has not been particularly successful, either in design terms or with respect to policy implementation.

REFERENCES

ELKE (2003), 'Invest in Greece Agency: Statistical Data', Research and Analysis Department (in Greek).

Georgantas and Psycharis (1999), 'Patterns of regional resource allocation in Greece', *Regional Studies Association International Conference*, Bilbao, Spain.

Ioannidis, Y. and Petrakos, G. (2000), 'Regional Convergence – Divergence Trends in Greece: An Assessment on The Basis of the Performance of Three Regions', *European Investment Bank Papers*, **5** (1), 31–58.

Konsolas, N., Papadaskalopoulos, A. and Plaskovitis, I. (2002), *Regional Development in Greece*, Springer.

Petrakos, G. and Saratsis, Y. (2000), 'Regional inequalities in Greece', *Papers in Regional Science*, **79**, 57–74.

Petrakos, G. and Psycharis, Y. (2003), '*Regional Development in Greece*', Athens: Kritiki Publishers (in Greek).

Petrakos, G. and Rodriguez-Pose, A. (2003), 'Regional inequality in the European Union', *TOPOS*, **18** (in Greek).

Petrakos, G., Karaveli, E. and Mardakis, P. (1999) 'Recent developments in the Greek system of urban centers', *Environment and Planning: B*, **26**, 2–13.

Psycharis, Y. (2006), 'Regional allocation of public investment in Greece', in *Regional Policy in Greece*, European Regional Science Association Greek Section (in Greek).

Psycharis, Y. (2004), 'Regional Programmes in Greece: evaluation and prospects', in *Regional Policy in Greece*, European Regional Science Association Greek Section (in Greek).

Index